Advances in Agronomy

Volume I

Advances in Agronomy
Volume I

Edited by **Jamie Hanks**

R CALLISTO REFERENCE

New York

Published by Callisto Reference,
106 Park Avenue, Suite 200,
New York, NY 10016, USA
www.callistoreference.com

Advances in Agronomy: Volume I
Edited by Jamie Hanks

International Standard Book Number: 978-1-63239-035-6 (Hardback)

Printed in the United States of America.

Contents

Permissions

List of Contributors

Preface

When the cave man turned civilized, the first thing he got into was agriculture. And since then, both of them have come a long way together. With the onset of modern days, the simple nature-dependent agriculture has also turned into an advanced field of occupation. The extensive use of science and technology in farming, to produce crops and plants then using those for food, fiber and other uses, is known as agronomy.

The scope and purpose of agronomy is to offer humanity a better, healthier and more environment-friendly ways of farming. The experts of the field called agronomists study various aspects of the process such as soil classification, plant breeding, enhancing plant and seeds fertility, irrigation & drainage control, benefits of pesticides and so on.

Agronomy goes beyond the realm of studying plant as a source of food. It studies plant as source of fuel, fiber for cloth, and other modern age purposes. It is in reality a dynamic combination of chemistry, botany, economics, ecology and genetics. Agriculture and allied industries are an integral part of most of the economies around the world. Not only it employs a significant percentage of global population, and changes in its framework, whether positive or negative, have the power to affect the entire world.

The Green Revolution that saved millions in India and Mexico from starvation is a perfect example of agronomy. The research and development on different types of renewable energy to be sourced from plants also branches out from agronomy.

This book is an honest attempt to focus on the various facets of agronomy and highlight its importance and contribution to the world at large.

Editor

Psidium guajava and *Piper betle* Leaf Extracts Prolong Vase Life of Cut Carnation (*Dianthus caryophyllus*) Flowers

M. M. Rahman, S. H. Ahmad, and K. S. Lgu

Department of Crop Science, Faculty of Agriculture, Universiti Putra Malaysia, Selangor, 43400 Serdang, Malaysia

Correspondence should be addressed to S. H. Ahmad, hajar@agri.upm.edu.my

Academic Editors: C. Dell and T. Takamizo

The effect of leaf extracts of *Psidium guajava* and *Piper betle* on prolonging vase life of cut carnation flowers was studied. "Carola" and "Pallas Orange" carnation flowers, at bud stage, were pulsed 24 hours with a floral preservative. Then, flowers were placed in a vase solution containing sprite and a "germicide" (leaf extracts of *P. guajava* and *P. betle*, 8-HQC, or a copper coin). Flowers treated with 8-HQC, copper coin, and leaf extracts had longer vase life, larger flower diameter, and higher rate of water uptake compared to control (tap water). The leaf extracts of *P. guajava* and *P. betle* showed highest antibacterial and antifungal activities compared to the other treatments. Both showed similar effects on flower quality as the synthetic germicide, 8-HQC. Therefore, these extracts are likely natural germicides to prolong vase life of cut flowers.

1. Introduction

The prospects for the cut flower industry in Malaysia are very promising due to the increase in the household and overseas markets. Per capita consumption of cut flowers per household had increased from 0.71 MYR in 1990 to 2.50 MYR in 2010. This trend is expected to increase further with greater consumer prosperity and aesthetic values for fresh cut flowers. There is also an increasing demand for cut flowers in the overseas market. This is expected to continue at a growth rate of 6% per annum. This value has reached 20 MYR and 36 MYR billion in the years 2000 and 2010, respectively.

Normally, dealers and consumers in cut flowers will put on view or store cut flowers in vases containing tap water plus sugar as the preservative or vase solution. When the bud is cut off from the mother plant, the bud no longer obtains its food from the mother plant. Sugar is used to provide food and more energy for bud opening and further flower development. Most preservative solutions contain two basic components: sugar and germicide. The sugar provides a respiratory substrate, whereas the germicide controls harmful bacteria and prevents plugging of the conducting tissues. The germicidal quality of 8-hydroxyquinoline citrate (8-HQC) ensures that bacteria do not grow in the vase

solutions which may result in a blockage of vascular tissues and inhibition of water uptake by the flowers. According to Jones and Truett [1], germicides extended the vase life of *Gloriosa rothschildiana* mainly by improving solution uptake. The extracts from mature leaves of *Psidium guajava* and *Piper betle* have antimicrobial properties [2]. The leaf extracts of *P. guajava* contain phytochemicals, which act as antimicrobial, astringent, and bactericide, while the leaf extract of *P. betle* has antifungal, antiseptic, and anthelmintic activity. The high levels of five propenylphenols (chavicol, chavibetol, allylpyrocatechol, chavibetol acetate, and allylpyrocatechol diacetate) in *P. betle* leaf extracts showed favourable response towards fungicidal and nematocidal activity.

A major form of deterioration in cut flowers is the blockage of xylem vessels by air and microorganisms that cause xylem occlusion [3]. The 8-HQS is a tremendously beneficial germicide in preservatives used in the floral industry [4] and acts as an antimicrobial agent [5] which can increase water uptake [6]. The application of 8-HQS increased the vase life as well as the fresh weight (percentage of initial) of the cut flowers. The 8-HQS treatment also prevented growth of microorganisms in xylem vessels of the cut flower stems and maintained water uptake. The combination treatment of 8-HQS and sucrose improved the postharvest quality of gladiolus spikes [7]. In *Dendrobium* cut flowers, holding solutions containing 8-HQS + sucrose extended the vase life and improved flower quality, water consumption, fresh weight, flower freshness, and reduced respiration rate and weight loss [8].

Synthetic germicides such as silver thiosulphate (STS), silver nitrate (AgNO₃), and 8-hydroxyquinoline citric (HQC) are expensive and not easily available in the local market. Furthermore, synthetic germicides, which contained silver, can pollute the environment due to its high phytotoxicity potential with harmful heavy-metal environmental contaminant [9]. The objectives of this research were to evaluate the effects of leaf extracts of *P. guajava* and *P. betle* on prolonging the vase life of cut carnation (*Dianthus caryophyllus*) flowers by controlling the microbial growth in the vase solution.

2. Materials and Methods

The experiment was conducted in the Postharvest Laboratory, Department of Crop Science, Universiti Putra Malaysia (UPM). The opened-bud stages of "Carola" and "Pallas Orange" carnation cut flowers were obtained directly from a supplier in the Cameron Highlands, Pahang. Cut carnations with firm buds and leaves that were free from damage, decay, and pest and diseases were selected. Each flower stalk was recut to a length of 45 cm. Leaves from the lower 15 cm of the stems were removed to avoid contamination in the vase solution. Then, the flower stalks were given pretreatments of sugar by submerging 10 cm of the stem ends in a vase solution containing 10% sucrose, 250 mg/L citric acid, and 250 mg/L 8-HQC for 12 h.

A 250 mL bottle was lined with a plastic tube (baby bottle system). This is to prevent contamination between tests and avoid the need for bottle washing. Sprite, a sterile-carbonated drink of pH 3.5 containing 26 grams of total sugar, 43 Kilocalories (Kcal) of energy, 3% citric acid, was used as the main vase solution into which either 8-HQC (250 mg/L), one copper coin, or leaf extracts (2 mL/L) of *P. guajava* or *P. betle* was added. Then, the pulsed flower stalk was divided randomly and placed into each tube, leaving the basal end about 1 cm away from the base of the bottle.

The flowers were evaluated at room temperature (25 ± 2°C) under continuous white fluorescent light at 1.2 klux. Vase life (day), flower diameter (cm), water uptake (mL), and petal colour were measured every other day. Vase life was measured as the longevity of flowers in the vase solution. Vase life was ended when the flower heads started to slump, followed by discolouration and abscission of petals, or when 30% of general appearance is no longer attractive. Flower diameter opening was measured by taking two measurements, which crossed at the centre of the flower, and the mean was calculated. Water uptake was measured daily by weighing the vase without the flower:

Water uptake (mL) on day N = Total weight (g) on day N − Total weight (g) on day N_{-1}.

Rate of water uptake (mL) = Amount of water taken up (mL/day).

Total water uptake = Water uptake on day $N + N_{+1} + N_{+2} \cdots N_{+n}$.

Petal colour was measured on an alternate day for 9 days. The measurement was made using a chromometer (Model CR-300b, Minolta, Ramsey, NJ, USA) in the lightness (L^*), chromaticity (C^*), and hue ($h°$) colour notation system (C illuminant, calibrated with standard white plate and 0° viewing angle).

2.1. Microbial Growth. The evaluation of microbes was conducted to determine the effectiveness of the leaf extracts of *P. guajava* and *P. betle* in controlling the microbial growth in the vase solution of the cut carnation flowers. Fourteen gram of ready mixed nutrient agar (NA) and 20 g of potato dextrose agar (PDA) was each diluted in 500 mL of distilled water. These two liquid media were sterilized at 121°C at 15 psi for 15 minutes in an autoclave. Then, 20 mL of each medium was poured separately into a 100 mm sterile petri dish and cooled for a few hours to 30°C under a laminar flow. The flowers were left in the vase solutions for 6 days. Then, by using an Eppendorf micropipette, 100 µL of the vase solution was taken from each treatment. Using an inoculation loop, a drop of the vase solution was taken and struck onto each of the NA and PDA media. Then, the petri dishes were incubated at 30°C in an incubator for 24 and 72 h, for bacterial and fungal growth, respectively. All the plates were evaluated for microbial growth. Then, growth of each microorganism was evaluated based on scores of 0–5: None = 0 colony/plate, Few = 1-2 (<30 colonies/plate), Moderate = 3-4 (30–300 colonies/plate), and Abundant = 5 (>300 colonies/plate) as shown in Table 1.

2.2. Leaf Extracts of P. guajava and P. betle. Leaves were extracted according to the methods of Rahman et al. [10] with some modifications. Fresh and mature leaves of *P. guajava* and *P. betle* were collected and sliced into small pieces (0.5 cm × 3 cm). The sliced leaves were dried at room temperature (27 ± 2°C) for 3 to 4 days and then grounded into a powder form. Then, the leaf was macerated with 80% ethanol for 24 hours at room temperature. Each extract was filtered and then evaporated with a rotary evaporator (Model CA-1310 Eyela, Tokyo Rikakikai Co., Ltd., Japan) under reduced pressure at a temperature not exceeding 50°C. The dark brown viscous residue was reconstituted in ethanol (5 mL). Two millilitre of each extract was mixed into 1 L of sprite to act as the vase solution.

3. Results

3.1. Flower Diameter. The Carola carnation treated with 8-HQC, copper coin, and the leaf extracts of *P. guajava* and *P. betle* showed a similar trend in flower opening as indicated by the increase in flower diameter (Figure 1). All flowers were partially opened on day 3. From days 3–5, there were rapid increases in diameter of flowers treated with 8-HQC, copper coin, and leaf extracts of *P. guajava* and *P. betle*. On day 5, control flowers had already attained the maximum diameter opening. Similarly, flowers treated with 8-HQC, copper coin,

TABLE 1: Scheme for scoring growth of microorganism based on colony.

Microorganism growth	Colony	Score
Abundant	(>300 per plate)	5/5
Moderate	(30–300 per plate)	3/5
Few	(<30 per plate)	1/5
None	0	0

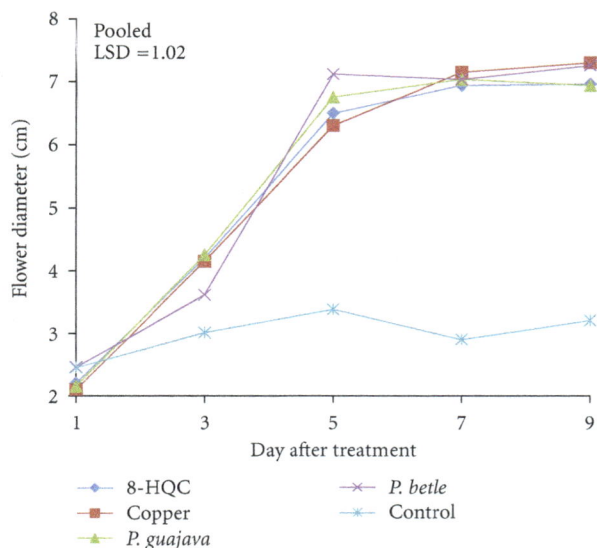

FIGURE 1: Effect of 8-HQC, copper coin, and leaf extracts of *P. guajava* and *P. betle* and control on flower diameter of "Carola" cut carnation flowers.

leaf extracts of *P. guajava* and *P. betle* obtained maximum diameter opening on day 5 and the diameter opening level off until day 9, by which time the flowers began to senesce. Similar results were found for the Pallas Orange flowers (data not included). After 9 days, flowers treated in tap water lost their turgidity and started to turn black and wilt.

On day 5, flowers treated with leaf extract of *P. betle* had significantly larger flower diameter compared to flowers treated with 8-HQC and copper coin (Figure 1). By day 9, diameters of flowers treated with 8-HQC, copper coin, and leaf extracts of *P. guajava* and *P. betle* were not significantly different from one another. The control flowers had the smallest diameter compared to flowers in all the other four treatments, throughout the evaluation period of the flowers. The cloudiness of the control vase solution (tap water) indicated that there was an abundance of microbes. These microbes clogged the xylem vessels of the flower stem.

3.2. Water Uptake. There were no significant differences in water uptake of the Carola carnation when treated with leaf extracts of *P. guajava* and *P. betle*, copper coin, 8-HQC, and control from days 3–5 after treatment (Figure 2). However, flowers treated with 8-HQC, copper coin, and *P. Betle* leaf extracts had no significant differences in water uptake throughout the study. On days 7–9, flowers treated

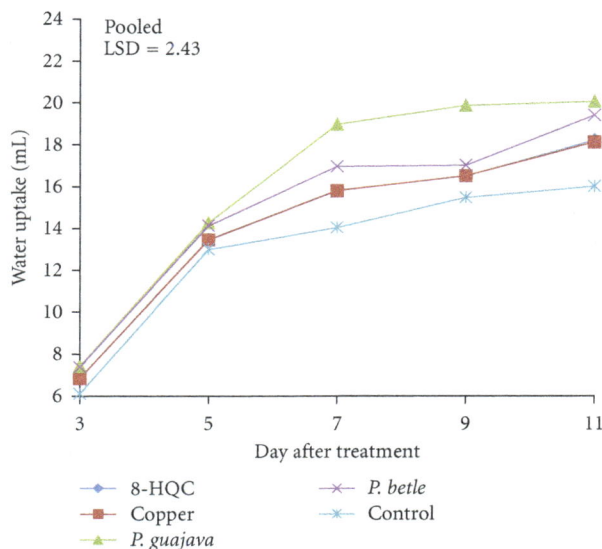

FIGURE 2: Effect of 8-HQC, copper coin, and leaf extracts of *P. guajava* and *P. betle* and control on water uptake of "Carola" cut carnation flowers.

with leaf extracts of *P. guajava* had significantly higher water uptake compared to the other treated flowers. But by day 11, flowers in all the four treatments were not significantly different in their water uptake. Water uptake of the control flowers was at the lowest compared to all the four treated flowers on day 11, due to flower senescence. However, water was still taken up for leaf survival.

The flowers treated with copper coin, 8-HQC, and leaf extracts of *P. guajava* and *P. betle* had gradual increases in water uptake from days 5–11. Similar results were found for Pallas Orange carnation flowers (data not included). Throughout the period of the evaluation, it was observed that water uptake of the control flowers were lowest after days 9–11 compared to flowers in leaf extracts of *P. guajava* and *P. betle*. In general, flowers treated with leaf extracts of *P. guajava* and *P. betle* had the higher water uptake compared to the control. Both 8-HQC and copper coin had a similar trend of water uptake from days 3–11.

3.3. Vase Life. There was no significant difference in the vase life of Carola carnation treated with copper coin and leaf extracts of *P. guajava* and *P. betle* throughout the evaluation period (Table 2). This showed that both of the leaf extracts could prolong the vase life of the flowers. Similarly, vase life of the Pallas Orange carnation treated with copper coin and leaf extracts of *P. guajava* and *P. betle* were significantly different from the control flowers (Table 2). Both types of flowers placed in the vase solution containing either 8-HQC, copper coin, or leaf extracts of *P. guajava* and *P. betle* had 5–7 days longer vase life compared to control flowers. In general, the vase life of flowers was terminated when 30% of the petals were rolled in and had senesced.

TABLE 2: Effect of 8-HQC, copper coin, and leaf extracts of *P. guajava* and *P. betle* and control on vase life of "Carola" and "Pallas Orange" cut carnation flowers.

Treatment	Vase life (days)	
	"Carola"	"Pallas Orange"
Sprite + 8-HQC	9.6 b^z	10.6 b^z
Sprite + copper coin	11.0 a	12.3 a
Sprite + *P. guajava*	10.3 ab	11.1 ab
Sprite + *P. betle*	10.4 ab	11.5 ab
Tap water (control)	4.8 c	5.1 c

^{z}Comparison of means within columns by LSD ($P = 0.05$). Means are averages for 8 replications with one flower stalk per replicate.

3.4. Petal Colour. Larger L^* values indicate a lighter colour compared to smaller L^* values that indicate a darker colour (0 = black to 100 = white). The L^* values of pink Carola flowers in all the treatments increased from days 1–9 except for flowers in the control treatment, which decreased after day 5 (Figure 3(a)). The L^* value of control flowers decreased until day 9. For the treated flowers, L^* value continued to increase, indicating that the flowers had become brighter in colour. However, all the Pallas Orange carnation flowers showed a decreasing trend of L^* values from days 1–5 (Figure 3(b)). From days 5–9, the L^* for the control flowers showed a rapid increase and was significantly different from the treated flowers. This indicated that petal of the control flowers became lighter than those treated with 8-HQC, copper coin, and leaf extracts of *P. guajava* and *P. betle*.

There were initial rapid declines in C^* values for Carola carnation in all the treatments from days 1–5 (Figure 4(a)) followed by a levelling off of C^* values until day 9. These indicated the loss of vividness or saturation of colour as the flowers senesced. The Pallas Orange carnation flowers showed a similar trend of decrease in C^* values throughout the evaluation period, except for days 3–5, during which flowers treated with *P. guajava* extract and copper coin were significantly different in C^* values from flowers treated with leaf extract of *P. guajava*, 8-HQC, and control (Figure 4(b)). After day 5, flowers treated with 8-HQC, leaf extract of *P. betle,* and the control showed a decline in C^* value. In general, by day 9, the flowers treated with leaf extract of *P. guajava* had the highest C^* value, while flowers treated with leaf extract of *P. betle* had the lowest C^* value.

From days 1–5 after treatment, Carola carnation flowers treated with 8-HQC, copper coin, and leaf extracts of *P. guajava* and *P. betle* had a similar trend of $h°$ values throughout the evaluation period (Figure 5(a)). After day 5, the $h°$ value of the control flowers showed a rapid increase indicating that the flowers had changed colour and senesce. From days 5–9, $h°$ values of flowers treated with leaf extracts of *P. guajava* and *P. betle* and 8-HQC remained constant. Similarly, the colour of the Pallas Orange carnation flowers remained constant from days 1–5 (Figure 5(b)). After day 5, $h°$ value increased rapidly indicating colour changes of flowers from orange to reddish orange. Flowers treated with

the copper coin had significantly higher $h°$ values from days 3–9 compared to flowers treated with leaf extracts of *P. guajava*.

3.5. Microbial Growth. The control (tap water only) vase solution contained a significantly higher bacterial colony compared to the vase solutions containing 8-HQC, copper coin, and leaf extracts of *P. guajava* and *P. betle* on day 3 after streaking (Table 3). It was observed that the control solution had an abundant growth of bacteria which started on day 2 (Figure 6(A)). Similarly, on day 3 after streaking, leaf extracts of *P. betle* and *P. guajava*, copper coin, and 8-HQC showed the highest antibacterial activities as proven by a few bacterial colonies on the nutrient agar media (Figure 7(A)).

There were significant differences in the score of microbial growth on potato dextrose agar with the control vase solution compared with the vase solutions containing 8-HQC, copper coin, and leaf extracts of *P. guajava* and *P. betle* (Table 3). There was an abundant growth of fungi on the control solution. Leaf extracts of *P. guajava* showed the highest antifungal activities (Figure 7(B)). Other germicides like 8-HQC, copper coin, and leaf extract of *P. betle* did not appear to be as good antifungal germicides. All these microorganisms growth was evaluated based on a score of 0–5, where None = 0 colony/plate, Few = 1-2 (<30 colonies/plate), Moderate = 3-4 (30–300 colonies/plate), and Abundant = 5 (>300 colonies/plate).

4. Discussion

In the present study, treated flowers had almost 30% larger flower diameter opening compared to control. Similarly, cut lotus (*Nelumbo nucifera*) flowers fail to open, and petal blackening occurs due to carbohydrate depletion in the leaves and sink activity of the flowers [11, 12]. Thus, it is not practical to use tap water alone, without any food source, as a vase solution for carnation flowers. All holding solutions must contain essentially two components: sugar and germicides. The sugar provides a respiratory substrate, while the germicides control harmful bacteria and prevent plugging of the water-conducting tissues. Among all the different types of sugars, sucrose has been found to be the most commonly used sugar in prolonging vase life of cut flowers. Sucrose added to the vase solution supplies cut flowers with substrates that are needed for respiration, thus enabling harvested buds to open into flowers [13]. The water uptake of tap water depends on factors like acidity (pH), total dissolved solids, and the presence of specific toxic ions [14]. The most toxic ion in tap water to carnation flower development is fluoride, and flowers were injured by 1 ppm fluoride [15]. A similar observation was found where decline in water uptake in rose was caused by bacteria ranging from 10^7 to 10^8/mL in the vase solution [14]. At 3×10^9/mL, the first sign of wilting would appear in an hour. Additionally, decreases in water uptake, which closely followed an increase in the tonoplast permeability, caused senescence in cut carnation flowers [16]. In the present study, initial increase of 20–30% water uptake rate indicated that water was taken up

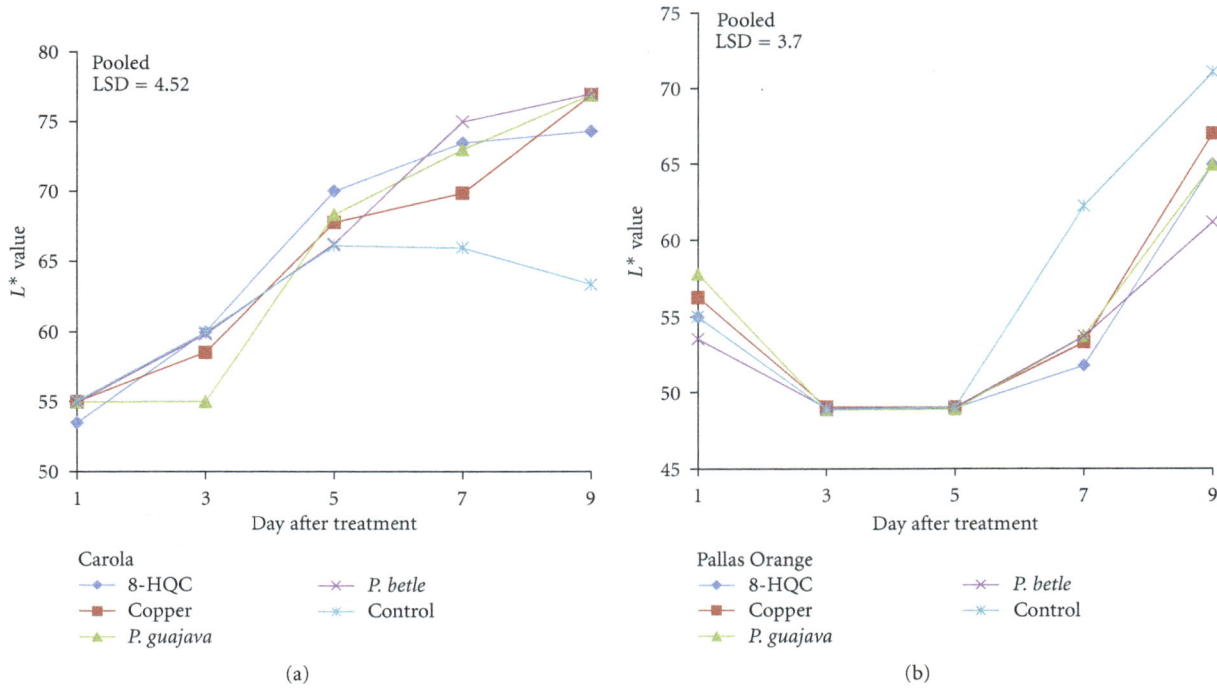

FIGURE 3: Effect of 8-HQC, copper coin, and leaf extracts of *P. guajava* and *P. betle* and control on L^* value of (a) "Carola" and (b) "Pallas Orange" cut carnation flowers.

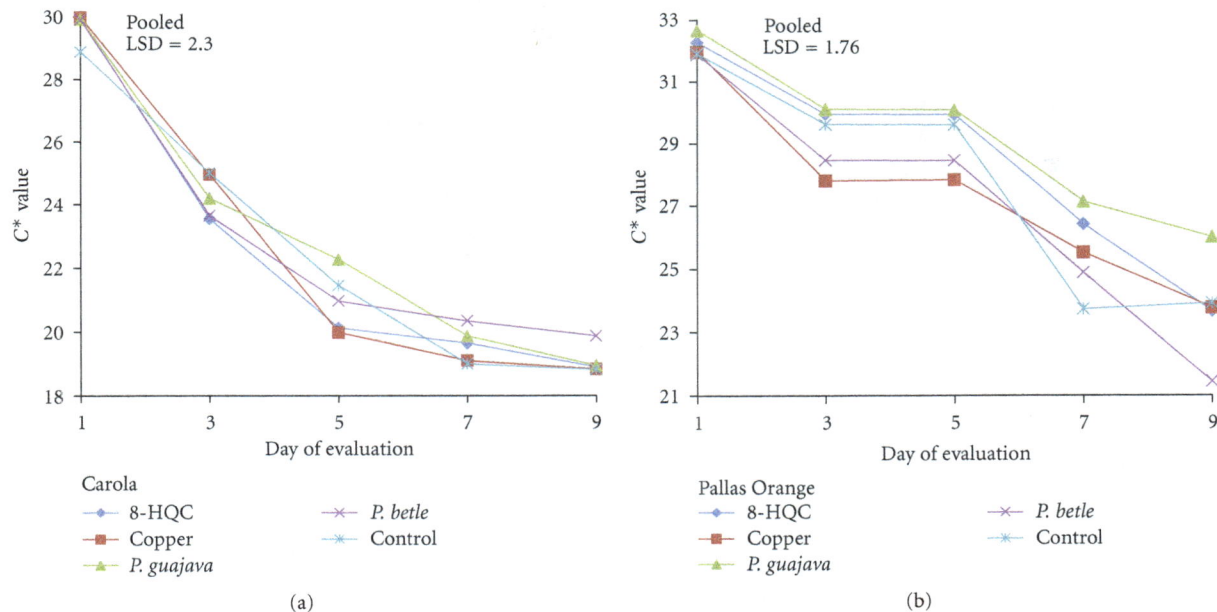

FIGURE 4: Effect of 8-HQC, copper coin, and leaf extracts of *P. guajava* and *P. betle* and control on C^* value of (a) "Carola" and (b) "Pallas Orange" cut carnation flowers.

through the stems to supply adequate food and nutrition for the flower development. A rapid decline of water uptake for the control flowers after day 5 was associated with senescence and abscission of the older leaves or flowers in the final phase of development [17, 18]. Cloudiness of the vase solution due to the activities of microbes could contribute to early senescence of the flowers. A similar observation was found whereby carnation flowers only lasted for 5-6 days in tap water [19]. The reasons for the short vase life of the control flowers were associated with the presence of microbes in the

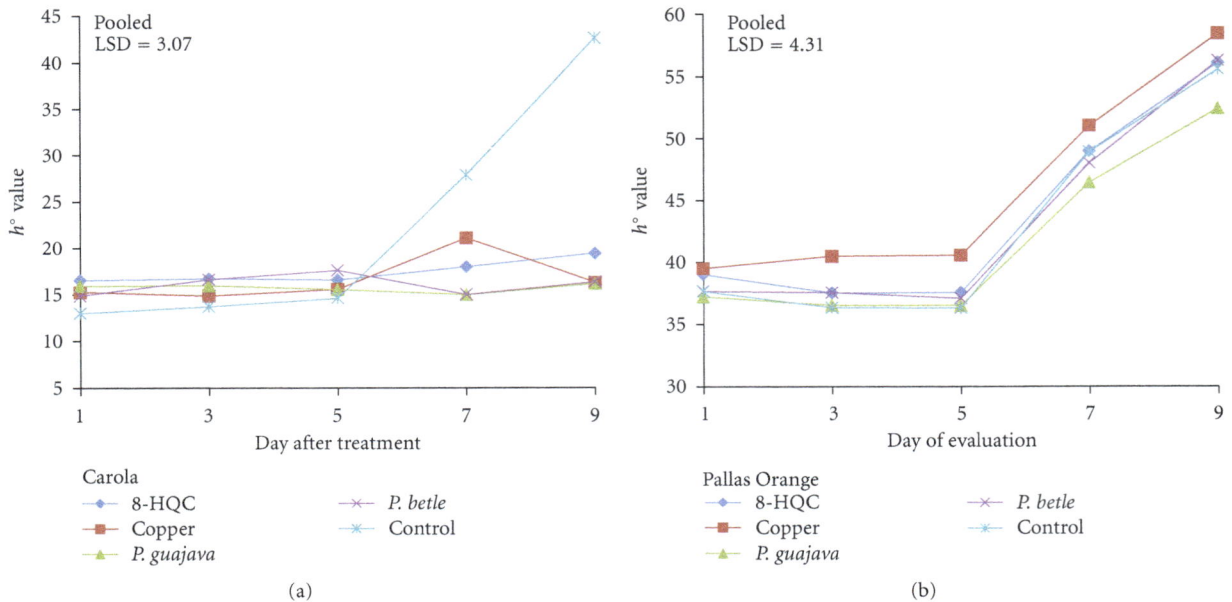

FIGURE 5: Effect of 8-HQC, copper coin, and leaf extracts of *P. guajava* and *P. betle* and control on $h°$ value of (a) "Carola" and (b) "Pallas Orange" cut carnation flowers.

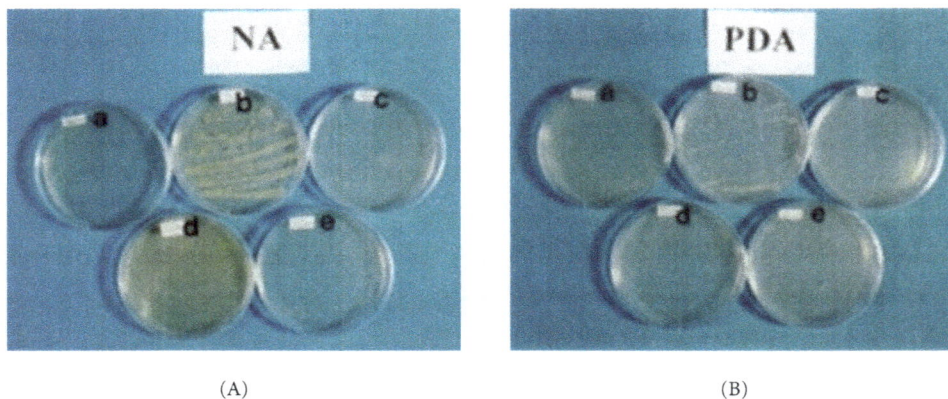

FIGURE 6: Growth of bacteria and fungi on (A) nutrient agar and (B) potato dextrose agar media plated on petri dish which contain copper coin (a), tap water (b), 8-HQC (c), *P. betle* leaf extract (d), and *P. guajava* leaf extract (e) on both media after day 2 of streaking vase solution of cut carnation flowers on day 9.

vase solution. Metabolites produced by certain bacteria also reduced longevity and water conductivity in carnation [20].

Senescence in carnation flowers was mainly associated with sensitivity of the carnation flowers to ethylene and carbohydrate metabolism. Ethylene was produced during normal metabolism in flowers and conditions created by infection, wound, and stress, like wilting, which accelerated its biosynthesis [21]. Other factors like microbial infection and lack of food supply are very important in affecting vase life. Pretreatment with sucrose (10%) could prolong the vase life of cut flowers. Pulsing with concentrations of sucrose > 10% for 20 h prolonged the vase life of the spikes of cut *Liatris spicata* [22]. Furthermore, recutting

stem ends could help in prolonging the vase life of cut flowers. Recutting the stem end promotes rehydration of the stems, speeds up flower opening, quickly revives wilted flowers, and increases the lasting qualities of the cut flowers [19]. Pretreatment with sugar improved colour and sizes of petals and the vase life of a number of cut flower species [23]. This requirement for sugar explained why the control flowers that were treated in tap water alone had poor colour. There were no significant differences in L^* values among flowers treated with 8-HQC, copper coin, and leaf extracts of *P. guajava* and *P. betle*, except that the control in both types of cut carnation flowers Pallas Orange carnation had orange petals, while the Carola carnations had pink

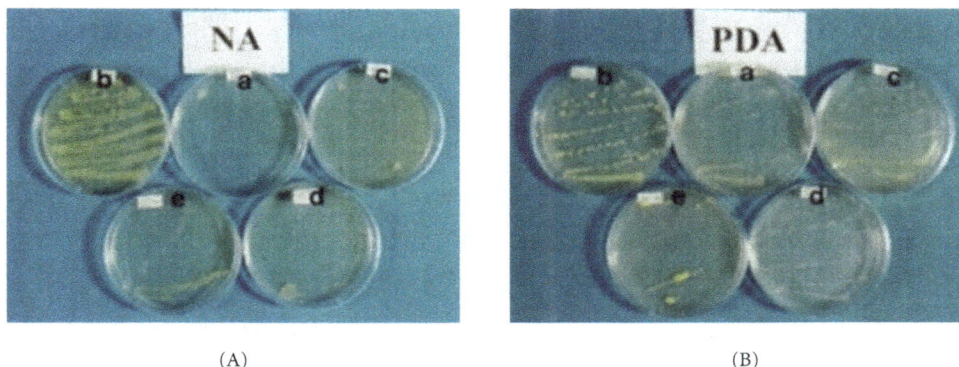

(A)

(B)

FIGURE 7: Growth of bacteria and fungi on (A) nutrient agar and (B) potato dextrose agar media plated on petri dish which contain copper coin (a), tap water (b), 8-HQC (c), *P. betle* leaf extract (d), and *P. guajava* leaf extract (e) on both media after day 3 of streaking vase solution of cut carnation flowers on day 9.

TABLE 3: Effect of vase solution containing 8-HQC, copper coin, leaf extracts of *P. guajava*, and *P. betle* and control at seven days after treatment on the microorganism growth after three days of inoculation. Media used were nutrient agar (NA) and potato dextrose agar (PDA).

Treatment	Score of microorganism growth	
	NA	PDA
Sprite + 8-HQC	1.4 b^z	3.4 b^z
Sprite + copper coin	1.0 b	2.4 c
Sprite + *P. guajava*	1.4 b	2.2 c
Sprite + *P. betle*	1.2 b	3.4 b
Tap water (control)	5.0 a	4.8 a

^{z}Comparison of means within columns by LSD ($P = 0.05$). Means are averages for 5 replications.

flowers. Initially on day 1, the Carola carnation flowers were light pink, with low L^* and $h°$ values but high C^* values. From the result, it is apparent that the optimum aesthetic values of the flowers were between days 4 and 6. During this period, the diameter of the germicide-treated flowers were about 65–95% larger than control flowers. The larger flower diameter indicated a normal opening behavior of the flowers. The diameter opening of control flowers was not normal throughout the evaluation period. This abnormal opening resulted in lower L^* and higher $h°$ values of control flowers compared to the germicide-treated flowers, by day 9. As for the Pallas Orange carnation, the initial petals had low L^* and $h°$ but with high C^* colour values. The results indicated that, by day 7, color saturation of control flowers had decreased, resulting in fading. Generally, when colour saturation of flowers decreased, the colour of the petals became darker, leading to a loss of flower colour. According to Salunkhe et al. [19], senescing flowers showed discoloration or fading of colour due to significant changes in carotenoid and anthocyanin pigments, which are responsible for different colours of flowers. NA of pH 7.4 and with a formulation was the favourite culture media for the growth

of heterotrophic bacteria [24]. The control solution was full of bacteria cells since no germicide was used to control the microbial growth. Germicides controlled microbial growth and partially decreased the resistance to water uptake [14].

Hence, leaf extracts of *P. guajava* appeared to be an excellent, natural antimicrobial agent. It showed antibacterial activity against the bacteria tested, in the most cases, with activity stronger than 50 μg streptomycin. Polyphenolic compounds like guaijaverin, quercetin, and avicularin are the active antimicrobial components in guava leaf [2]. Leaf extracts of *P. betle* contain two phenols, betel-phenol (chavibetol) and chavicol. Leaf extracts of *P. betle* had antifungal activities [2]. The high levels (1% of leaf fresh weight) of five propenylphenols (chavicol, chavibetol, allylpyrocatechol, chavibetol acetate, and allylpyrocatechol diacetate) were considered responsible for the fungicidal and nematocidal activities. Leaf extracts of *P. betle* have been shown to be effective against pathogens that cause collar rots, such as *Pyricularia oryzae* Cav., *Cochliobolus miyabeanus*, *Rhizoctonia solani* Kuhn, and *Thanatephorus cucumeris* (Frank) Donk [25].

5. Conclusions

It is not reasonable to use tap water alone as a vase solution to preserve carnation flowers. The response of Carola and Pallas Orange carnation flowers to tap water was a lower water uptake, shorter vase life, and smaller flower diameter compared to treated carnation flowers. In addition, tap water of pH 7.2 provided a favourable environment for bacteria and fungi to multiply in the vase solution.

Sprite, in combination with leaf extracts of *P. betle* or *P. guajava*, was able to prolong the vase life of Carola and Pallas Orange carnation flowers. The high concentrations of sugar, 10%, in the sprite had become the main source of energy for maintaining the biochemical and physiological processes after the flowers were separated from the mother plant. The presence of sodium benzoate and carbon dioxide in the sprite inhibited ethylene action while the leaf extracts of *P. guajava* and *P. betle* acted as germicides.

When treated with leaf extracts of *P. guajava* and *P. betle,* longevity of the Carola and Pallas Orange carnation flowers doubled when compared to the flowers placed in the control solution. These results indicated that leaf extracts of *P. guajava* and *P. betle* were able to prolong the vase life of carnations flowers. Conversely, carnation flowers that were pulsed overnight in high concentrations of sucrose (10%) together with 250 mg/L 8-HQC and 250 mg/L citric acid could extend longevity of the carnation flowers. However, longevity and colour of flowers in the vase solution depend much more on the variety of the carnation flowers. In this experiment, a slightly longer vase life and gradual decrease in colour saturation (C^* values) were observed in the Pallas Orange than Carola carnation flowers.

Flowers treated with leaf extracts of the *P. guajava* and *P. betle* showed similar results in diameter opening, water uptake, and vase life as the other synthetic germicides like 8-HQC and copper coin. Therefore, they have potential as natural germicides for antibacterial activities. The leaf extracts of *P. guajava* and *P. betle* contain phenolic compounds, either bactericidal or bacteriostatic, depending on the concentration used. Chavicol and chavibetol are compounds found in *P. betle* leaf extracts. They are responsible for the antifungal activities. However, further investigation should be carried out to determine the actual concentration of leaf extracts of *P. guajava* and *P. betle* to inhibit the growth of bacteria and fungi in the vase solution.

Acknowledgments

This study was supported by the Malaysian Government Research Grants through E-science project no. 5450473. Graduate Research Assistance (GRA) from University Putra Malaysia is acknowledged for providing scholarship to the first author for his doctorate program.

References

[1] R. B. Jones and J. R. Truett, "Postharvest handling of cut *Gloriosa rothschildiana* O'Brien (Liliaceae) flowers," *Journal of the American Society for Horticultural Science,* vol. 177, pp. 442–445, 1992.

[2] M. Suhaila, E. F. P. Henie, and H. Zaiton, "Bacterial membrane disruption in food pathogens by *Psidium guajava* leaf extracts," *International Food Research Journal,* vol. 16, no. 3, pp. 297–311, 2009.

[3] R. E. Hardenburg, "The commercial storage of fruits, vegetables and florist and nursery stock," in *Agricultural Handbook,* J. M. Lutz, Ed., no. 66, p. 130, Department of Agriculture. Agricultural Research Service, USA, 1968.

[4] J. Nowak and R. M. Rudnicki, *Postharvest Handling and Storage of Cut Flower, Florist, Greens and Potted Plants,* Timber Press, 1990.

[5] S. Ketsa, Y. Piyasaengthong, and S. Prathuangwong, "Mode of action of AgNO$_3$ in maximizing vase life of *Dendrobium* "Pompadour" flowers," *Postharvest Biology and Technology,* vol. 5, no. 1-2, pp. 109–117, 1995.

[6] B. S. Reddy, K. Singh, and A. Singh, "Effect of sucrose, citric acid and 8- hydroxyquinoline sulphate on the postharvest physiology of tuberose cv. single," *Advances in Agricultural Research,* vol. 3, pp. 161–167, 1996.

[7] S. Beura, S. Ranvir, S. Beura, and R. Sing, "Effect of sucrose pulsing before storage on postharvest life of Gladiolus," *Journal of Ornamental Horticulture,* vol. 4, pp. 91–94, 2001.

[8] M. Dineshbabu, M. Jawaharlal, and M. Vijayakumar, "Influence of holding solutions on the postharvest life of *Dendrobium* hybrid Sonia," *South Indian Horticulture,* vol. 50, pp. 451–457, 2002.

[9] J. W. Damunupola and D. C. Joyce, "When is a vase solution biocide not, or not only, antimicrobial?" *Journal of the Japanese Society for Horticultural Science,* vol. 77, no. 3, pp. 211–228, 2008.

[10] M. Rahman, S. H. Ahmad, M. T.M. Mohamed, M. Zaki, and A. Rahman, "Extraction of *Jatropha curcas* fruits for antifungal activity against anthracnose (*Colletotrichum gloeosporioides*) of papaya," *African Journal of Biotechnology,* vol. 10, no. 48, pp. 9796–9799, 2011.

[11] R. B. Jones, R. McConchie, W.G. van Doorn, and M. S. Reid, "Leaf blackening in cut *protea* flowers," *Horticultural Reviews,* vol. 17, pp. 173–201, 1995.

[12] W. G. van Doorn, "Leaf blackening in protea flowers: recent developments (Proceedings: International Protea Research Symposium)," *Acta Horticulturae,* vol. 545, pp. 197–204, 2001.

[13] U. K. Pun and K. Ichimura, "Role of sugars in senescence and biosynthesis of ethylene in cut Flowers," *Japan Agricultural Research Quarterly,* vol. 37, no. 4, pp. 219–224, 2003.

[14] R. B. Jones and M. Hill, "The effect of germicides on the longevity of cut flowers," *Journal of the American Society for Horticultural Science,* vol. 118, pp. 350–354, 1993.

[15] V. I. Lohr and C. H. Pearson-Mims, "Damage to cut roses from fluoride in keeping solutions varies with cultivar," *HortScience,* vol. 25, pp. 215–216, 1990.

[16] R. Battelli, L. Lombardi, H. J. Rogers, P. Picciarelli, R. Lorenzi, and N. Ceccarelli, "Changes in ultrastructure, protease and caspase-like activities during flower senescence in *Lilium longiflorum,*" *Plant Science,* vol. 180, no. 5, pp. 716–725, 2011.

[17] R. E. Brevedan and D. B. Egli, "Short periods of water stress during seed filling, leaf senescence, and yield of soybean," *Crop Science,* vol. 43, no. 6, pp. 2083–2088, 2003.

[18] S. Mahajan and N. Tuteja, "Cold, salinity and drought stresses: an overview," *Archives of Biochemistry and Biophysics,* vol. 444, no. 2, pp. 139–158, 2005.

[19] D. K. Salunkhe, N. R. Bhat, and B. B. Desai, *Postharvest Biotechnology of Flowers and Ornamental Plants,* Springer, New York, NY, USA, 1990.

[20] S. Mayak and E. Accati-Garibaldi, "The effect of microorganisms on susceptibility to freezing-damage in petals of cut rose flowers," *Scientia Horticulturae,* vol. 11, no. 1, pp. 75–81, 1979.

[21] R. Nichols, "A description model of the senescence of the carnation (*Dianthus caryophyllius*) induced by ethylene," *Planta,* vol. 130, pp. 47–52, 1977.

[22] S. H. Susan, "Role of sucrose in bud development and vase life of cut *Liatris spicata* (L) wild," *Journal of Horticultural Science,* vol. 27, pp. 1198–1200, 1992.

[23] K. Ichimura and T. Hiraya, "Effect of silver thiosulfate complex (STS) in combination with sucrose on the vase life of cut sweet pea flowers," *Plant Growth Regulation,* vol. 28, pp. 117–122, 1999.

[24] J. P. Michael, E. C. S. Cha, and R. K. Noel, *Microbiology: Concepts and Applications*, McGraw–Hill, New York, NY, USA, 1st edition, 1993.

[25] M. Suhaila, M. Suzana, S. H. El-Sharkawy, A. M. Ali, and S. Muid, "Antimycotic screening of 58 Malaysian plants against plant pathogens," *Pesticide Science*, vol. 47, no. 3, pp. 259–264, 1996.

2

Temporal Downscaling of Crop Coefficient and Crop Water Requirement from Growing Stage to Substage Scales

Songhao Shang

State Key Laboratory of Hydroscience and Engineering, Department of Hydraulic Engineering, Tsinghua University, Beijing 100084, China

Correspondence should be addressed to Songhao Shang, shangsh@tsinghua.edu.cn

Academic Editor: Enrico Porceddu

Crop water requirement is essential for agricultural water management, which is usually available for crop growing stages. However, crop water requirement values of monthly or weekly scales are more useful for water management. A method was proposed to downscale crop coefficient and water requirement from growing stage to substage scales, which is based on the interpolation of accumulated crop and reference evapotranspiration calculated from their values in growing stages. The proposed method was compared with two straightforward methods, that is, direct interpolation of crop evapotranspiration and crop coefficient by assuming that stage average values occurred in the middle of the stage. These methods were tested with a simulated daily crop evapotranspiration series. Results indicate that the proposed method is more reliable, showing that the downscaled crop evapotranspiration series is very close to the simulated ones.

1. Introduction

Crop water requirement, the essential data for agricultural water management and regional water resources planning, is defined as the amount of water needed to compensate the water consumed by plant transpiration and soil evaporation from cropped field under nonrestricting soil conditions [1]. The values of crop water requirement in any growing periods are equal to crop evapotranspiration (ET_c) under nonrestricting soil conditions in the same period. Therefore, crop water requirement was usually measured or calculated through ET_c [2].

In 1998, Food and Agriculture Organization (FAO) recommended an updated procedure to calculate ET_c from reference evapotranspiration (ET_0) and crop coefficient (K_c) [2]. ET_0 is mainly influenced by meteorological factors and can be calculated by the Penman-Monteith equation, and K_c is mainly influenced by the crop type, its growing stage, and climate factors. Allen et al. [2] used a piecewise line to depict the variation of K_c with crop development and gave tabulated values of K_c in the initial and midseason stages and at the end of the late growing stage for main crops. However, this generalized crop coefficient curve and these tabulated values of K_c represent general crop conditions and cannot fully consider differences of crop varieties and local environment.

On the other hand, crop coefficient and crop water requirement had been studied for main crops in many places over the world. For example, an irrigation experiment network with over 100 stations had studied crop water requirement in China during the past several decades, and results of crop coefficient and crop water requirement for main crops were available at the temporal scale of growing stage [3]. These results can be spatially interpolated to obtain crop coefficient and crop water requirement of growing stages in unmeasured sites. However, in agricultural water management and regional water resources planning, crop water requirement in different time scales is usually necessary, such as monthly, weekly, or even daily scales. Therefore, it is necessary to downscale or disaggregate crop water requirement from growing stages to values in shorter periods.

Since crop evapotranspiration is usually measured over specified growing periods, it belongs to flow variables [4]. Temporal disaggregation methods for flow variables have been widely used in many disciplines, such as hydrology

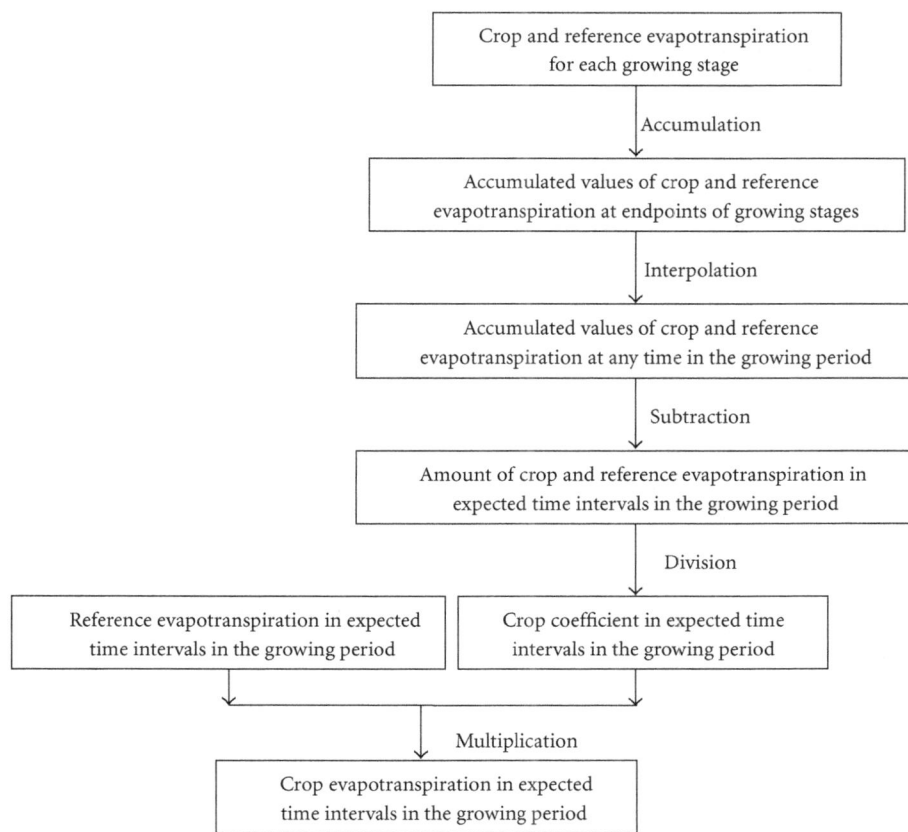

FIGURE 1: Proposed procedure to temporally downscale crop coefficient and crop water requirement.

[5, 6], meteorology [7, 8], and economics [9, 10]. Main temporal disaggregation methods include pure mathematical methods without auxiliary information [11], time series analysis models [5, 12], regression models [13], and dynamic models [14]. *Eurostat* [5], *Feijoó* et al. [9] and *Bojilova* [10] compared main methods for temporal disaggregation. However, these methods usually require a large number of data [12] or depend on assumptions on the smoothness of the flow variables and/or some plausible minimizing criteria [11].

In this paper, we proposed a method to downscale crop coefficient and crop water requirement from growing stage to substage scales using crop evapotranspiration data in growing stages and reference evapotranspiration data in both stages and substages. This method was compared with two direct interpolation methods and tested with a simulated crop evapotranspiration series of winter wheat at Xiaohe Station in North China.

2. Method to Temporally Downscale Crop Coefficient and Crop Water Requirement

For flow variables whose values are measured over specified periods, interpolation methods are not applicable directly to estimate their values over unknown periods or points. To use direct interpolation method, one straightforward method is to assume that stage average values occurred in the middle of the stages [7]. However, this assumption is valid for linear case and usually invalid for nonlinear cases. For crop evapotranspiration, direct interpolation of crop evapotranspiration or crop coefficient can be used following the above linear assumption.

To avoid the unrealistic linear assumption, an alternative method to temporally downscale crop coefficient and crop water requirement was proposed, which was based on the interpolation of accumulated crop and reference evapotranspiration. This method used the interpolation of accumulated evapotranspiration rather than evapotranspiration itself. Considering that the accumulated crop evapotranspiration is a stock variable, it can be used for direct interpolation to approximate accumulated crop evapotranspiration at any time of the growing period. Then values of crop evapotranspiration in any expected periods can be estimated. Considering the match of crop and reference evapotranspiration in calculating the crop coefficient, reference evapotranspiration at growing stage scale is also disaggregated similarly. Procedure of this method is shown in Figure 1.

Suppose that the whole or part crop growing period is divided into n stages, the duration of stage i ($i = 1, 2, \ldots, n$) is Δt_i, and measured or simulated crop evapotranspiration (crop water requirement) during stage i is ET_{ci}. Reference evapotranspiration in both growing stages (ET_{0i}) and sub-stages can be calculated by the Penman-Monteith equation [2] with monitored meteorological data, which is used as an auxiliary variable to downscale crop coefficient and crop evapotranspiration. The accumulated crop and reference evapotranspiration at the beginning and end of growing stages can be calculated from crop and reference evapotranspiration in growing stages with (1) and (2), respectively:

$$\text{AET}_c(t_0) = 0,$$

$$\text{AET}_c(t_i) = \sum_{k=1}^{i} \text{ET}_{ck} = \text{AET}_c(t_{i-1}) + \text{ET}_{ci}, \quad i = 1, 2, \ldots, n,$$

(1)

$$\text{AET}_0(t_0) = 0,$$

$$\text{AET}_0(t_i) = \sum_{k=1}^{i} \text{ET}_{0k} = \text{AET}_0(t_{i-1}) + \text{ET}_{0i}, \quad i = 1, 2, \ldots, n,$$

(2)

where $\text{AET}_c(t_i)$ and $\text{AET}_0(t_i)$ are accumulated crop and reference evapotranspiration at time t_i, and t_i is defined as

$$t_0 = 0, \qquad t_i = \sum_{k=1}^{i} \Delta t_k = t_{i-1} + \Delta t_i, \quad i = 1, 2, \ldots, n. \quad (3)$$

Interpolation polynomials or piecewise polynomials [15], $S_c(t)$ and $S_0(t)$, can be constructed to approximate accumulated crop and reference evapotranspiration using interpolation conditions of (1) and (2), respectively. When choosing appropriate type of interpolation polynomials or piecewise polynomials, the number of interpolation nodes ($n + 1$) and the characters of the accumulated evapotranspiration should be considered. In general, polynomials are appropriate for smaller n, and piecewise polynomials for larger n. Moreover, interpolation functions for accumulated crop and reference evapotranspiration should be strictly increasing functions since values for crop and reference evapotranspiration are always positive. Therefore, the monotone piecewise cubic interpolation method proposed by Fritsch and Carlson (1980) [16] is appropriate for larger n.

From $S_c(t)$ and $S_0(t)$, the amount of crop and reference evapotranspiration over expected time interval $[t_u, t_v]$ ($0 \leq t_u < t_v \leq t_n$) in the growing period can be estimated with (4) and (5), respectively:

$$\text{ET}_c^i(t_u, t_v) = S_c(t_v) - S_c(t_u), \quad 0 \leq t_u < t_v \leq t_n, \quad (4)$$

$$\text{ET}_0^i(t_u, t_v) = S_0(t_v) - S_0(t_u), \quad 0 \leq t_u < t_v \leq t_n, \quad (5)$$

where superscript i refers to interpolated values. Specially, daily crop and reference evapotranspiration in the day of t_u ($t_u = 1, 2, \ldots, n$) can be estimated with (6) and (7), respectively:

$$\text{ET}_c^i(t_u) = S_c(t_u) - S_c(t_u - 1), \quad t_u = 1, 2, \ldots, n, \quad (6)$$

$$\text{ET}_0^i(t_u) = S_0(t_u) - S_0(t_u - 1), \quad t_u = 1, 2, \ldots, n. \quad (7)$$

Since only crop and reference evapotranspiration in limited growing stages were available for interpolation, fluctuations of evapotranspiration in higher frequencies were smoothed out. Therefore, the disaggregated evapotranspiration varies smoothly over time, which represents the trend of evapotranspiration processes.

Using interpolated crop and reference evapotranspiration in substage periods, crop coefficient can be calculated as the ratio of crop to reference evapotranspiration, that is,

$$K_c^d = \frac{\text{ET}_c^i}{\text{ET}_0^i}, \quad (8)$$

where K_c^d is the downscaled crop coefficient in a substage period, and ET_c^i and ET_0^i are interpolated crop and reference evapotranspiration in the same substage period.

Owing to the smoothness of disaggregated crop and reference evapotranspiration, the variation of downscaled crop coefficient is also smooth over time. To reflect the fluctuation of crop evapotranspiration over growing period, downscaled crop evapotranspiration (ET_c^d) can be calculated with the product of actual reference evapotranspiration and the downscaled crop coefficient with (9):

$$\text{ET}_c^d = K_c^d \text{ET}_0. \quad (9)$$

3. Data and Evaluation Criteria for the Downscaling Method

To evaluate the effectiveness of the proposed method to temporally downscale crop coefficient and crop water requirement, a daily crop evapotranspiration series without soil water stress to crop simulated with a soil water balance model [17] was used. This series includes 110 values of daily crop evapotranspiration of winter wheat from March 6 (greening) to June 23 (harvesting) in 2003 at Xiaohe Experiment Station in Shanxi province in North China.

These 110 values were aggregated to 4 values of crop evapotranspiration in four stages (Table 1), which were then downscaled to daily scale to evaluate the performance of different disaggregation methods.

Besides the proposed method, direct interpolation of crop evapotranspiration and crop coefficient by assuming that stage average values occurred in the middle of the stage were also used for comparison (Table 2).

When actual values of crop evapotranspiration in substage periods are available from measurement or simulation, the scatter plot of original (ET_c^o) versus downscaled (ET_c^d) crop evapotranspiration can be used to evaluate the performance of the downscaling method through visual inspection and regression analysis. If most of the scatter points are close to the $1:1$ line, then the agreement between original and downscaled crop evapotranspiration is more likely to be good. The linear regression equation between ET_c^o and ET_c^d can be expressed as

$$\text{ET}_c^d = a + b\text{ET}_c^o + e, \quad (10)$$

where a and b are the intercept and slope of the regression line, respectively, and e is the residual. The slope b,

TABLE 1: Values of reference (ET_0) and crop (ET_c) evapotranspiration and crop coefficient (K_c) in growing stages of winter wheat after greening in 2003.

Growing stage	Starting date	Ending date	Duration (d)	ET$_0$ (mm)	ET$_c$ (mm)	K_c
Greening	March 6	April 4	30	67.6	27.4	0.40
Jointing	April 5	May 10	36	143.9	145.6	1.01
Heading	May 11	May 31	21	102.3	127.6	1.25
Milking	June 1	June 23	23	115.8	109.0	0.94
Greening to harvest	March 6	June 23	110	429.6	409.6	0.95

TABLE 2: Disaggregation methods for crop coefficient (K_c) and crop evapotranspiration (ET$_c$).

No.	Disaggregation method	Input	Output
1	Interpolation of crop evapotranspiration with the linear assumption	ET$_c$ in growing stages	ET$_c$ in substages
2	Interpolation of crop coefficient with the linear assumption	K_c in growing stages and ET$_0$ in substages	ET$_c$ and K_c in substages
3	Interpolation of accumulated crop and reference evapotranspiration	ET$_c$ and ET$_0$ in growing stages and ET$_0$ in substages	ET$_c$ and K_c in substages

the intercept a, and the Pearson product moment correlation coefficient r can be calculated from

$$b = \frac{\sum_{j=1}^{N}\left(\mathrm{ET}_{c,j}^{o} - \overline{\mathrm{ET}_{c}^{o}}\right)\left(\mathrm{ET}_{c,j}^{d} - \overline{\mathrm{ET}_{c}^{d}}\right)}{\sum_{j=1}^{N}\left(\mathrm{ET}_{c,j}^{o} - \overline{\mathrm{ET}_{c}^{o}}\right)^2}, \quad (11)$$

$$a = \overline{\mathrm{ET}_{c}^{d}} - b\overline{\mathrm{ET}_{c}^{o}}, \quad (12)$$

$$r = \frac{\sum_{j=1}^{N}\left(\mathrm{ET}_{c,j}^{o} - \overline{\mathrm{ET}_{c}^{o}}\right)\left(\mathrm{ET}_{c,j}^{d} - \overline{\mathrm{ET}_{c}^{d}}\right)}{\left[\sum_{j=1}^{N}\left(\mathrm{ET}_{c,j}^{o} - \overline{\mathrm{ET}_{c}^{o}}\right)^2 \sum_{j=1}^{N}\left(\mathrm{ET}_{c,j}^{d} - \overline{\mathrm{ET}_{c}^{d}}\right)^2\right]^{1/2}}, \quad (13)$$

where N is the number of data points, and $\overline{\mathrm{ET}_{c}^{o}}$ and $\overline{\mathrm{ET}_{c}^{d}}$ are average values of the original and downscaled crop evapotranspiration, respectively. For perfect downscaling results, a would be 0, and b and r would be 1.

Relative volume error (RVE) and root mean squared error (RMSE) are also two widely used indexes for model evaluation. RVE represents the volume conservative characteristic of a disaggregation method, while RMSE represents the level of overall agreement between the original and downscaled crop evapotranspiration. They are defined as

$$\mathrm{RVE} = \frac{\sum_{j=1}^{N}\left(\mathrm{ET}_{c,j}^{o} - \mathrm{ET}_{c,j}^{d}\right)}{\sum_{j=1}^{N}\mathrm{ET}_{c,j}^{o}}, \quad (14)$$

$$\mathrm{RMSE} = \left[\frac{1}{N}\sum_{j=1}^{N}\left(\mathrm{ET}_{c,j}^{d} - \mathrm{ET}_{c,j}^{o}\right)^2\right]^{1/2}. \quad (15)$$

Both RVE and RMSE would be 0 for perfect downscaling results.

4. Results and Discussions

4.1. Crop Coefficient. Crop coefficients in substage periods can be estimated with methods 2 and 3 in Table 2. For method 3, the accumulated values of crop and reference evapotranspiration were calculated from their values in four growing stages and used to construct interpolation polynomials. Considering that the numbers of accumulated values were five, quartic polynomials with constants equaling to 0 were used. Interpolation polynomials (Figure 2) for accumulated crop and reference evapotranspiration were (16) and (17), respectively:

$$\mathrm{AET}_{c}^{i} = -918.135t'^{4} + 1346.967t'^{3} - 52.342t'^{2} + 33.099t', \quad (16)$$

$$\mathrm{AET}_{0}^{i} = -24.764t'^{4} - 127.654t'^{3} + 445.714t'^{2} + 136.390t, \quad (17)$$

where $t' = t/110$ is the relative time, and t is days after greening. These two interpolation polynomials are appropriate since they are both strictly increasing functions.

From Figure 2, reference evapotranspiration tends to increase from greening (early March) to harvesting (late June), while crop evapotranspiration increases slowly in greening and milking stages and quickly in jointing and heading stages. Consequently, the increase of accumulated reference evapotranspiration is superlinear, while the variation of accumulated crop evapotranspiration follows a sigmoid pattern. The interpolation polynomials in Figure 2 were used to calculate reference and crop evapotranspiration in any time intervals during the crop growing period using (4) and (5). Then crop coefficients in these intervals were calculated with (8), and downscaled daily crop coefficient is shown in Figure 3.

TABLE 3: Evaluation of different disaggregation methods.

Method	Total ET_c (mm)	REV	RMSE (mm/d)	a	b	r
1	406.5	0.0075	1.12	0.939	0.740	0.865
2	404.3	0.0129	0.22	0.112	0.957	0.996
3	406.8	0.0068	0.16	−0.010	0.996	0.998
Perfect value	409.6	0	0	0	1	1

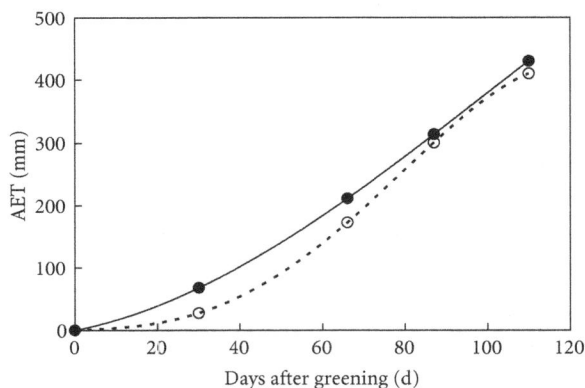

FIGURE 2: Accumulated values of reference (dot) and crop (circle) evapotranspiration and corresponding interpolation polynomials (solid and dashed lines) from greening to harvesting of winter wheat in 2003.

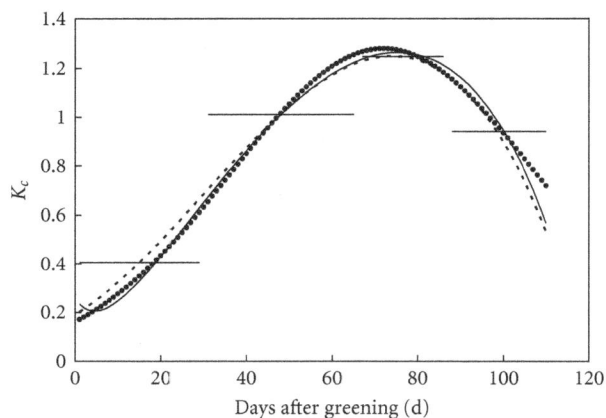

FIGURE 3: Daily crop coefficient (dot), average ones at different growing stages (thick solid line), and disaggregated ones with methods 2 (dashed line) and 3 (solid line).

For method 2, crop coefficients in the middle of growing stages were assumed to be the stage average values, and crop coefficients in substage periods were estimated with the corresponding interpolation polynomial (Figure 3).

Disaggregated daily crop coefficients were compared with the original and stage average values (Figure 3). In early growing periods, method 2 tends to overestimate the crop coefficient. In final growing periods, both methods 2 and 3 underestimate the crop coefficient. In general, disaggregated daily crop coefficients are both acceptable compared with the original ones, and the results of method 3 are slightly superior to those of method 2.

4.2. Crop Evapotranspiration. Crop evapotranspiration can be disaggregated with all three methods in Table 2. Figure 4 shows the variations of original and disaggregated crop evapotranspiration on a daily basis. Disaggregated results of methods 1 vary smoothly over time as expected, which represents the smoothed trend of original daily evapotranspiration. On the other hand, disaggregated results of methods 2 and 3 are very close. Moreover, their fluctuations are very similar to original simulated ones, since daily reference evapotranspiration is used in these two disaggregation methods.

Values of evaluation criteria for these three disaggregation methods are listed in Table 3. Interpolation of crop evapotranspiration with the linear assumption (Method 1) gave the worst results for all evaluation criteria except REV, while interpolation of accumulated crop and reference evapotranspiration (method 3) gave the best results. Therefore,

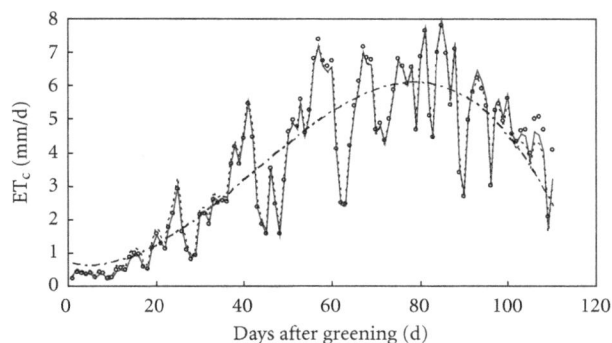

FIGURE 4: Original daily crop evapotranspiration (circle) and disaggregated ones with methods 1 (dash-dotted line), 2 (dashed line), and 3 (solid line).

method 3 is the most appropriate method to disaggregate crop coefficient and crop evapotranspiration from stage to substage scales. Interpolation of crop coefficient with the linear assumption (Method 2) is also acceptable.

5. Conclusions

We proposed a method to estimate crop coefficient and crop water requirement in substage scales from crop evapotranspiration in growing stages and reference evapotranspiration in growing stages and substages. In this method, accumulated values of crop and reference evapotranspiration calculated from their values in growing stages were used to estimate crop and reference evapotranspiration in substage periods through interpolation, which were then used to obtain the crop coefficient in substage periods. The crop water requirement in substage periods was estimated with downscaled crop coefficient and the reference evapotranspiration at substage scale.

The method was tested with a simulated daily crop evapotranspiration series. Results indicate that the variation of crop coefficient is close to the simulated one, and the downscaled crop evapotranspiration series is very close to the original one and superior to two direct interpolation methods.

Since crop coefficient and crop water requirement in growing stages for main crops are available in many irrigation experiment stations [3], they can be used to obtain crop coefficient and crop water requirement in shorter time interval using the present downscaling method. The downscaled values are more useful in agricultural water management and regional water resources planning.

Acknowledgments

This work was supported by the National Natural Science Foundation of China (Grant no. 50939004) and the National Key Technology R&D Program of China (Grant no. 2011BAD25B05).

References

[1] L. S. Pereira and I. Alves, "Crop water requirements," in *Encyclopedia of Soils in the Environment*, D. Hillel, Ed., pp. 322–334, Elsevier, Oxford, UK, 2005.

[2] R. G. Allen, L. S. Pereira, D. Raes, and M. Smith, *Crop Evapotranspiration—Guidelines for Computing Crop Water Requirements*, FAO, Rome, Italy, 1998.

[3] Research Group for Isograms of Crop Water Requirement for Main Crops in China, *Study on Isograms of Crop Water Requirement for Main Crops in China*, China Agriculture Press, Beijing, China, 1993.

[4] A. Ford, *Modeling the Environment*, Island Press, Washington, DC, USA, 2nd edition, 2009.

[5] E. K. Bojilova, "Disaggregation modelling of spring discharges," *International Journal of Speleology*, vol. 33, pp. 65–72, 2004.

[6] J. Zhang, R. R. Murch, M. A. Ross, A. R. Ganguly, and M. Nachabe, "Evaluation of statistical rainfall disaggregation methods using Rain-Gauge information for West-Central Florida," *Journal of Hydrologic Engineering*, vol. 13, no. 12, pp. 1158–1169, 2008.

[7] J. D. Wu and F. T. Wang, "Study on the creation of daily variation scenarios with a stochastic weather generator and various interpolations," *Quarterly Journal of Applied Meteorology*, vol. 11, no. 2, pp. 129–136, 2000.

[8] B. Debele, R. Srinivasan, and J. Yves Parlange, "Accuracy evaluation of weather data generation and disaggregation methods at finer timescales," *Advances in Water Resources*, vol. 30, no. 5, pp. 1286–1300, 2007.

[9] Eurostat, *Handbook on Quarterly National Accounts*, Office for Official Publications of the European Communities, Luxembourg, 1999.

[10] S. Rodríguez Feijoó, A. Rodríguez Caro, and D. D. Dávila Quintana, "Methods for quarterly disaggregation without indicators; a comparative study using simulation," *Computational Statistics and Data Analysis*, vol. 43, no. 1, pp. 63–78, 2003.

[11] G. Gudmundsson, "Estimation of continuous flows from observed aggregates," *Journal of the Royal Statistical Society Series D*, vol. 50, no. 3, pp. 285–293, 2001.

[12] W. W. S. Wei and D. O. Stram, "Disaggregation of time series models," *Journal of the Royal Statistical Society B*, vol. 52, pp. 453–467, 1990.

[13] G. Chow and A. L. Lin, "Best linear unbiased interpolation, distribution and extrapolation of time series by related series," *The Review of Economics and Statistics*, vol. 53, pp. 372–375, 1971.

[14] T. Di Fonzo, *Temporal Disaggregation of Economic Time Series: Towards a Dynamic Extension*, Office for Official Publications of the European Communities, Luxembourg, 2003.

[15] J. L. Chen and Q. Y. Li, "Interpolation," in *Computation and Numerical Analysis*, Editorial Board for Modern Applied mathematics, Ed., pp. 24–85, Tsinghua University Press, Beijing, China, 2005.

[16] F. N. Fritsch and R. E. Carlson, "Monotone piecewise cubic interpolation," *SIAM Journal on Numerical Analysis*, vol. 17, pp. 238–246, 1980.

[17] S. Shang and X. Mao, "Application of a simulation based optimization model for winter wheat irrigation scheduling in North China," *Agricultural Water Management*, vol. 85, no. 3, pp. 314–322, 2006.

3

Effect of Growth Stage on the Efficacy of Postemergence Herbicides on Four Weed Species of Direct-Seeded Rice

Bhagirath Singh Chauhan and Seth Bernard Abugho

Crop and Environmental Sciences Division, International Rice Research Institute, Los Baños, Philippines

Correspondence should be addressed to Bhagirath Singh Chauhan, b.chauhan@irri.org

Academic Editor: David W. Archer

The efficacy of bispyribac-sodium, fenoxaprop + ethoxysulfuron, and penoxsulam + cyhalofop was evaluated against barnyard-grass, Chinese sprangletop, junglerice, and southern crabgrass when applied at four-, six-, and eight-leaf stages. When applied at the four-leaf stage, bispyribac-sodium provided greater than 97% control of barnyardgrass, junglerice, and southern crabgrass; however, it was slightly weak (74% control) on Chinese sprangletop. Irrespective of the weed species, fenoxaprop + ethoxysulfuron provided greater than 97% control when applied at the four-leaf stage. At the same leaf stage, penoxsulam + cyhalofop controlled 89 to 100% barnyardgrass, Chinese sprangletop, and junglerice and only 54% of southern crabgrass. The efficacy of herbicides was reduced when applied at the eight-leaf stage of the weeds; however, at this stage, fenoxaprop + ethoxysulfuron was effective in controlling 99% of Chinese sprangletop. The results demonstrate the importance of early herbicide application in controlling the weeds. The study identified that at the six-leaf stage of the weeds, fenoxaprop + ethoxysulfuron can effectively control Chinese sprangletop and southern crabgrass, penoxsulam + cyhalofop can effectively control Chinese sprangletop, and bispyribac-sodium can effectively control junglerice.

1. Introduction

Rice is an important crop in Asia, where 90% of this crop is grown and consumed. Rice is traditionally grown by transplanting seedlings into puddled soil. Recently, due to high labor cost and less availability of water, there has been a trend to shift from transplanting to direct-seeded rice (DSR) in many Asian countries [1]. However, weeds are a greater problem in DSR than in transplanted rice because of the absence of the crop seedling size advantage and standing water at the time of crop emergence [2].

Many weeds, including barnyardgrass (*Echinochloa crus-galli* (L.) Beauv.), Chinese sprangletop (*Leptochloa chinensis* L.), junglerice (*Echinochloa colona* (L.) Link), and southern crabgrass (*Digitaria ciliaris* (Retz.) Koel.) are problematic grass species in DSR [3, 4]. Barnyardgrass and junglerice are examples of "crop mimicry" as they closely resemble rice at the seedling stage. By the time these weeds can be easily recognized by farmers, crop yield losses may already be inevitable [5]. In DSR, junglerice and southern crabgrass

seedling emergence was greater in a no-till system compared with a conventionally tilled system [6]. In another study, rice residue of up to $4\,\mathrm{Mg\,ha^{-1}}$ was not able to reduce the growth of barnyardgrass [7], suggesting that the crop residue at this amount, as a mulch on the soil surface, may not provide suppression of this weed. Barnyardgrass at a density of $9\,\mathrm{plants\,m^{-2}}$ can reduce rice yield by more than 50% [8] and heavy infestation of this weed can remove up to 80% of the nitrogen (N) from the soil [9]. In a recent study, Chinese sprangletop grown under rice interference responded with increased leaf area ratio (amount of leaf area per unit plant dry biomass) and specific leaf area (amount of leaf area per unit leaf biomass), reflecting changes in the distribution of leaf biomass under competition [10]. Furthermore, Chinese sprangletop was not a prevalent and dominant weed in rice fields of Malaysia while transplanting of rice was the usual establishment method, but it became widespread with the shift to DSR [11]. These studies suggest the importance of these weed species in DSR.

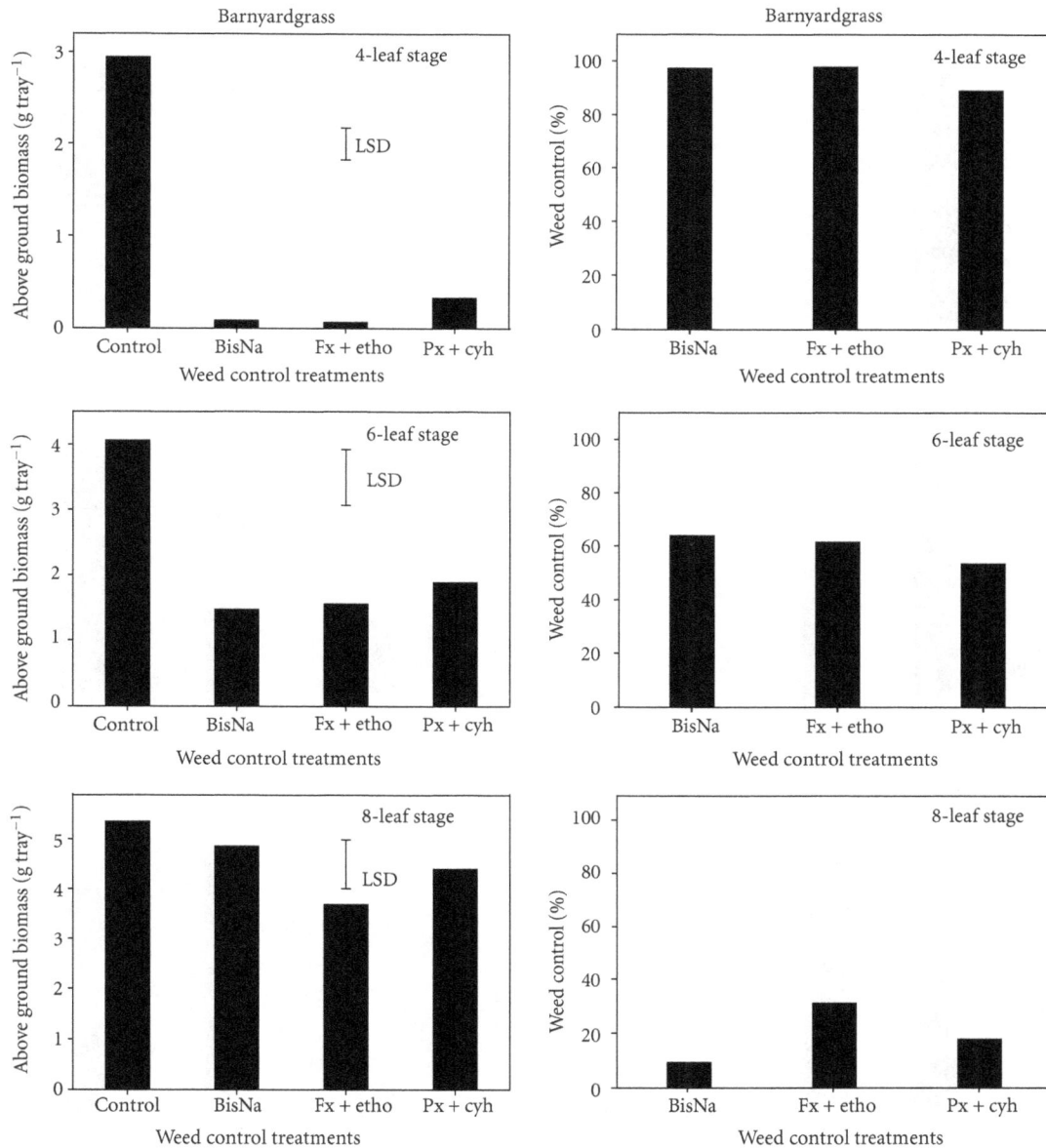

FIGURE 1: Effect of postemergence herbicides (BisNa, bispyribac-sodium; Fx + etho, fenoxaprop-p-ethyl + ethoxysulfuron; Px + cyh, penoxsulam + cyhalofop) on aboveground dry biomass (g) and control (%) of barnyardgrass when sprayed at its four-, six-, and eight-leaf stages. Mean comparisons were performed based on least significant difference test at 5%.

Postemergence herbicides are a major tool used to control weeds in DSR. The growth stage of weed species may have an effect on herbicide efficacy by influencing uptake and metabolism of herbicides [12]. Diclofop, for example, was more effective on green foxtail (*Setaria viridis* (L.) Beauv.) and wild oat (*Avena fatua* L.) when applied at an early growth stage [13]. Conversely, trifloxysulfuron was more effective on yellow nutsedge (*Cyperus esculentus* L.) at late application stages [12]. Generally, the herbicide efficacy is lower when applied on bigger weeds. The herbicide degradation rate may be faster in big plants, and herbicide rates may need to be increased to achieve the same level of control [12]. In addition, reliance on a single herbicide may result in evolution of herbicide resistance in weeds and shift in weed flora. In Sri Lanka, for example, continuous use of bispyribac-sodium to control propanil-resistant barnyardgrass has resulted in a shift to dominance by Chinese sprangletop in rice [14]. There are reports from India that Chinese sprangletop is poorly controlled by bispyribac-sodium [15]. Therefore, optimum time of herbicide application and range of herbicides may help control these weeds effectively.

The objective of this study was to evaluate the efficacy of different postemergence herbicides on barnyardgrass, Chinese sprangletop, junglerice, and southern crabgrass when applied at their different growth stages.

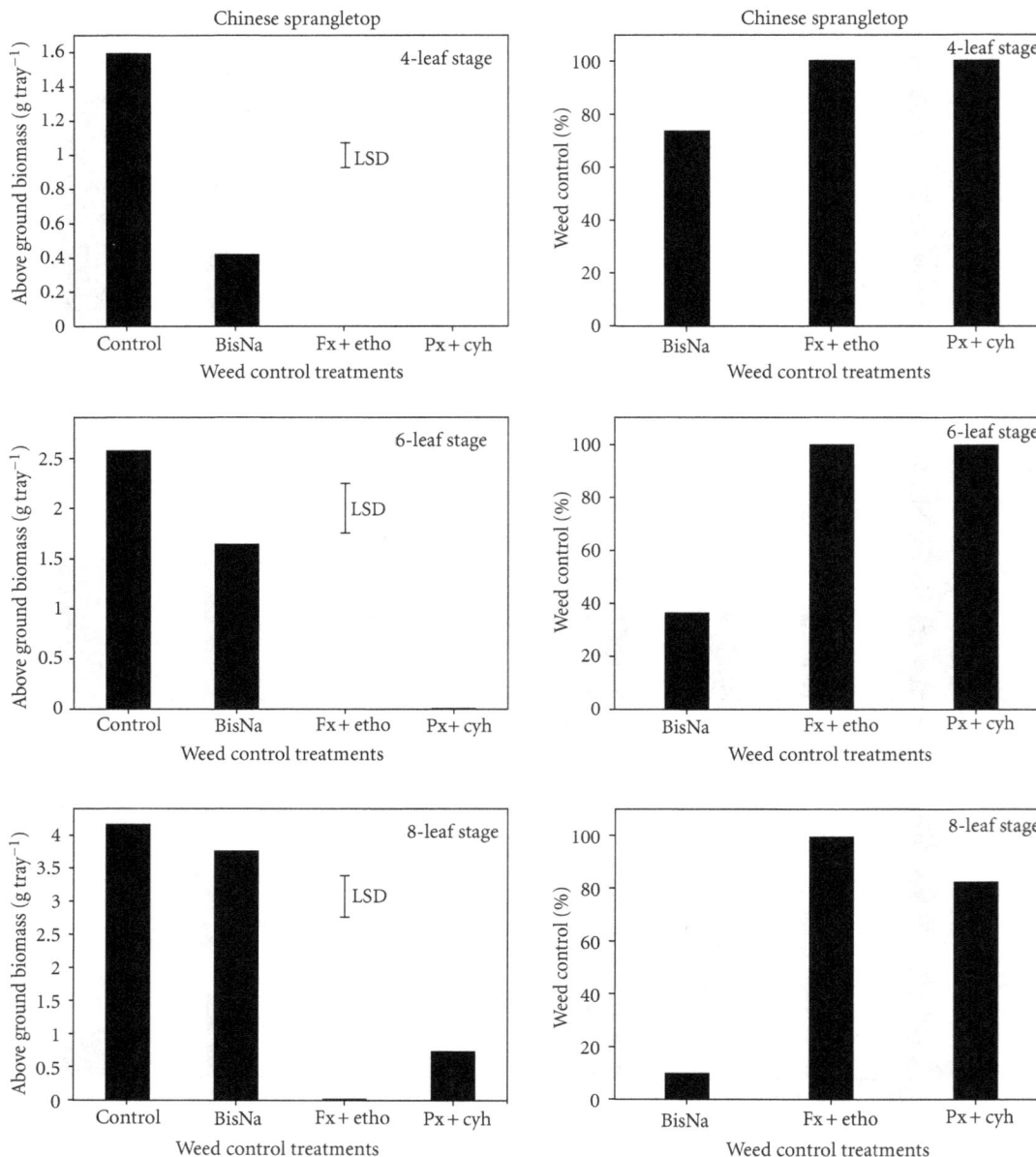

FIGURE 2: Effect of postemergence herbicides (BisNa, bispyribac-sodium; Fx + etho, fenoxaprop-p-ethyl + ethoxysulfuron; Px + cyh, penoxsulam + cyhalofop) on aboveground dry biomass (g) and control (%) of Chinese sprangletop when sprayed at its four-, six-, and eight-leaf stages. Mean comparisons were performed based on least significant difference test at 5%.

2. Materials and Methods

2.1. Experimental Details. Greenhouse experiments were conducted at the International Rice Research Institute, Los Baños, Philippines. Twenty-five seeds, each of barnyardgrass, Chinese sprangletop, junglerice, and southern crabgrass were planted on the soil surface in plastic trays (8 cm by 8 cm by 5 cm). Soil used in this study was collected from upland rice fields and it had a pH of 6.6 with 31% sand, 37% silt, and 32% clay. Seedlings were thinned to 10 plants per tray immediately after emergence. No fertilizer was applied to the weeds.

Plants were sprayed at the four-, six-, and eight-leaf stages using a research track sprayer that delivered $210 \, L \, ha^{-1}$ spray solution at a spray pressure of 140 kPa. Flat nozzles (Teejet 80015) were used in the sprayer. The postemergence herbicide treatments included bispyribac-sodium at $30 \, g \, ai \, ha^{-1}$ and commercial mixtures of fenoxaprop-p-ethyl + ethoxysulfuron at $45 \, g \, ai \, ha^{-1}$, and penoxsulam + cyhalofop at $72 \, g \, ai \, ha^{-1}$. There were control treatments for each leaf stage and weed species in which herbicides were not sprayed.

Plants were watered daily such that there was no water stress in the plants. Seedling survival was determined 14 d

TABLE 1: Effect of postemergence herbicides on seedling survival (%) of barnyardgrass, Chinese sprangletop, junglerice, and southern crabgrass when sprayed at their four-, six-, and eight-leaf stage.

Weed control treatment	Rate	Seedling survival		
		Four-leaf	Six-leaf	Eight-leaf
	$g\,ha^{-1}$		%	
Barnyardgrass				
Untreated control	0	100	100	100
Bispyribac-sodium	30	19	94	100
Fenoxaprop-p-ethyl + ethoxysulfuron	45	13	56	80
Penoxsulam + cyhalofop	72	63	83	100
$LSD_{0.05}$		17	22	12
Chinese sprangletop				
Untreated control	0	100	100	100
Bispyribac-sodium	30	91	99	100
Fenoxaprop-p-ethyl + ethoxysulfuron	45	0	1	1
Penoxsulam + cyhalofop	72	0	2	61
$LSD_{0.05}$		13	3	17
Junglerice				
Untreated control	0	100	100	100
Bispyribac-sodium	30	4	0	48
Fenoxaprop-p-ethyl + ethoxysulfuron	45	6	41	89
Penoxsulam + cyhalofop	72	16	53	89
$LSD_{0.05}$		11	20	18
Southern crabgrass				
Untreated control	0	100	100	100
Bispyribac-sodium	30	9	82	100
Fenoxaprop-p-ethyl + ethoxysulfuron	45	0	7	65
Penoxsulam + cyhalofop	72	91	100	100
$LSD_{0.05}$		14	15	17

after herbicide application with the criterion of at least one green leaf on the plant. Aboveground biomass was measured after drying plant samples in an oven at 70°C for 72 h and expressed as percent control.

2.2. *Statistical Analyses.* The experiments with each weed species were conducted twice using a randomized complete block design with five replicates. Because there was no significant interaction between treatments and experiments, the data from the repeated experiments were pooled ($n = 10$) before being subjected to ANOVA (GenStat 8.0 2005). The data were analyzed separately for each leaf stage by using one-way ANOVA. Mean comparisons were performed based on least significant difference test at 0.05 probability.

3. Results and Discussion

3.1. *Barnyardgrass.* All herbicides had a phytotoxic effect on plant survival when applied at the early growth stage of barnyardgrass. Delayed herbicide application after the four-leaf stage increased the number of surviving plants, and all plants of barnyardgrass survived when bispyribac-sodium and penoxsulam + cyhalofop were applied at its eight-leaf stage (Table 1). Irrespective of the growth stage of barnyardgrass, there was no difference in efficacy of

these herbicides and the biomass and percent weed control decreased with the progress in weed growth (Figure 1). The application of all herbicides gave 89–98% weed control of barnyardgrass when applied at the four-leaf stage. However, delayed application from the four-leaf to the six-leaf stages reduced the control to 53–64%. Similarly, in a previous study in India, fenoxaprop + ethoxysulfuron ($150 + 18\,g\,ai\,ha^{-1}$) sprayed 21 d after sowing in dry-seeded rice gave 68% control of barnyardgrass [16]. In our study, very poor weed control (9–31%) of barnyardgrass (3.7–4.9 g biomass tray^{-1} in herbicide treatments versus 5.4 g biomass tray^{-1} in the untreated control) was achieved when these herbicides were applied at their eight-leaf stage.

3.2. *Chinese Sprangletop.* No (or very few) seedlings of Chinese sprangletop survived when fenoxaprop + ethoxysulfuron and penoxsulam + cyhalofop were applied at its four- and six-leaf stages (Table 1). However, when penoxsulam + cyhalofop were applied at the eight-leaf stage, more than 60% Chinese sprangletop seedlings survived (with at least one green leaf). Chinese sprangletop sprayed with bispyribac-sodium had more surviving plants than those sprayed with the other two herbicides, and this was true at all leaf stages. Bispyribac-sodium at any stage of Chinese sprangletop was not able to kill more than 10% seedlings (Table 1).

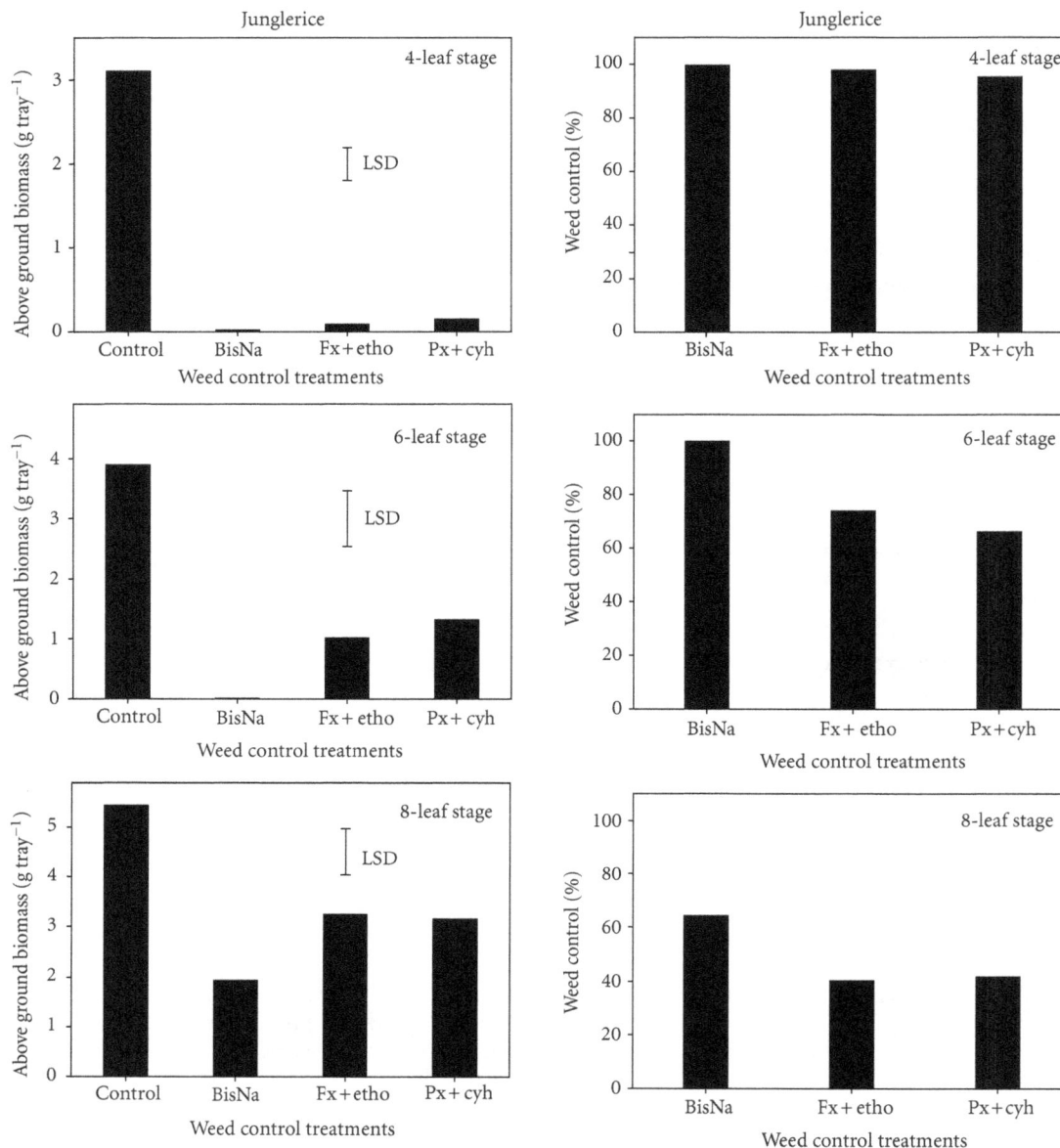

FIGURE 3: Effect of postemergence herbicides (BisNa, bispyribac-sodium; Fx + etho, fenoxaprop-p-ethyl + ethoxysulfuron; Px + cyh, penoxsulam + cyhalofop) on aboveground dry biomass (g) and control (%) of junglerice when sprayed at its four-, six-, and eight-leaf stages. Mean comparisons were performed based on least significant difference test at 5%.

Penoxsulam + cyhalofop gave more than 99% weed control when applied at the four-to-six-leaf stage (Figure 2). The control of Chinese sprangletop with this herbicide was slightly reduced (82% control) when application was delayed to the eight-leaf stage. Irrespective of leaf stage, fenoxaprop + ethoxysulfuron was very effective in controlling (>99%) Chinese sprangletop (Figure 2). On the other hand, the effect of bispyribac-sodium on this weed was poor and control decreased with progress in weed growth (74, 36, and 10% control at the four-, six-, and eight-leaf stages, resp.). This is consistent with reports from India where bispyribac-sodium was shown to be weak on Chinese sprangletop [15]. Penoxsulam alone was also not effective in controlling

Chinese sprangletop [15]; however, its commercial mixture with cyhalofop was able to control this weed effectively.

3.3. Junglerice. As observed in barnyardgrass, all herbicides had a phytotoxic effect on junglerice survival when applied at the early stage; however, delayed herbicide application increased the number of surviving plants (Table 1). As compared with the untreated control, the herbicides used in our study reduced more than 95% junglerice biomass when applied at its four-leaf stage (Figure 3). Delayed application of fenoxaprop + ethoxysulfuron and penoxsulam + cyhalofop from the four-to-six-leaf stage reduced junglerice control to 66–74%; however, bispyribac-sodium still gave greater

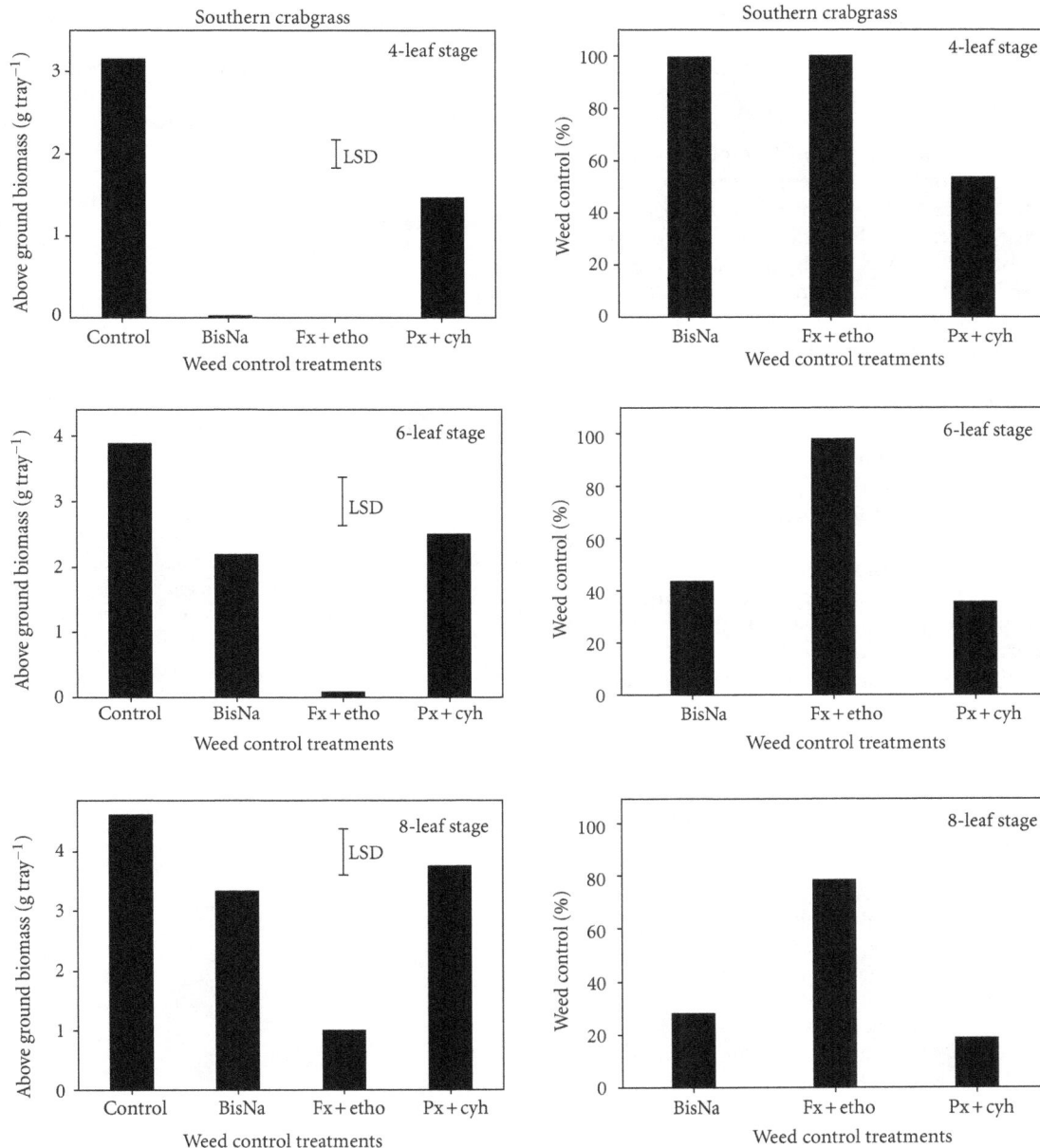

FIGURE 4: Effect of postemergence herbicides (BisNa, bispyribac-sodium; Fx+etho, fenoxaprop-p-ethyl + ethoxysulfuron; Px+cyh, penoxsulam + cyhalofop) on aboveground dry biomass (g) and control (%) of southern crabgrass when sprayed at its four-, six-, and eight-leaf stages. Mean comparisons were performed based on least significant difference test at 5%.

than 99% junglerice control at this stage. Recently, this herbicide has been widely used by farmers in India and Sri Lanka to control junglerice. Weed control by all herbicides was reduced to 40–64% when applied at the eight-leaf stage. Earlier, fenoxaprop + ethoxysulfuron (150 + 18 g ai ha^{-1}) sprayed 21 d after rice sowing gave 83% control of junglerice; however, the study did not mention the leaf stage of the weed [16].

3.4. Southern Crabgrass. Only a few plants (<10%) of southern crabgrass survived when fenoxaprop + ethoxysulfuron were applied at the four- or six-leaf stages and greater

than 90% plants survived when penoxsulam + cyhalofop were applied at these same stages (Table 1). Bispyribac and fenoxaprop + ethoxysulfuron gave more than 99% control of southern crabgrass when applied at the four-leaf stage (Figure 4). However, the efficacy of bispyribac was reduced markedly when application was delayed from four-to-six-leaf stage. Fenoxaprop + ethoxysulfuron was the most effective herbicide in controlling southern crabgrass at the six- (98%) and eight-leaf stage (78%). Penoxsulam + cyhalofop, on the other hand, did not effectively control southern crabgrass at any leaf stage: 54, 36, and 19% control at the four-, six-, and eight-leaf stage, respectively (Figure 4). This could be due to lower herbicide uptake translocation or to faster metabolism [12].

4. Conclusions

The results demonstrate the importance of early herbicide application in controlling the weeds. Generally, the effectiveness of herbicides was low on weeds when applied at their eight-leaf stage. The herbicide degradation rate or metabolism could be faster in big plants, thus herbicide rates may need to be increased to achieve the same level of control [12]. In situations such as rains, farmers may not be able to apply postemergence herbicides at the early stage. Our study identified that at the six-leaf stage of the weeds, fenoxaprop + ethoxysulfuron can effectively control Chinese sprangletop and southern crabgrass, penoxsulam + cyhalofop can effectively control Chinese sprangletop, and bispyribac-sodium can effectively control junglerice. It has been suggested that the control of barnyardgrass is improved with the addition of urea ammonium nitrate to bispyribac [17]. However, it is also possible that the interaction between herbicide and fertilizer may influence the amount of crop injury [3]. Fenoxaprop may cause injury on rice plants. Further research is therefore needed to understand the interaction of these postherbicides and nitrogen fertilizer on rice growth and yield.

References

[1] S. Pandey and L. Velasco, "Trends in crop establishment methods in Asia and research issues," in *Rice Is Life: Scientific Perspectives for the 21st Century*, Japan International Research Center for Agricultural Sciences, pp. 178–181, International Rice Research Institute and Tsukuba, Los Baños, Philippines, 2005.

[2] B. S. Chauhan and D. E. Johnson, "The role of seed ecology in improving weed management strategies in the tropics," *Advances in Agronomy*, vol. 105, pp. 221–262, 2009.

[3] B. S. Chauhan and D. E. Johnson, "Growth response of direct-seeded rice to oxadiazon and bispyribac-sodium in aerobic and saturated soils," *Weed Science*, vol. 59, no. 1, pp. 119–122, 2011.

[4] B. S. Chauhan, V. P. Singh, A. Kumar, and D. E. Johnson, "Relations of rice seeding rates to crop and weed growth in aerobic rice," *Field Crops Research*, vol. 121, no. 1, pp. 105–115, 2011.

[5] L. G. Holm, D. L. Plucknett, J. V. Pancho, and J. P. Herberger, *The World's Worst Weeds: Distribution and Biology*, The University Press of Hawaii, Malabar, Fla, USA, 1991.

[6] B. S. Chauhan and D. E. Johnson, "Influence of tillage systems on weed seedling emergence pattern in rainfed rice," *Soil and Tillage Research*, vol. 106, no. 1, pp. 15–21, 2009.

[7] B. S. Chauhan and D. E. Johnson, "Row spacing and weed control timing affect yield of aerobic rice," *Field Crops Research*, vol. 121, no. 2, pp. 226–231, 2011.

[8] M. A. Maun and S. C. H. Barrett, "The biology of Canadian weeds. 77. Echinochloa crus-galli (L.) Beauv," *Canadian Journal of Plant Science*, vol. 66, pp. 739–759, 1986.

[9] L. G. Holm, D. L. Plucknett, J. V. Pancho, and J. P. Herberger, *The World's Worst Weeds: Distribution and Biology*, University of Hawaii Press, Honolulu, Hawaii, USA, 1977.

[10] B. S. Chauhan and D. E. Johnson, "Ecological studies on Echinochloa crus-galli and the implications for weed management in direct-seeded rice," *Crop Protection*, vol. 30, no. 11, pp. 1385–1391, 2011.

[11] M. Azmi, D. V. Chin, P. Vongsaroj, and D. E. Johnson, "Emerging issues in weed management of direct-seeded rice in Malaysia, Vietnam, and Thailand," in *Rice Is Life: Scientific Perspectives for the 21st Century*, Japan International Research Center for Agricultural Sciences, pp. 196–198, International Rice Research Institute and Tsukuba, Los Baños, Philippines, 2005.

[12] S. Singh and M. Singh, "Effect of growth stage on trifloxysulfuron and glyphosate efficacy in twelve weed species of citrus groves," *Weed Technology*, vol. 18, no. 4, pp. 1031–1036, 2004.

[13] H. A. Friesen, P. A. O'Sullivana, and W. H. Vanden Born, "HOE 23408, a new selective herbicide for wild oats and green foxtail in wheat and barley," *Canadian Journal of Plant Science*, vol. 56, pp. 567–578, 1976.

[14] B. Marambe, "Emerging weed problems in wet-seeded rice due to herbicide use in Sri Lanka," in *Proceedings of the International Rice Congress*, p. 430, Beijing, China, 2002.

[15] R. Gopal, R. K. Jat, R. K. Malik et al., "Direct dry seeded rice production technology and weed management in rice-based systems," Tech. Rep., International Maize and Wheat Improvement Center, New Delhi, India, 2010.

[16] S. Singh, L. Bhushan, J. K. Ladha, R. K. Gupta, A. N. Rao, and B. Sivaprasad, "Weed management in dry-seeded rice (*Oryza sativa*) cultivated in the furrow-irrigated raised-bed planting system," *Crop Protection*, vol. 25, no. 5, pp. 487–495, 2006.

[17] C. H. Koger, D. M. Dodds, and D. B. Reynolds, "Effect of adjuvants and urea ammonium nitrate on bispyribac efficacy, absorption, and translocation in barnyardgrass (*Echinochloa crus-galli*). I. Efficacy, rainfastness, and soil moisture," *Weed Science*, vol. 55, no. 5, pp. 399–405, 2007.

4

Determination of Effective Factors on Power Requirement and Conveying Capacity of a Screw Conveyor under Three Paddy Grain Varieties

Ezzatollah Askari Asli-Ardeh[1] and Ahmad Mohsenimanesh[2]

[1] Department of Agricultural Machinery, College of Agriculture, University of Mohaghegh Ardabili, Ardabil, Iran
[2] Agricultural Engineering Research Institute, Shahid Fahmideh Boulevard, P.O. Box 31585-845, Karaj, Iran

Correspondence should be addressed to Ahmad Mohsenimanesh, mohsenimanesh@yahoo.com

Academic Editors: I. Watanabe and A. Welle

An experiment was conducted to investigate the effect of screw speed, inclination angle and variety on the required power, and conveying capacity of a screw conveyor. The experiment was designed with four levels of screw speed (600, 800, 1000, and 1200 rpm), five levels of inclination angle (0, 20, 40, 60, and 80°), and three levels of variety (Alikazemi, Hashemi, and Khazar). The Length, diameter, and pitch of screw were 2, 0.78, and 0.5 m, respectively. The experimental design was a randomized complete block (RCB) with factorial layout. Maximum and minimum power requirements of tested screw conveyor were 99.29 and 81.16 Watt corresponding to conveying capacity of 3.210 and 1.975 ton/hour obtained for khazar and Alikazemi varieties, respectively. The results indicated that as screw inclination angle increased from 0 to 80°, the conveying capacity decreased significantly from 3.581 to 0.932 t/h. It can be concluded that the most conveying capacity was 4.955 t/h at tests with khazar variety and conveyor inclination angle zero degree.

1. Introduction

Rice is one of the commonly consumed cereals and food staples for more than half of the world's population. It is an important source of energy, vitamins, mineral elements, and rare amino acids. World rice production increased from 520 million tons in 1990 to 605 million tons in 2004, while Iran's rice production increased from 1.3 million tons in 1980 to 3.4 million tons in 2004 [1].

Application of engineering principles for reducing energy requirement in the form of mechanical and electrical power is necessary to reduce cost of production (http://www.crri .nic.in/Research/Divisions/Engineering.htm). Efficient rice conveying implements are to be used to achieve energy savings. Screw Conveyers (augers) are used to convey free flowing materials such as grain to more difficult fibrous materials such as straw and alfalfa. At gain elevators and on the farmstead, augers are used to move grain to and from storage.

In these applications, the conveyor consists of a rotating shaft which carries a helicoid flighting through a stationary tube. Factors affecting capacity include auger dimen-sions (diameter, auger geometry), shear-plane flighting orientation, auger speed, angle of inclination, commodity being conveyed, and entrance-opening configuration [2].

Nicolai et al. [2] investigated the capacity, volumetric effi-ciency, and power requirements for a 20 cm and a 25 cm di-ameter conveyor and operating in a speed range of 250 to 1100 rpm at inclination angles of 13, 20, and 30°. They found that increasing the conveyor speed increased the capacity up to a maximum value, and further increases in speed caused a decrease in capacity. The screw speed for maximum capacity averaged 823 rpm for both 20 cm (8 in) and 25 cm (10 in) diameter of the conveyor. Volumetric efficiency decreased an average of 3% for every 100 rpm increase in conveyor screw speed. Changing the conveyor inclination angle did not affect the volumetric efficiency relative to the screw speed.

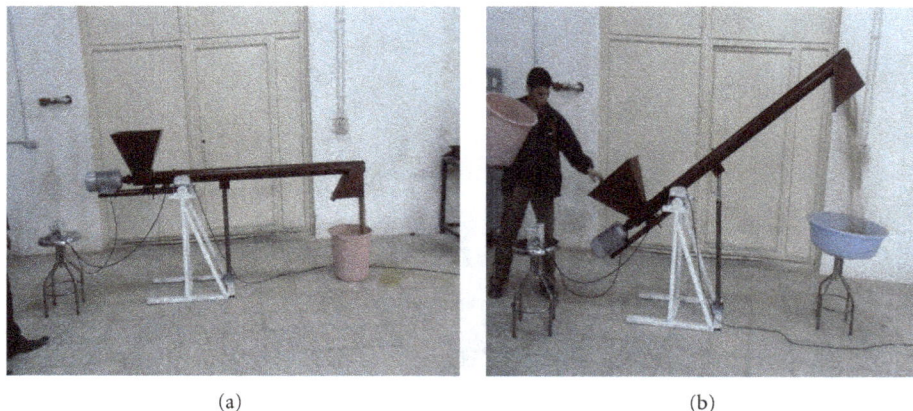

FIGURE 1: Tested screw conveyor at two conditions of (a) horizontal and (b) inclination angle of 40°.

Power requirements per conveyor screw speed were effected by inclination angle up to 20° but not for inclination angles greater than 20°.

Zareiforoush et al. [3, 4] designed and evaluated a screw conveyor for one rice variety, namely, Hashemi. They found that the specific power requirement of the screw conveyor increased with increasing the screw clearance and screw rotational speed. The net power requirement of the conveyor increased with increasing the screw rotational speed; whilst the value decreased with increasing the screw clearance. As the rotational speed of the screw conveyor increased, the actual volumetric capacity increased up to a maximum value and further increases in speed caused a decrease in capacity. With increasing the screw clearance and screw rotational speed, the volumetric efficiency of the screw conveyor decreased.

In recent times, performance-test procedure develops for screw conveyors designed for different agricultural grains products which characterize conveying capacity, volumetric efficiency, and power requirement. Since grain property variation is wide, especially when considering variety difference, rice cannot be considered to have uniform properties [5]. In other words, variation in rice-milling quality could be as a result of variation in its physical properties [6]. So, mechan-ical and physical properties of different rice varieties are important in optimum design of machinery for handling. However, a few screw conveyor concern different rice varieties type. Clearly, conveying capacity and requirement power of screw conveyor with respect to conveying materials characteristics are quite different from those of agricultural grains products. There are still improvements to be made for higher accuracy and reliability of screw conveyor. Therefore, an experiment was developed with the following objectives:

(1) to evaluate the effects of variety, screw rotational speed, and inclination angle on the requirement power and conveying capacity;

(2) to optimize conveying capacity and power requirement of the screw conveyor with respect to the conveying materials characteristics.

2. Materials and Methods

An experiment was conducted at Agricultural Machinery Department, Ardabil (Iran), a facility of University of Mohaghegh Ardabili, to investigate the effect of screw speed, inclination angle and variety on the required power and conveying capacity of a screw conveyor. The experiment was designed with four levels of screw speed (600, 800, 1000 and 1200 rpm), five levels of inclination angle (0, 20, 40, 60, and 80°), and three levels of variety (Alikasemi, Hashemi, and Khazar). The experimental design was a randomized complete block (RCB) with factorial layout at four replications. For means comparison, Duncan's Multiple Range Test was used.

The length of screw, screw pitch, and diameter were 2000, 50, and 78 mm, respectively. The clearance between screw and tube is 6 mm, which was developed by Iranian Ashtad cooperative company (Figure 1). A three-phase electromotor (3 HP) was used for driving the screw shaft. Moisture content of paddy Grain was 12 to 13% (w.b.) in all varieties.

An inverter, a rotational speed gauge, and a digital wattmeter were used for continuously determining the variable transmission of screw shaft, the screw rotational speed, and the electromotor electrical power, respectively. The wattmeter was connected to a personal computer for recording and displaying the power data during the tests so that power consumption variations were graphically visible on monitor during the experiments. Each test was started by pouring 30 kg paddy grain into the input section. After a few seconds of unloading grain at outlet section, a container was directed towards the discharge, and then discharged grain mass was weighted. The time of purring of paddy grain into the container was between 10 and 12 seconds. Conveying capacity data was recorded every second for future analyzing.

3. Results and Discussion

3.1. Power Requirement. The analysis of variance of power requirement and conveying capacity is presented in Table 1. The results showed that the main and interactions' effects

Determination of Effective Factors on Power Requirement and Conveying Capacity of a Screw Conveyor
under Three Paddy Grain Varieties

25

TABLE 1: Variance analysis of power requirement and convening capacity.

Variation sources	Freedom degree	Power requirement (watt)		Conveying capacity	
		Square mean	F ratio	Square mean	F ratio
Replication	3	153	2.930ns	0.113	1.274ns
Variety (V)	2	8635	165.451**	30.88	387.70**
Error	6	52		0.089	
Inclination angle (α.)	4	23626	260.637**	68.11	1275.37**
Interactions (V × α)	8	1762	19.43**	0.053	56.22**
Error	36	90		5.833	
Screw speed (N)	3	56790	610.08**	0.203	129.8**
Interactions (V × N)	6	390	4.196**	0.914	4.62**
Interactions (α × N)	12	1942	20.86**	0.515	20.34**
Interactions (V × α × N)	24	1575	16.92**	0.045	11.44**
Error	135	93			
Total	239	376298			

** Significant at less than 1% probability levels, ns: not significant.

TABLE 2: The mean comparisons of variety, screw speed, and inclination angle on power requirement and conveying capacity.

	Varieties		Screw speed (rpm)		Inclination angle (degree)	
	Alikazemi	81.16 b	600	46.93 d	0	127.8 a
	Hashemi	81.41 b	800	81.56 c	20	87.28 b
Power requirement (Watt)	Khazar	99.29 a	1000	118.0 a	40	18.59 e
			1200	102.7 b	60	28.11 d
					80	54.42 c
	Alikazemi	1.975 c	600	2.102 c	0	3.581 a
	Hashemi	2.484 b	800	2.621 b	20	3.532 a
Conveying capacity (t/h)	Khazar	3.210 a	1000	2.803 a	40	3.062 b
			1200	2.697 b	60	1.672 c
					80	0.932 d

Dissimilar letters show different significantly at probability level %5.

of variety, screw rotational speed, and inclination angle were significant ($P < 1\%$).

Means comparison results of the main effects revealed that power requirement of Khazar variety was more than others (Table 2). The reason may be due to the differences among the bulk density and friction coefficients of varieties. This is in agreement with the work of Ray [7] and Roberts [8], who found that power requirement varies with physical properties of agricultural grains. By increasing the screw speed from 600 to 1000 rpm, the power requirement was increased and a further increase up to 1200 rpm decreased the power requirement. The reason was that at speeds higher than 1000, centrifugal force prevents axial movement of the grains, and therefore the power requirement is reduced [9]. With increasing of inclination angle up to 40°, the power requirement decreased significantly because of the decrease in the conveying capacity. As inclination angle increased from 40 to 80°, power requirement increased. The reason may be due to paddy grains returning and compacting into the tube. Therefore, friction force and even the amount of grains damage increases. A similar result was obtained by Zareiforoush et al. [5].

Mean comparison results of interaction effects of variety and inclination angle are shown in Figure 2. With increasing inclination angle from 0 to 40° at tests with Alikazemi and Hashemi varieties, power requirement has been increased significantly. But with further increase in inclination angle up to 80°, power requirement decreased significantly. With increasing inclination angle from 20 to 60° at tests with Khazar variety, power requirement mean was decreased.

Mean comparison results of interaction effects of variety and screw speed are showed in Figure 3. At testes with all varieties, power requirement increased as screw speed increased up to 1000 rpm. The maximum and minimum power requirements were 123.52 and 38.22 W, which was obtained at tests with Khazar variety at screw speed 1000 rpm and with Alikazemi at screw speed 600 rpm, respectively.

Mean comparison results of interaction effects of inclination angle and screw speed are showed in Figure 4. For each

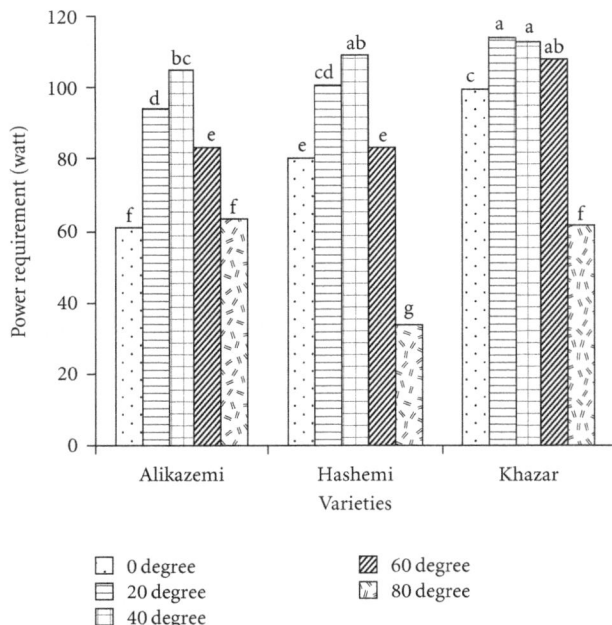

FIGURE 2: Mean comparison results of interaction effects of variety and inclination angle on power requirement (dissimilar letters show different significantly at probability level %5).

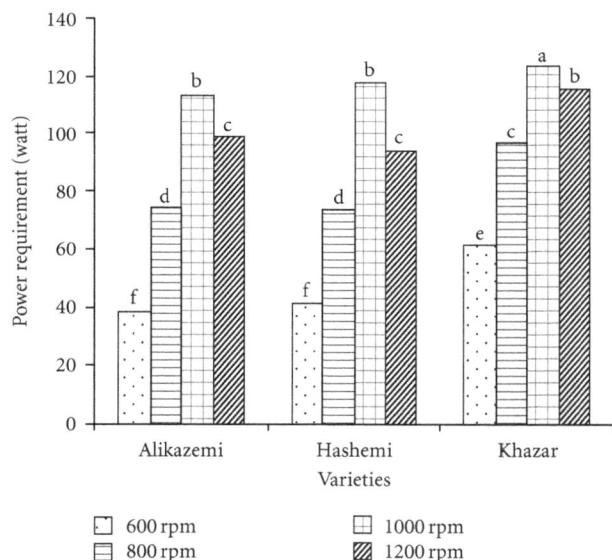

FIGURE 4: Mean comparison results of interaction effects of inclination angle and screw rotational speed on power requirement (dissimilar letters show different significantly at probability level %5).

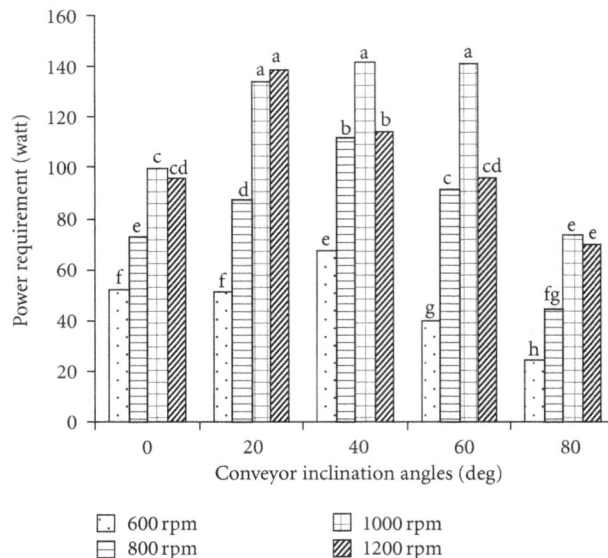

FIGURE 3: Mean comparison results of interaction effects of variety and screw speed on power requirement (dissimilar letters show different significantly at probability level %5).

angle, power requirement was significantly increased with increasing screw speed up to 1000 rpm. The lowest power requirement was obtained at 80° inclination angle for all levels of the screw rotational speed due to the lowest amount of conveying. A similar result was obtained by Srivastava et al. [9].

Mean comparison results of interactions between three factors revealed that the most power requirement mean

was 156.88 W at tests with Khazar variety, screw speed of 1000 rpm, and inclination angle of 60°.

3.2. Conveying Capacity. Analysis of the variance of data obtained from measuring conveying capacity was presented in Table 1. The results showed that the main and interactions effects of variety, screw rotational speed, and inclination angle were significant ($P < 1\%$). These results conform to the obtained findings of the researchers in the study of screw conveyor performance [7, 9]. Theoretical capacity of a screw conveyor has confirmed a direct relationship with screw rotational speed. The other factors affecting the conveying capacity are grain bulk density and internal and external friction that may have different values in the paddy grain varieties. By increasing of conveyor inclination angle, weight force of grain mass prevents from the axial movement of grain mass in tube, so the conveying capacity results are reduced. Such trends in the conveying capacity have also been observed in a study of conveyor inclination angle [8].

Means comparison results of variety effects on conveying capacity was significant ($P < 5\%$). The highest and lowest conveying capacity was allocated to Khazar and Alikazemi varieties, respectively. By increasing screw speed from 600 to 1000 rpm, conveying capacity was increased significantly and then decreased at test with screw speed 1200 rpm. Similarly, this case had happened for power requirement. By increasing screw inclination angle from 20 to 80°, conveying capacity decreased significantly from 3.352 to 0.932 t/h, respectively. In horizontal state, the screw conveyor had been the greatest amount of 3.581 t/h of conveying capacity. Similar results were obtained by Ray [7] and Srivastava et al. [9].

Determination of Effective Factors on Power Requirement and Conveying Capacity of a Screw Conveyor under Three Paddy Grain Varieties

27

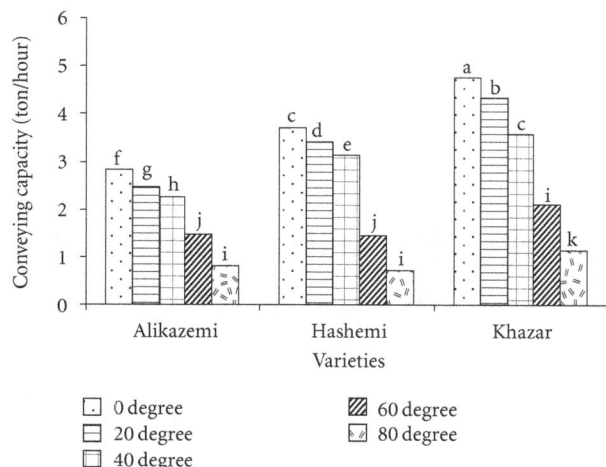

FIGURE 5: Mean comparison results of interaction effects of variety and conveyor inclination angle on the conveying capacity (dissimilar letters show different significantly at probability level %5).

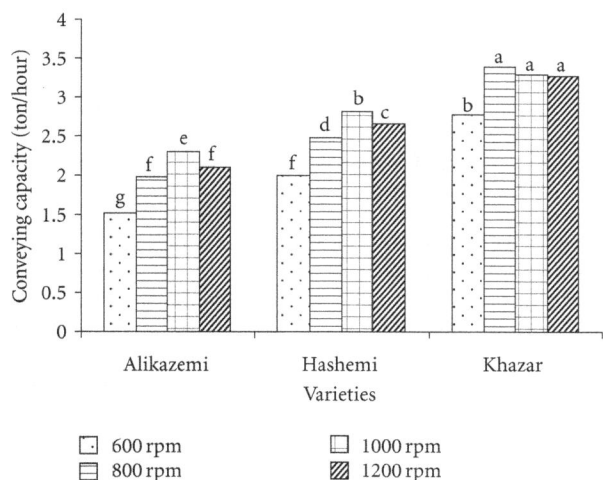

FIGURE 6: Mean comparison results of interaction effects of variety and screw rotational speed on the conveying capacity (dissimilar letters show different significantly at probability level %5).

Mean comparison results of interaction variety and conveyor inclination angle revealed that at tests with all varieties, conveying capacity reduced significantly by changing of conveyor inclination angle from 0 to 80°. In tests with Khazar variety, by changing screw speed from 800 to 1200 rpm, conveying capacity was not significantly changed (Figure 5).

Mean comparison results of interaction effects of variety and screw speed were shown in Figure 6. At testes with all varieties, conveying capacity has been increased by increasing screw speed up to 1000 rpm except in Khazar variety. The maximum and minimum conveying capacities were 3.40 and 1.52 t/h, which were obtained at tests with Khazar variety at screw speed 800 rpm and with Alikazemi at screw speed 600 rpm, respectively.

Mean comparison results of interaction effects of three factors showed that the most conveying capacity of 4.955 t/h was obtained at tests with Khazar variety, conveyor inclination angle 0 degree, and screw rotational speed 1000 rpm.

4. Conclusions

(1) The main effects of variety, conveyor inclination angle, and screw rotational speed and their interactions were significant on power requirement and conveying capacity. Maximum and minimum power requirements of tested screw conveyor were 99.29 and 81.16 Watt corresponding to conveying capacity of 3.210 and 1.975 t/h obtained for khazar and Alikazemi varieties, respectively.

(2) The most and least power requirement, and conveying capacity were obtained at screw rotational speeds of 1000 rpm and conveyor inclination angle of 40°, respectively. By increasing the conveyor inclination angle from 0 to 80°, conveying capacity was decreased significantly from 3.581 to 0.932 t/h. Generally, the most conveying capacity was 4.955 t/h at tests with khazar variety and zero inclination angle degree.

References

[1] FAOSTAT, "Rice production," 2005, http://faostat.fao.org/.

[2] R. Nicolai, J. Ollerich, and J. kelly, "Screw auger power and throuput analysis," in *Proceedings of the ASABE Annual International Meeting*, 046134, Ottawa, ON, Canada, 2004.

[3] H. Zareiforoush, M. H. Komarizadeh, M. R. Alizadeh, and M. Mosoomi, "Screw conveyors power and throughput analysis during horizontal handling of paddy grains," *Journal of Agricultural science*, vol. 2, no. 2, pp. 147–157, 2010.

[4] H. Zareiforoush, M. H. Komarizadeh, and M. R. Alizadeh, "Performance evaluation of a 15.5 cm screw conveyor during handling process of rough rice (Oriza Sativa L.) grains," *Nature and Science*, vol. 8, no. 6, pp. 66–74, 2010.

[5] H. Zareiforoush, M. H. Komarizadeh, and M. R. Alizadeh, "Effect of moisture content on some physical properties of paddy grains," *Research Journal of Applied Sciences, Engineering and Technology*, vol. 1, no. 3, pp. 132–139, 2009.

[6] H. A. Araghi, M. Sadeghi, and A. Hemmat, "Physical properties of two rough rice varieties affected by moisture content," *International Agrophysics*, vol. 24, no. 2, pp. 205–207, 2010.

[7] T. K. Ray, *Mechanical Handling of Materials*, Asian Books Private, New Delhi, India, 2004.

[8] A. W. Roberts, "The influence of granular vortex motion on the volumetric performance of enclosed screw conveyors," *Powder Technology*, vol. 104, no. 1, pp. 56–67, 1999.

[9] A. K. Srivastava, C. E. Goering, R. P. Rohrbach, and D. R. Buckmasrer, *Engineering Principals of Agricultural Machines*, American Society of Agricultural and Biological Engineers (ASABE), St. Joseph, Mich, USA, 2nd edition, 2006.

5

Deployment of Municipal Solid Wastes as a Substitute Growing Medium Component in Marigold and Basil Seedlings Production

Nikos Tzortzakis,[1] Sofia Gouma,[1] Costas Paterakis,[2] and Thrassyvoulos Manios[1]

[1] *Department of Organic Greenhouse Crops and Floriculture, School of Agricultural Technology,*
Technological Educational Institute of Crete, 71004 Heraklion, Greece
[2] *Inter-Municipal Enterprise for the Management of Solid Wastes-IMEMSW, 73100 Chania, Greece*

Correspondence should be addressed to Nikos Tzortzakis, ntzortzakis@staff.teicrete.gr

Academic Editor: Tadashi Takamizo

The possible use of municipal solid waste compost (MSWC) in the production of marigold and basil seedlings examined. Six medium prepared from commercial peat (CP) and MSWC (0, 15, 30, 45, 60, and 100% v/v). There was not any plant growth when MSWC used alone (100%). The addition of MSWC in low content (15% and 30%) improved seed emergence for marigold and basil respectively, while greater content revealed opposed impacts. Mean emergence time delayed as MSWC content increased into substrates. Addition of MSWC (especially in content greater than 30%) into CP reduced (from 34 to 64%) plant height, leaf number and stem diameter as a consequence reduced plant fresh weight (plant biomass) for both species. The number of lateral stems decreased (up to 81%) in basil when MSWC added into substrate mixtures. Chlorophyll b content decreased (up to 58%) in substrates with MSWC content greater than 15% or 30% while similar reduction observed in content of Chlorophyll a and total carotenoids for basil with MSWC > 60%. However, Chlorophyll a and total carotenoids content increased as MSWC content increased for marigold. K and Na leaf content increased but P equivalent decreased as MSWC content increased. Nursery-produced basil and marigold seedlings grown in 15% MSWC; displayed quality indices similar to those recorded for conventional mixtures of peat and may act as component substitute.

1. Introduction

Transplants, compared with direct sowing, are a more reliable method of ensuring the proper establishment of a range of commercial horticultural crops with great economic value. The production of container-grown flowers and aromatic plants is a highly competitive business; uniform and rapid seed emergence is essential prerequisites to increase yield, quality, and profits in crops. Use of good crop substrates is therefore critical [1].

It is well known that peat, a wide used substrate, is a nonrenewable resource, and diminishing availability is prompting price increases. The extensive use of peat as a substrate has led growers to consider its replacement in the medium to long term [1] with alternative candidates achieving attention. Numerous studies have shown that organic residues such as urban solid wastes, sewage sludge, pruning

waste, and even green wastes following composting process can be used with very good results as growth media instead of peat [2, 3]. Additionally, composting has positive environmental impacts towards organic residues. The introduction and interest of compost into nursery plant production increased nowadays. It has been found that mixtures of compost with perlite (20–50% MSWC) may be used as substrates without the need for additional mineral fertilizer [4, 5]. There are, however, certain limitations on some composts use, including the increase in salt content to levels which might affect the growth of sensitive crops, heavy metal toxicity, low overall porosity, and a marked variation in physical/chemical properties [6, 7].

Municipal solid waste compost (MSWC) as an organic soil additive when applied in field trials suggested that it can be used in agricultural production, improving soil physicochemical properties, increasing water retention as well as

Deployment of Municipal Solid Wastes as a Substitute Growing Medium Component in Marigold
and Basil Seedlings Production

29

supply with considerable amount of essential nutrients [8, 9]. However, little information is available regarding the use of MSWC as a peat alternative for nursery production of horticultural crops. Indeed, most studies have focused on ornamental potted plants, woody shrubs, and trees [10]. Each particular compost has to find the best amounts for particular plant growth as there is no one standard growing medium recommended for all container crops under all growing conditions.

A previous study showed that the growth and development of nursery-produced tomato seedlings using a peat and MSWC mixture was similar to that obtained with the standard peat mixture [5]. The present study sought to evaluate the effect of varying the proportion of MSWC mixed with conventional peat substrates, as a growth medium in the nursery production of marigold and basil plants.

2. Material and Methods

2.1. Seed and Municipal Solid Waste Compost Source. Seeds of marigold (*Tagetes erecta* L. cv Erecta) and basil (*Ocimum basilicum* L. cv sweet basil) were purchased from Agrospito company (Goldsmith seeds Ltd, CA, USA) and Agrimore (Agrimore SA, Thessaloniki, Greece), respectively.

Municipal solid waste compost was punctuated by Inter-Municipal Enterprise for the Management of Solid Wastes (IMEMSW), based in Chania. The compost used was made from the organic fraction of selectively collected urban waste. Following electromagnetic separation, manual sorting and use of an 80 mm trommel screen to remove as many bulking agents as possible, organic material was arranged in piles of 5 m wide of 2.5 m high of 45 m long, which were regularly turned and watered over a 140-day period to ensure appropriate composting conditions (turned windrow system). This material was then passed through a densimetric table and a 15 mm trommel screen to remove the largest particles. The composting procedure lasted for 5-6 months. The 60% of compost consisted of particles with <4 mm size.

2.2. Germination and Plant Growth Studies in Nursery Tests. Marigold and basil seeds were used for nursery tests. A mix of commercial compost peat (Professional peat, Gebr. Brill Substrate GmbH & Co.KG, Georgsdorf, Germany), perlite (Perloflor, Protectivo EPE, Athens, Greece), and MSWC were used to create six treatments which were (% v/v) (1) peat : MSWC (100 : 0) as control, (2) peat : MSWC (85 : 15), (3) peat : MSWC (70 : 30), (4) peat : MSWC (55 : 45), (5) peat : MSWC (40 : 60), and (6) peat : MSWC (0 : 100). In each substrate medium was added 10% of perlite. Perlite physicochemical properties were reported in previous studies [11].

Seeds were sown (0.5 cm depth; 1.0–1.5 cm between seeds in plastic seedling trays (5 seeds per well; 4 wells per replication; 5 replications per treatment, 40 cm^3 well capacity)) on top of the surface of the media. The experiment was carried out in a completely randomized design in an unheated glasshouse with a north-south orientation at the School of Agricultural Technology at Heraklion, Crete,

Greece, located at the latitude of 35′ 35°N, longitude 24′ 02°E, and 8 m altitude (temperature: 25.7 ± 6.8°C max, 15.1 ± 5.1°C min; RH (%): 93.5 ± 1.9 max, 74.8 ± 4.1 min) with alternate-day watering by mist system (initially with 1 min/2 h and then up to 1 min/5 h). Over the growth-period in the nursery, no fertilizer was applied; seedling nutritional requirements were thus met entirely by the substrates. Daily observations were recorded for seed germination (seeds recorded as emerged when the hypocotyls appeared above the surface of substrate medium). After 15 days seedlings were thinned to single plant, maintaining 4-5 cm distance among seedlings. Mean germination time (MGT) was calculated as follows, according to Labouriau [12]:

$$t = \frac{(\Sigma ni.ti)}{\Sigma n} \text{ (days)}, \tag{1}$$

where t is the mean germination time, ti is the given time interval, ni is the number of germinated seeds during a given time interval, and n is the total number of germinated seeds.

After 48 days, seedling growth was assessed by harvesting six individuals/treatment. Seedlings were harvested above substrate, the leaf number and height (cm) per seedling, measured from substrate surface, stem diameter (mm) measured below the cotyledon node, number of stems (for basil), number of flowers (for marigold), upper fresh weight (g), total dry matter content (%), leaf fluoresces, content (μg/g fresh weight) of chlorophyll a (Chla), chlorophyll b (Chlb), and total carotenoids (Car) determined. Leaf elemental analysis for K, Na (photometric), P (spectrophotometric), and N (Kjeldahl) was determined.

2.3. Statistical Analysis. The experiment was carried out twice. Percentage data were log-transformed before analysis. Data were tested for normality and then subjected to analysis of variance (ANOVA). Significant differences between mean values were determined using Duncan's Multiple Range test following One-Way ANOVA. Various correlations were also calculated. Statistical analyses were performed using SPSS (SPSS Inc., Chicago, Ill, USA).

3. Results and Discussion

3.1. Compost Properties. The main physicochemical characteristics of compost (dry weight: dwt) were pH, 7.7; electrical conductivity (EC), 18.2 mS/cm; ashes, 50.6% dwt; organic matter, 49.4% dwt; Carbon, 27.5% dwt; N, 1.9% dwt; and ratio C/N, 7.2, under low limits for heavy metal content. The C/N ratio is widely used as an indicator of the maturity and stability of organic matter. The low values recorded here for the C/N ratio in MSWC suggest that composts were stable and mature. Davidson et al. [13] reported that composts with a C/N ratio of less than 20 are ideal for nursery plant production. Ratios above 30 may be toxic, causing plant death [14].

3.2. Seed Germination and Emergence Time In Vivo. The first germination was observed after two and six days of sowing for marigold and basil, respectively, while the first

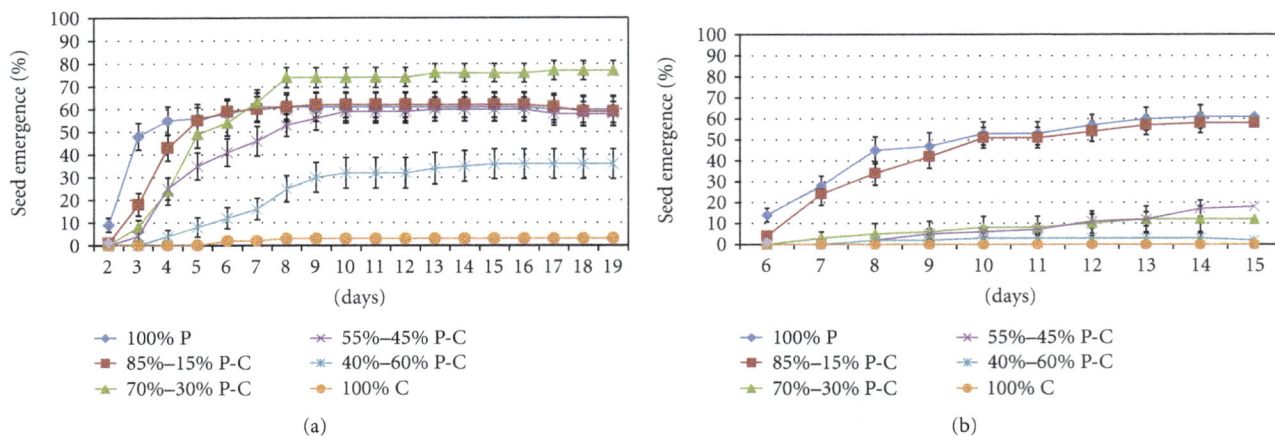

FIGURE 1: Influence of substrate medium (commercial peat : CP; municipal solid waste compost : MSWC) at different ratios on cumulative seedling emergence of marigold and basil seeds germinated in greenhouse nursery. Values represent mean ($\pm$SE) of measurements made on 5 independent replication (4 wells per replication; 5 seeds per well) per treatment.

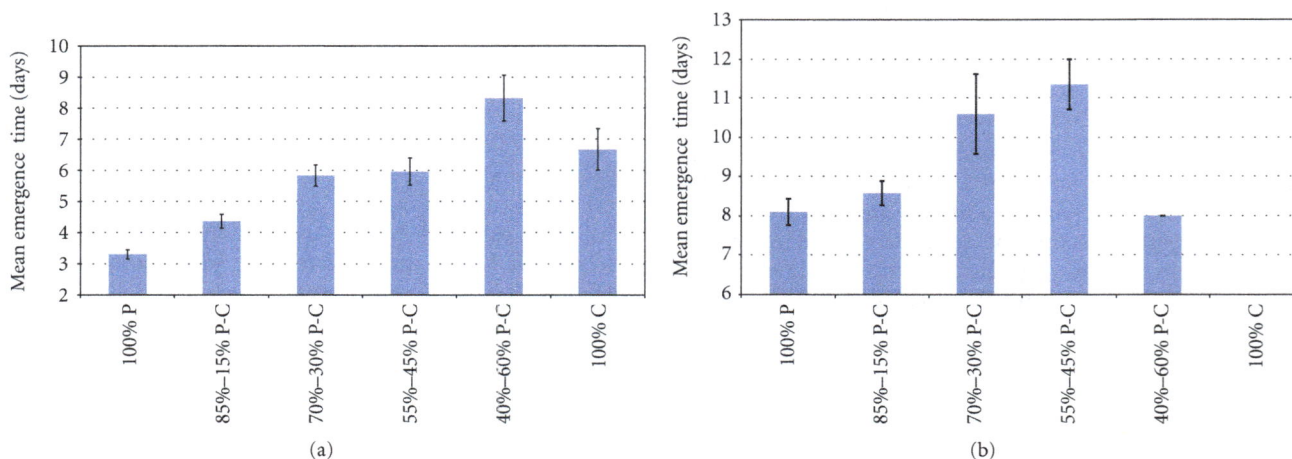

FIGURE 2: Mean emergence time for marigold and basil in substrate medium (commercial peat : CP; municipal solid waste compost : MSWC) at different ratios under nursery condition. Values represent mean ($\pm$SE) of measurements made on 5 replication (4 wells per replication; 5 seeds per well) per treatment.

true-leaf was emerged after five and nine days, respectively. The ratios of mix when peat and MSWC were used affected seed germination/emergence (Figure 1). Marigold treated with the 70 : 30 CP : MSWC ratios increased (up to 22%) seed emergence comparing with the control (PC) being in accordance with previous studies in melon seedlings [9]. In contrast, high content (>45%) of MSWC into the substrate reduced seed emergence comparing with the control. In basil seedlings, there were no differences with low (15%) MSWC content comparing with control (see Figure 1). Thus, increased MSCW content into the substrate revealed reduction in seed emergence. The worst overall emergence rates were recorded for 100% MSWC, only 3% emergence for marigold by day 8 after which emergence ceased, while no emergence was observed in case of basil.

The ratios of mix when CP and/or MSWC were used affected seed MGT (Figure 2). For marigold seeds, increased

MSWC content resulted in MGT reduction. For example, 45% of MSWC revealed 3 days delay in MGT comparing with the control. For basil seeds, MGT delayed as MSWC content increased (up to 45% MSWC content), while in 60% content, MGT did not differ comparing with the control. It is worthwhile to mention that seeds sowed in pure MSWC (100%) did not germinate/emerge at all.

Adding MSWC into the substrate in low content benefits the seed germination/emergence, possible due to the fact that MSWC provided nutritional value as organic material and/or improved substrate medium properties. The stimulation of several presowing treatments (hydropriming, halopriming, osmopriming, thermopriming, solid matrix priming, and biopriming as reported by Ashraf and Foolad [15]) of seed comparing with untreated seeds might be due to altered physiology of embryos and activation of enzymes, so that developmental processes occur more rapidly after sowing

Deployment of Municipal Solid Wastes as a Substitute Growing Medium Component in Marigold
and Basil Seedlings Production

31

TABLE 1: Influence of substrate medium (commercial peat: CP; municipal solid waste compost: MSWC) at different ratios on seedling height (cm/plant), number of leaf produced, stem diameter (mm), number of flower produced and opened, and number of lateral shoots on marigold and basil plant grown in greenhouse nursery.

| | Marigold | | | | | Basil | | | |
	Height	Leaf no.	Stem diameter	Flower no.	Flower open	Height	Leaf no.	Stem diameter	Shoots no.
P-C (100:0)	13.57 a[Y]	7.33 a	2.07 a	0.67 c	0.00 c	10.91 a	9.33 a	2.47 a	7.00 a
P-C (85:15)	13.87 a	7.33 a	1.65 b	1.83 a	0.50 a	11.58 a	8.33 a	2.49 a	3.00 bc
P-C (70:30)	12.82 a	7.33 a	1.99 a	1.67 a	0.50 a	7.25 b	6.51 b	1.65 b	3.33 bc
P-C (55:45)	9.6 b	5.67 b	1.41 b	1.00 b	0.17 b	8.75 b	7.50 ab	1.71 b	4.00 b
P-C (40:60)	4.8 c	4.50 c	1.51 b	0.50 c	0.00 c	3.96 c	4.66 c	0.89 c	1.33 c
P-C (0:100)	—	—	—	—	—	—	—	—	—

[Y] Values (n = 6) in columns followed by the same letter are not significantly different P < 0.05.

TABLE 2: Influence of substrate medium (commercial peat: CP; municipal solid waste compost: MSWC) at different ratios on seedling fresh weight (g/plant) and dry matter content (%) on marigold and basil plant grown in greenhouse nursery.

| | Marigold | | Basil | |
	Fresh weight	Dry matter	Fresh weight	Dry matter
P-C (100:0)	1.92 a[Y]	13.88 b	2.56 a	8.46 b
P-C (85:15)	2.38 a	12.37 c	1.98 a	8.45 b
P-C (70:30)	2.28 a	18.90 a	1.48 ab	8.45 b
P-C (55:45)	0.82 b	11.27 d	1.28 b	8.48 a
P-C (40:60)	0.57 c	9.47 d	0.45 c	8.44 b
P-C (0:100)	—	—	—	—

[Y] Values (n = 6) in columns followed by the same letter are not significantly different, P ≤ 0.05.

[16] and this is possible with the seed germination under MSWC enrichment.

Although there is no single, ideal growth medium for nursery-produced horticultural crops [17], most greenhouse-grown species display better growth at slight acid pH values (5.2–7.0); peat mixtures approached these values but MSWC did not. Like pH levels, the highest initial substrate EC values were recorded for mixtures containing MSWC. Ribeiro and Santos [18] reported that substrates with high EC values reduce water retention, negatively affecting the imbibing process, and may delay seed emergence rates.

3.3. Seedling Growth In Vivo. Seedling growth failed when 100% MSWC was used as substrate due to seed emergence failure. The highest values for all growth parameters were recorded in substrate without MSWC (control), which in most cases differed significantly from the other substrates (Tables 1 and 2). Analyses of variance showed that the addition (>15% for basil and >30% for marigold) of MSWC in commercial peat significantly reduced leaf number (between 20% and 50%), seedling height (between 20% and 65%), stem diameter (between 27% and 64%), and fresh weight (between 50% and 82%), for basil and marigold seedlings (Table 1) which are in agreement with previous studies in cucumber and melon seedlings [9, 19]. Seedlings dry matter reduced as MSWC content increased into the substrate. Thus, seedlings grown in the MSWC mixtures in high content displayed worse quality and suitability for transplanting, possible due to increased EC and/or alternated medium physicochemical properties. Seedling resistance to transplant stress is directly related to dry matter content, which improves seedling establishment in the soil or growth substrate [20]. In case of basil, stem number reduced as MSWC content increased, and this reduction fluctuated between 57 to 83% comparing with the control treatment, CP. In case of marigold, low content of MSWC into the substrate mixture accelerated the number of flowers produced as well as the number of open flowers (considered open when the yellow colour of petals appeared).

In marigold seedlings, significant increases in Chla and total carotenoids were observed in substrates with >60% MSWC and >45% MSWC, respectively, while Chlb decreased when the content of MSWC was >45% (Table 3). In basil seedlings, significant increases in Chla and total carotenoids were observed in CP:MSWC (55:45) but decreased in substrate CP:MSWC (40:60). However, Chlb decreased when the content of MSWC was >30%. No differences were observed in leaf fluoresce among the different substrates.

Leaf elemental content for marigold revealed that K and Na increase (up to 33% and 83%, resp.) with the addition of MSWC into the substrate, being in agreement with melon seedling production with the same MSWC [9], while P and N content reduced in substrates with >30% and >60% MSWC content, respectively, compared with the control (Table 4). In case of basil, K content increased with the addition of 15–45% MSWC but reduced when MSWC content was greater than 60%. Phosphorus content was reduced as MSWC content increased into the substrate and this might be due to high pH value of MSWC. Nitrogen content decreased in substrates with >60% MSWC content. Thus, considerable nutritive value was marked due to MSWC addition into the substrates.

MSW compost was found to be an ideal component of mixed-peat substrates for marigold and basil seedlings, provided that it accounts for less than 15% of the mixture. These proportions reduce the negative effects of high pH and

TABLE 3: Influence of substrate medium (commercial peat: CP; municipal solid waste compost: MSWC) at different ratios on leaf fluoresces, Chlorophyll a (Chla; μg/g fw), Chlorophyll b (Chlb; μg/g fw), and total carotenoids (Car; μg/g fw) on marigold and basil plant grown in greenhouse nursery.

	Marigold				Basil			
	Fluoresce	Chla	Chlb	Car	Fluoresce	Chla	Chlb	Car
P-C (100:0)	0.67 a[Y]	53.79 b	67.57 a	19.67 c	0.80 a	57.09 c	29.68 a	28.35 c
P-C (85:15)	1.83 a	55.98 b	70.16 a	20.42 c	0.81 a	57.68 bc	25.32 a	30.68 abc
P-C (70:30)	1.67 a	61.72 ab	60.36 ab	27.92 bc	0.78 b	58.46 b	16.78 b	30.60 b
P-C (55:45)	1.00 a	61.58 ab	50.35 b	30.87 b	0.79 b	58.96 a	15.74 b	31.91 a
P-C (40:60)	0.50 a	72.96 a	28.02 c	40.09 a	0.66 ab	38.11 d	18.89 ab	19.10 d
P-C (0:100)	—	—	—	—	—	—	—	—

[Y]Values ($n = 6$) in columns followed by the same letter are not significantly different, $P \leq 0.05$.

TABLE 4: Influence of substrate medium (commercial peat: CP; municipal solid waste compost: MSWC) at different ratios on leaf elemental (K, P, Na, N) concentration (mg/g fresh weight) on marigold and basil seedlings grown in greenhouse nursery. Values represent mean ($\pm$SE) of measurements made on 3 replication (3 seedlings mixed per replication) per treatment.

	Marigold				Basil			
	N	K	P	Na	N	K	P	Na
P-C (100:0)	10.91 a[Y]	0.177 c	0.020 a	0.015 d	8.58 ab	0.096 b	0.017 a	0.029 b
P-C (85:15)	11.02 a	0.231 ab	0.018 ab	0.050 b	7.41 abc	0.120 a	0.013 b	0.036 a
P-C (70:30)	12.64 a	0.232 ab	0.016 b	0.136 a	9.06 a	0.084 b	0.009 b	0.029 b
P-C (55:45)	11.71 a	0.266 a	0.013 c	0.032 c	8.21 b	0.116 ab	0.008 bc	0.039 a
P-C (40:60)	8.93 b	0.228 b	0.017 b	0.122 a	7.33 c	0.048 c	0.005 c	0.017 c
P-C (0:100)	—	—	—	—	—	—	—	—

[Y]Values ($n = 3$) in columns followed by the same letter are not significantly different, $P \leq 0.05$.

EC on seedling growth and provide a seedling comparable to that obtained using standard peat-based mixtures, being in agreement with previous study in melon seedling production [9]. Thus, the mixture of CP (85%) and MSWC (15%) provides an ideal substrate for nursery production of marigold and basil seedlings, yielding quality indices similar to those provided by conventional peat. Similarly, nursery-produced tomato seedlings grown in peat with MSWC (30%) displayed good quality indices [5, 6]. This is in all probability due to a correct balance between nutrient supply from the MSW compost and the physical characteristics of CP, particularly substrate porosity and aeration.

There is a growing public concern about the environmental impact of industrial development and population expansion in recent decades. Improved methods of selective waste collection and compost processing will enable increasingly widespread use of this renewable organic compost, as an alternative to high-quality sphagnum peat, which—because they are nonrenewable—are less available and more expensive for growers.

Acknowledgments

This work was funded by the Greek National Research program XM EOX: EL0031. The authors would also like to thank C. Saridakis, M. Papamichalaki, and E. Dagianta for their technical support in plant analysis.

References

[1] S. B. Sterrett, "Compost as horticultural substrates for vegetable transplant production," in *Compost Utilization in Horticultural Cropping Systems*, P. J. Stoffella and B. A. Kahn, Eds., pp. 227–240, Lewis Publication, Boca Raton, Fla, USA, 2001.

[2] H. I. Siminis and V. I. Manios, "Mixing peat with MSW compost," *BioCycle*, vol. 31, no. 11, pp. 60–61, 1990.

[3] M. Benito, A. Masaguer, R. De Antonio, and A. Moliner, "Use of pruning waste compost as a component in soilless growing media," *Bioresource Technology*, vol. 96, no. 5, pp. 597–603, 2005.

[4] O. Kostov, Y. Tzvetkov, N. Kaloianova, and O. Van Cleemput, "Production of tomato seedlings on composts of vine branches and grape prunings, husks and seeds," *Compost Science and Utilization*, vol. 4, no. 2, pp. 55–61, 1996.

[5] J. E. Castillo, F. Herrera, R. J. López-Bellido, F. J. López-Bellido, L. López-Bellido, and E. J. Fernández, "Municipal Solid Waste (MSW) compost as a tomato transplant medium," *Compost Science and Utilization*, vol. 12, no. 1, pp. 86–92, 2004.

[6] C. Vavrina, "Municipal solid waste materials as soilless media for tomato transplant," *Proceedings of the Florida State Horticultural Society*, vol. 108, pp. 232–234, 1995.

[7] T. M. Spiers and G. Fietje, "Green Waste Compost as a Component in Soilless Growing Media," *Compost Science and Utilization*, vol. 8, no. 1, pp. 19–23, 2000.

[8] D. B. McConnell, A. Shiralipour, and W. H. Smith, "Compost application improves soil properties," *BioCycle*, vol. 34, no. 4, pp. 61–63, 1993.

Deployment of Municipal Solid Wastes as a Substitute Growing Medium Component in Marigold and Basil Seedlings Production

33

[9] N. Tzortzakis, E. Dagianta, G. Daskalakis, V. Manios, C. Paterakis, and T. Manios, "Municipal solid waste compost: a growing medium component for melon seedling production," in *Proceedings of the 5th European Bioremediation Conference in Chania*, 2011.

[10] G. E. Fitzpatrick, E. R. Duke, and K. A. Klock-Moore, "Use of compost products for ornamental crop production: research and grower experiences," *HortScience*, vol. 33, no. 6, pp. 941–944, 1998.

[11] N. G. Tzortzakis and C. D. Economakis, "Shredded maize stems as an alternative substrate medium. Effect on growth, flowering and yield of tomato in soilless culture," *Journal of Vegetation Science*, vol. 11, pp. 57–70, 2005.

[12] L. G. Labouriau, *Seed Germination*, Organization of American, Washington, DC, USA, 1983.

[13] H. Davidson, R. Mecklenburg, and C. Peterson, *Nursery Management: Administration and Culture*, Prentice Hall, Englewood Cliffs, NJ, USA, 3rd edition, 1994.

[14] F. Zucconi, A. Pera, M. Forte, and M. De Bertoldi, "Evaluating toxicity of immature compost," *BioCycle*, vol. 22, no. 2, pp. 54–57, 1981.

[15] M. Ashraf and M. R. Foolad, "Pre-sowing seed treatment-a shotgun approach to improve germination, plant growth, and crop yield under saline and non-saline conditions," *Advances in Agronomy*, vol. 88, pp. 223–271, 2005.

[16] K. N. Kattimani, Y. N. Reddy, and B. Rajeswar Rao, "Effect of pre-sowing seed treatment on germination, seedling emergence, seedling vigour and root yield of Ashwagandha (*Withania somnifera* Daunal.)," *Seed Science and Technology*, vol. 27, no. 2, pp. 483–488, 1999.

[17] G. J. Bugbee, "Growth of Rhododendron, Rudbeckia and Thujia and the leaching of nitrates as affected by the pH of potting media amended with biosolids compost," *Compost Science and Utilization*, vol. 4, no. 1, pp. 53–59, 1996.

[18] H. M. F. Ribeiro and J. Q. e Santos, "Utiliçao de residuos sólidos urbanos comportados na formulaçao de sustratos: efeito nas propiedades físicas e químicas dos sustrato," Actas de Horticultura, vol. 18-II. Cong-reso Iberoamericano e III Congreso Ibérico de Ciencias Hortícolas, Vilamoura, Portugal, 1997.

[19] Y. Mami and G. Peyvast, "Substitution of municipal solid waste compost for peat in cucumber transplant production," *Journal of Horticulture and Forestry*, vol. 2, pp. 157–160, 2010.

[20] F. Pimpini and G. Gianquinto, "Primi resultati sulle modalita di allevamento inviaio di piantina di pomodoro da industria. Riflessi su aecrescimento e produzion e in campo," 2 Convego Nazionale "Il vivaismo Octicolo", Foggia, Italy, 1991.

6

Water Use Efficiency and Physiological Response of Rice Cultivars under Alternate Wetting and Drying Conditions

Yunbo Zhang,[1,2] Qiyuan Tang,[1] Shaobing Peng,[3] Danying Xing,[2] Jianquan Qin,[1] Rebecca C. Laza,[4] and Bermenito R. Punzalan[4]

[1] Crop Physiology, Ecology, and Production Center (CPEP), Hunan Agricultural University, Changsha, Hunan 410128, China
[2] Agricultural College, Yangtze University, Jingzhou, Hubei 434025, China
[3] Crop Physiology and Production Center (CPPC), MOA Key Laboratory of Crop Physiology, Ecology and Cultivation (The Middle Reaches of Yangtze River), Huazhong Agricultural University, Wuhan, Hubei 430070, China
[4] Crop and Environmental Sciences Division, International Rice Research Institute (IRRI), DAPO Box 7777, Metro Manila, Philippines

Correspondence should be addressed to Qiyuan Tang, cntqy@yahoo.com.cn

Academic Editors: D. W. Archer and M. Dumont

One of the technology options that can help farmers cope with water scarcity at the field level is alternate wetting and drying (AWD). Limited information is available on the varietal responses to nitrogen, AWD, and their interactions. Field experiments were conducted at the International Rice Research Institute (IRRI) farm in 2009 dry season (DS), 2009 wet season (WS), and 2010 DS to determine genotypic responses and water use efficiency of rice under two N rates and two water management treatments. Grain yield was not significantly different between AWD and continuous flooding (CF) across the three seasons. Interactive effects among variety, water management, and N rate were not significant. The high yield was attributed to the significantly higher grain weight, which in turn was due to slower grain filling and high leaf N at the later stage of grain filling of CF. AWD treatments accelerated the grain filling rate, shortened grain filling period, and enhanced whole plant senescence. Under normal dry-season conditions, such as 2010 DS, AWD reduced water input by 24.5% than CF; however, it decreased grain yield by 6.9% due to accelerated leaf senescence. The study indicates that proper water management greatly contributes to grain yield in the late stage of grain filling, and it is critical for safe AWD technology.

1. Introduction

Rice (*Oryza sativa* L.) is a major staple food for the world's population with about two-thirds of the total rice production grown under irrigation [1]. In the past 10 years, the growth of rice yield has dropped below 1% per year worldwide, but an increase of more than 1.2% per year is required to meet the growing demand for food [2]. Rice production in Asia is increasingly constrained by water limitation [3] and increasing pressure to reduce water use in irrigated production as a consequence of global water crisis [4]. Guerra [5] reported that 60% of the world's irrigated fields are in Asia, half of which are devoted to rice production. Irrigated lowland rice consumes more than 50% of total freshwater, and irrigated flooded rice requires two or three times more water than other cereal crops, such as wheat and maize [6]. In addition, rice production is facing increasing competition with rapid urban and industrial development in terms of freshwater resource [7]. The need for "more rice with less water" is crucial for food security, and irrigation plays a greater role in meeting future food needs than it has in the past [8].

Continuous flooding (CF) provides a favorable water and nutrient supply under anaerobic conditions. However, the conventional system consumes a large amount of water [9]. A number of water-saving irrigation (WSI) technologies to reduce water use, to increase water use efficiency, and to maintain or increase production for rice-based systems have been developed [10, 11]. One of the most commonly practiced WSI techniques is alternate wetting and drying

(AWD) irrigation [7, 12, 13]. In AWD, water is applied to irrigate the field depending on the weather condition or until some fine cracks appear on the soil surface.

Water use efficiency (WUE) is defined as the units of yield produced per unit of available water [14]. Crop WUE is especially an important consideration where available water resources are limited or diminishing. According to Fisher [15], WUE among plant varieties is essentially the same while gene improvement cannot change WUE [16]. However, WUE could be enhanced under restricted water by increasing transpiration and evaporation rate and by improving harvest index.

Nitrogen is one of the most important agricultural inputs to increase yield, and its use and uptake are affected by availability of water. Fertilizer application can improve both the crop yield and WUE. Hatfield [17] reported that the vital issue of nutrition is how to fertilize and improve WUE under restricted water conditions. Chlorophyll meter (SPAD) is a convenient tool to estimate leaf nitrogen concentration of rice plant. It is a simple, quick, and nondestructive method [18], and SPAD values are closely correlated with leaf N concentration [19]. Senescence is a genetically programmed process that involves remobilization of nutrients from vegetative tissues to grains [20, 21]. In China's super rice, too much use of nitrogen fertilizer leads to slow grain filling and low harvest index because leaves stay "green" for a too long time in the late stage of grain filling [22]. Water stress imposed during grain filling, especially at the early stage, usually results in a reduction in grain weight [23].

Water deficiency can accelerate plant senescence and lead to a faster and better remobilization of carbon from vegetative tissues to the grain [24]. Grain filling, which is an essential determinant of grain yield in cereal crops, is characterized by its duration and rate; these parameters correlated with other yield-related components of rice grain filling rate were more important than duration [25]. Grain filling rate was positively correlated with actual panicle weight and 100-grain weight and was negatively correlated with panicles m^{-2} [26].

In our current study, we compared hybrid rice varieties and inbred varieties under two N rates (low N, high N) and two water management treatments (AWD, CF). The objectives of this study were (1) to determine rice yield potential and water use efficiency under two N rates and two water management methods, (2) to identify the factors that contribute to increased yield and water productivity under these conditions, and (3) to determine if there exist interactions among N, water management, and varieties.

2. Materials and Methods

The field experiments were conducted for three consecutive seasons (2009 dry season (DS), 2009 wet season (WS), and 2010 DS) in the same field at the International Rice Research Institute (IRRI) farm, Los Baños (14°11′N, 121°15′E, and 21 m als), Philippines. The soil was an Aquandic Epiaquoll with pH 6.2; 20.0 g kg^{-1} organic C; 2.0 g kg^{-1} total N; 11.4 mg kg^{-1} Olsen P; 0.43 cmol kg^{-1} exchangeable K and

34.5 cmol kg^{-1} cation exchange capacity; and 58.3% clay, 34.0% silt, and 8.0% sand. The soil test was based on samples taken from the upper 20 cm of the soil before transplanting in 2010 DS.

The experimental design was split-split plot with four replications in the three seasons. The main plots were two water management treatments (AWD and CF). The subplots were two N treatments: low N rate (60 kg ha^{-1} in WS, 100 kg ha^{-1} in DS) and high N (120 kg ha^{-1} in WS, 200 kg ha^{-1} in DS). The sub-subplots were four rice varieties; they belong to two groups: hybrid rice (IR72 and PSBRc80) and inbred varieties (IR82372H and Mestizo7). SL8-H was replaced with Mestizo7 because of its disease susceptibility in the WS.

In the CF plots, ponded water was kept with a depth of 3–5 cm during the 7 days after transplanting until the 7 days before maturity. In the AWD plots, soil water potential was measured with two porous-cup tensiometers installed at 20 cm and 40 cm depth. The depth of groundwater table was monitored using piezometers in open-bottom PVC tubes installed at a depth of 100 cm. Holes were perforated on all sides of the tube. When the ponded water dropped to 15 cm below the soil surface, then irrigation was applied to reflood the field up to 5 cm in AWD treatment. This cycle was repeated throughout the season. The first AWD treatment was initiated in the 3 weeks after transplanting. The irrigated water of each plot was measured using a 90° boxed Weir connected to an irrigation outlet. Daily mean temperature and rainfall were recorded from the weather station adjacent to the experimental site. Total water input = the amount of irrigated water applied + rainfall. Water productivity = grain yield/total amount of water supplied.

Pregerminated seeds were sown in seedling trays to produce uniform seedings. Fourteen-day-old seedlings were manually transplanted on January 6, June 10, and January 14 for 2009 DS, 2009 WS, and 2010 DS, respectively. Four seedlings per hill were transplanted at a hill spacing of 20 cm × 20 cm. Insects, diseases, and weeds were intensively controlled by using approved pesticides to avoid biomass and yield loss. Fertilizers were manually broadcasted and incorporated during basal application: 30 kg P ha^{-1}, 40 kg K ha^{-1}, and 5 kg Zn ha^{-1} in the DS and 15 kg P ha^{-1}, 20 kg K ha^{-1}, and 2.5 kg Zn ha^{-1} in the WS. Nitrogen in the form of urea was applied. During DS, low N rate was supplied with 40, 20, 40, and 20 kg N ha^{-1} at basal, midtillering, panicle initiation, and booting, respectively. High N rate corresponded to 60-40-60-40 kg N ha^{-1}. During WS, low N rate was lowered to 20-10-20-10 kg N ha^{-1} while the rate for high N was reduced to 30-20-30-20 kg N ha^{-1}. In the dry season, total N rate was 120 and 200 kg ha^{-1} for the low and high N rates, respectively. In the wet season experiments, total N rate was 60 and 100 kg ha^{-1} for the low and high N rates, respectively.

The soil water content (SWC) of the soil was monitored when water was deficient in the AWD treatment in 2010 DS. In each plot, soil samples were taken every 2 days using a core sampler. Fresh weight of the soil samples was measured immediately. Dry weight was obtained after oven drying at 105°C for 24 h. The soil water content was calculated following the equation: SWC = 100 × (fresh weight − dry weight)/fresh weight. Three varieties (IR72, IR82372H, and SL-8H) were used to measure grain filling and SPAD value. At the onset of flowering, 150 panicles headed on the same day were initially tagged from the high N plots. Among these panicles, ten were taken every two days from heading until maturity. The SPAD value of its flag leaf was also measured before sampling. Dry weights of the spikelets were determined after oven drying at 70°C to constant weight.

For growth analysis, 12 hills were sampled from each plot at flowering to measure plant height, stem number, leaf area index, and aboveground total dry weight. Plant height was measured from the plant base to the tip of the highest leaf. Plants were separated into green leaves and stems. Green leaf area was measured with a leaf area meter (LI-3000, LI-COR, Lincoln, NE, USA) and expressed as leaf area index. The dry weight of each component was determined after oven drying at 70°C to constant weight. Total dry weight was the sum of the weights of green leaves and stems. At maturity, 12 hills were taken diagonally from a $5\,m^2$ area in each plot where grain yield was determined to measure the above ground total dry weight, harvest index, and yield components. Panicles of each hill were counted to determine the panicle number per m^2. Plants were separated into straw and panicles. Straw dry weight was determined after oven drying at 70°C to constant weight. Panicles of all 12 hills were hand threshed and filled spikelets were separated from unfilled spikelets by submerging them in tap water. Three subsamples each of $30\,g$ filled spikelets and $2\,g$ unfilled spikelets were taken to determine the number of spikelets. Dry weights of rachis and filled and unfilled spikelets were measured after oven drying at 70°C to constant weight. Aboveground total dry weight was the total dry matter of straw, rachis, and filled and unfilled spikelets. Spikelets per panicle, grain filling percentage ($100 \times$ filled spikelet number/total spikelet number), and harvest index ($100 \times$ filled spikelet weight/aboveground total dry weight) were calculated. Grain yield was determined from a $5\,m^2$ area in each plot and adjusted to the standard moisture content of $0.14\,g\,H_2O\,g^{-1}$ fresh weight. Grain moisture content was measured with a digital moisture tester (DMC-700, Seedburo, Chicago, IL, USA).

Data were analyzed following the analysis of variance (SAS Institute) and means were compared based on the least significant difference test (LSD) at the 0.05 probability level [27].

3. Results and Discussion

Average temperatures during the growing season in 2009 DS were 1.1–1.3°C higher than that in the 2009 WS (Figure 1). Seasonal mean values of maximum temperature were 29.9°C in 2009 DS, 31.2°C in 2009 WS, and 31.9°C in 2010 DS, whereas seasonal mean minimum temperatures were 23.6, 24.7, and 23.3°C for 2009 DS, 2009 WS, and 2010 DS, respectively. Higher daily minimum temperature and lower radiation were observed in the WS compared with the DS. No significant differences in daily maximum temperature between the two DS were observed. Seasonal mean radiation

was 15.3, 13.9, and $19.3\,MJ\,M^{-2}\,day^{-1}$ in 2009 DS, 2009 WS, and 2010 DS, respectively. The difference in radiation during the growing season between the DS and WS in 2009 was about 15% and about 10% between the two DS.

Total rainfall of each season was 349, 1079, and 92 mm in 2009 DS, 2009 WS, and 2010 DS, respectively (Table 1). There was about 67% difference in rainfall during the growing season between the DS and WS in 2009 and about 73% difference between the two DS. The total amount of water input (irrigation plus rainfall) in the AWD was 876, 1184, and 833 mm in 2009 DS, 2009 WS, and 2010 DS, which was 7.2%, 5.3%, and 24.5% less than the CF, respectively. CF greatly consumed more water than AWD, especially in the 2010 DS.

Soil water content during the growing period under AWD in 2010 DS was shown in Figure 2. Analysis showed that N rate and variety had no significant effect on soil water content; the soil water content at 0–10 cm depth was higher than that of 10–20 cm during the vegetative stage. However, it was lower during the late growth stage because several reirrigations will influence soil structure. In many previous studies [7, 28], the time of irrigation was determined by soil water potential, and 0–20 kpa in the root zone was defined as mild stress and 50–80 Kpa as severe stress. This study followed an irrigation scheme according to soil water content of the upper 20 cm soil in 2010 DS and took SWC of 40% and 30% as irrigation threshold at PI and grain filling stage, respectively. Compared with soil water potential, it is a more accurate and simpler method to measure soil water content in field.

Interaction effects of variety, water management, and N rate in all the three experiments were not significant. Grain yield was not significantly different between AWD and CF across the three seasons (Table 2). Varietal differences in grain yield were significant in the two DS experiments, but not significant in the 2009 WS (Table 3). Average yield of AWD was $7.22\,t\,ha^{-1}$ in 2009 DS, $5.07\,t\,ha^{-1}$ in 2009 WS, and $8.01\,t\,ha^{-1}$ in 2010 DS, respectively. Compared with CF, AWD reduced water input of 7.2%, 5.3%, and 24.5% and lost grain yield of 5.3%, 2.9%, and 6.9% in 2009 DS, 2009 WS, and 2010 DS, respectively. Cabangon [28] reported that mild stress AWD reduced irrigation water input by 8%–20% and severe stress by 19%–25% compared with CF. In this study, there was a large amount of rainfall in both 2009 DS and 2009 WS, which resulted in high water input particularly in 2009 WS. Earlier studies showed that even a 2%–70% water irrigation reduction would not lead to rice yield decrease [5, 7]. In this study, grain yield was not significantly different between AWD and CF in all the three experiments. CF produced a greater yield due to its higher grain weight.

Nitrogen rate had a significant effect on grain yield in all the three experiments. In this study a significant difference in water productivity between N treatments only in the normal dry season such as 2010 DS was observed. In the 2010 DS, CF received 16 irrigations from transplanting to maturity, while 10 irrigations were applied to AWD. The number of irrigation was reduced in 2009 WS, when 2 and 1 irrigations were applied to CF and AWD, respectively. The differences in water productivity between AWD and CF treatments were

(a)

(b)

(c)

—●— Maximum temperature
—○— Minimum temperature
—▲— Solar radiation

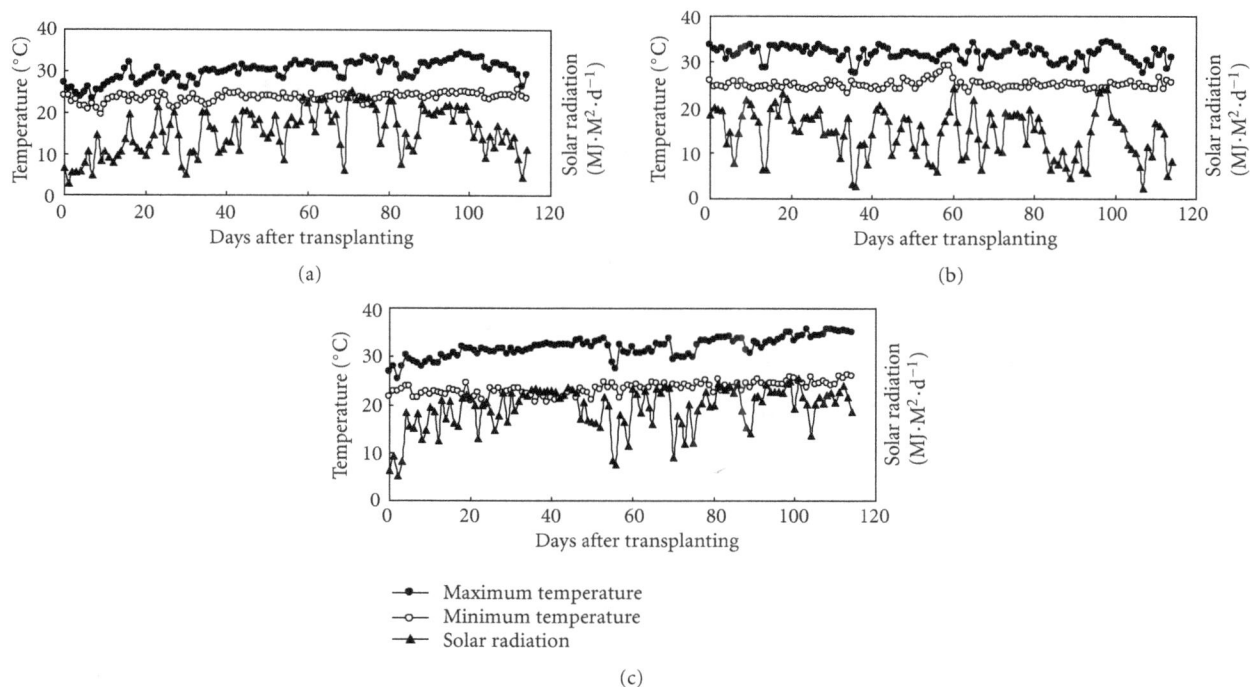

FIGURE 1: Daily maximum and minimum temperatures and solar radiation during rice-growing seasons at the IRRI farm in 2009 DS (a), 2009 WS (b), and 2010 DS (c).

TABLE 1: Rainfall and total water supply (irrigation plus rainfall) of the four rice varieties grown under two water management treatments and two N rates at IRRI farm in the three consecutive seasons.

N	Irrigation (mm)		Rainfall	Total water input (mm)		Reduction (%)
	AWD	CF		AWD	CF	
2009 DS						
LN	517	606	349	866	955	9.3
HN	537	584	349	886	933	5.0
2009 WS						
LN	107	168	1079	1186	1247	4.9
HN	102	172	1079	1181	1251	5.6
2010 DS						
LN	753	1024	92	845	1116	24.3
HN	729	998	92	821	1090	24.7

Data are the means across four varieties. Variety had insignificant effect on the amount of water supply.

TABLE 2: Analysis of variance for grain yield and water use efficiency (WUE) in the three consecutive seasons at IRRI farm, Philippines.

Year	2009 DS		2009 WS		2010 DS	
Source of variation	Yield	WUE	Yield	WUE	Yield	WUE
Water regime (W)	ns	ns	ns	ns	ns	*
Nitrogen (N)	*	ns	**	*	**	*
Variety (V)	*	*	*	ns	*	*
W × N	ns	ns	ns	ns	ns	ns
W × V	ns	ns	ns	ns	ns	ns
N × V	ns	ns	ns	ns	ns	ns
W × N × V	ns	ns	ns	ns	ns	ns

* Significance at the 0.05 level based on analysis of variance.
** Significance at the 0.01 level based on analysis of variance.
ns: denotes nonsignificance based on analysis of variance.

TABLE 3: Grain yield and water productivity of the four rice varieties grown under two water management treatments and two N rates at the IRRI farm for the three consecutive seasons.

Variety	Grain yield (t ha^{-1})				Water productivity (kg m^{-3})			
	LN		HN		LN		HN	
	AWD	CF	AWD	CF	AWD	CF	AWD	CF
2009 DS								
IR72	7.45^a	7.80^b	7.74^{ab}	7.75^b	0.86^a	0.82^{ab}	0.87^{ab}	0.83^b
PSBRc80	7.52^a	7.45^b	7.67^{ab}	8.29^{ab}	0.87^a	0.78^b	0.87^{ab}	0.89^a
IR82372H	7.66^a	8.16^{ab}	8.55^a	9.09^a	0.88^a	0.85^a	0.97^a	0.97^a
SL-8H	7.90^a	8.23^a	7.30^b	8.48^{ab}	0.91^a	0.86^a	0.82^b	0.91^a
Mean	7.63	7.91	7.82	8.40	0.88	0.83	0.88	0.90
2009 WS								
IR72	4.96^a	5.05^a	5.6^a	5.63^a	0.42^a	0.40^a	0.47^a	0.45^a
PSBRc80	5.11^a	4.92^a	5.03^a	5.29^a	0.43^a	0.39^a	0.43^a	0.42^a
IR82372H	4.73^a	4.97^a	5.30^a	5.47^a	0.40^a	0.40^a	0.45^a	0.44^a
Mestizo7	4.64^a	5.12^a	5.22^a	5.34^a	0.39^a	0.41^a	0.44^a	0.43^a
Mean	4.86	5.02	5.29	5.43	0.41	0.4	0.45	0.44
2010 DS								
IR72	7.51^a	7.69^b	8.93^a	8.74^b	0.89^a	0.69^b	1.09^a	0.80^b
PSBRc80	7.57^a	7.94^b	8.86^a	9.50^a	0.90^a	0.71^b	1.08^a	0.87^a
IR82372H	7.21^a	7.97^b	8.35^b	8.74^b	0.85^a	0.71^b	1.02^a	0.80^b
SL-8H	7.53^a	8.92^a	8.14^b	9.37^a	0.89^a	0.80^a	0.99^a	0.86^a
Mean	7.46	8.13	8.57	9.09	0.88	0.73	1.05	0.86

Data are the means across two N rates. Within a column for each season, means followed by the same letters are not significantly different according to LSD (0.05).

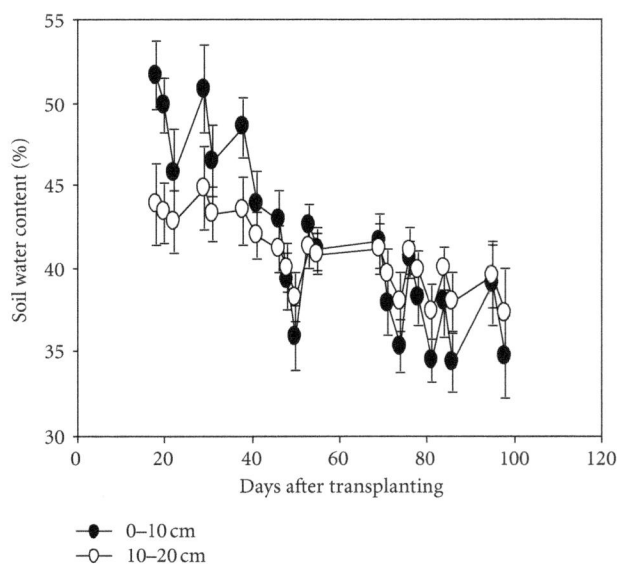

FIGURE 2: Change in soil water content during the growing season under AWD in 2010 DS at the IRRI farm. Data were the means across two rates and four varieties; N rate and variety had no significant effect on soil water content.

insignificant. Water productivity in the two DS ranged from 0.78 to 1.09, which was 2.0–2.4 times higher than in 2009 WS. No significant interactions were observed in terms of variety, water management and, N rate. Varieties with higher yield had greater WUE. AWD received higher WUE than CF due to the decrease in water input. Using high nitrogen fertilization and high yield varieties were the two ways to improved water productivity in this study, as discussed by Hatfield [17].

The difference in grain yield between the hybrid and inbred varieties was relatively slight, except in 2010 DS. Nitrogen rate had a significant effect on grain yield in all the three experiments. Significant differences in grain weight between the AWD and CF treatments were observed in 2010 DS. Panicles per m^2 and spikelets per m^2 were significantly higher in high N than low N. Among the four varieties, the hybrid ones had more spikelets number per m^2 compared with inbreds ones (Table 4). Hybrids had an average of 109 spikelets per panicle, which was 23% higher than the inbreds. IR72 had the highest panicles per m^2 among the varieties. In general, grain filling percentage was the lowest in IR82372H and the highest in IR72. Grain weight of the hybrid variety SL-8H was more than 26.0 mg in two DS. Spikelet number per m^2 was higher in the DS than in the WS and higher in 2010 DS than in 2009 DS.

The LAI at flowering was significantly higher in high N than low N in the two DS. LAI in hybrids was higher than in inbreds in 2009 WS and 2009 DS (Table 5). Differences in the total dry weight at maturity were significant in N treatments, but not significant in water treatments across the three seasons. Harvest index (HI) was significantly higher in hybrids than the inbreds in 2009 WS and 2009 DS. Both HI and LAI (leaf area index) at flowering were higher in the DS

TABLE 4: Yield components of the four rice varieties grown under two water management treatments and two N rates at the IRRI farm for the three consecutive seasons.

	Spikelets panicle^{-1}		Panicles m^2		Grain filling (%)		Grain weight (mg)	
	AWD	CF	AWD	CF	AWD	CF	AWD	CF
2009 DS								
IR72	80.2^c	83.4^c	412.3^a	441.2^a	87.9^a	85.5^a	23.2^d	23.0^d
PSBRc80	101.7^b	104.6^b	370.3^b	363.8^b	82.8^{ab}	81.6^b	23.9^c	23.6^c
IR82372H	120.0^a	117.4^a	330.2^c	344.8^b	78.3^b	77.1^c	24.4^b	24.5^b
SL-8H	119.8^a	124.8^a	228.3^d	301.1^c	81.3^b	83.0^{ab}	26.8^a	27.1^a
Mean	105.4	107.6	335.3	362.7	82.6	81.8	24.6	24.6
2009 WS								
IR72	79.6^b	81.0^b	351.1^a	351.6^a	74.0^a	75.9^{ab}	21.9^c	22.1^c
PSBRc80	94.3^a	96.9^a	295.6^b	296.1^b	74.9^a	76.7^a	22.7^b	22.5^b
IR82372H	98.5^a	99.6^a	307.8^b	298.7^b	66.6^b	71.5^b	23.2^a	23.4^a
Mestizo7	100.4^a	95.2^a	291.2^b	298.2^b	72.1^a	74.7^{ab}	23.4^a	23.5^a
Mean	93.2	93.2	311.4	311.2	71.9	74.7	22.8	22.9
2010 DS								
IR72	70.5^b	72.9^c	517.5^a	504.5^a	90.0^a	89.8^a	22.4^d	22.7^d
PSBRc80	97.4^a	98.8^b	424.0^b	420.1^b	81.9^{bc}	83.6^{ab}	23.0^c	23.1^c
IR82372H	108.9^a	106.6^{ab}	390.3^c	407.0^b	79.0^c	82.4^b	23.5^b	23.8^b
SL-8H	105.9^a	110.1^a	346.9^d	344.0^c	84.8^b	85.9^{ab}	26.2^a	26.5^a
Mean	95.7	97.1	419.7	418.9	83.9	85.4	23.8	24.0

Data are the means across two N rates. Within a column for each site, means followed by the same letters are not significantly different according to LSD (0.05).

TABLE 5: Growth duration, leaf area index (LAI) at flowering, harvest index, and total dry weight of the four rice varieties grown under two water management treatments and two N rates at the IRRI farm for the three consecutive seasons.

	Growth duration (days)		LAI at flowering		Total dry weight (g m^{-2})		Harvest index (%)	
	AWD	CF	AWD	CF	AWD	CF	AWD	CF
2009 DS								
IR72	104	104	5.49^b	5.35^c	1481^a	1557^a	45.5^b	46.4^c
PSBRc80	106	104	5.56^b	5.58^{bc}	1481^a	1459^b	50.1^a	50.1^b
IR82372H	100	100	6.63^a	6.26^{ab}	1441^b	1464^b	50.6^a	52.2^a
SL-8H	106	106	6.82^a	6.71^a	1478^a	1609^a	52.1^a	52.4^a
Mean	104	104	6.13	5.98	1470	1522	49.6	50.3
2009 WS								
IR72	101	102	3.16^b	3.31^a	1089^a	1117^a	41.7^b	42.7^c
PSBRc80	103	103	3.30^{ab}	3.53^a	1078^a	1119^a	43.8^b	44.3^{bc}
IR82372H	102^c	102	3.56^a	3.50^a	1078^a	1105^a	43.5^b	44.9^b
Mestizo7	100	100	3.55^a	3.71^a	1045^b	1055^a	47.2^a	47.1^a
Mean	102	102	3.39^c	3.51	1073	1099	44.1	44.8
2010 DS								
IR72	105	106	4.70^a	5.15^a	1545^b	1560^a	47.7^b	48.1^b
PSBRc80	107	107	4.63^a	5.40^a	1594^b	1578^a	48.6^{ab}	50.7^{ab}
IR82372H	100	100	4.72^a	5.05^a	1554^b	1631^a	50.4^a	50.0^a
SL-8H	111	111	5.04^a	5.01^a	1658^a	1667^a	49.1^{ab}	51.7^{ab}
Mean	106	106	4.77	5.15	1588	1609	49.0	50.1

Data are the means across two N rates. Within a column for each season, means followed by the same letters are not significantly different according to LSD (0.05).

(a)

(b)

AWD
CK

(c)

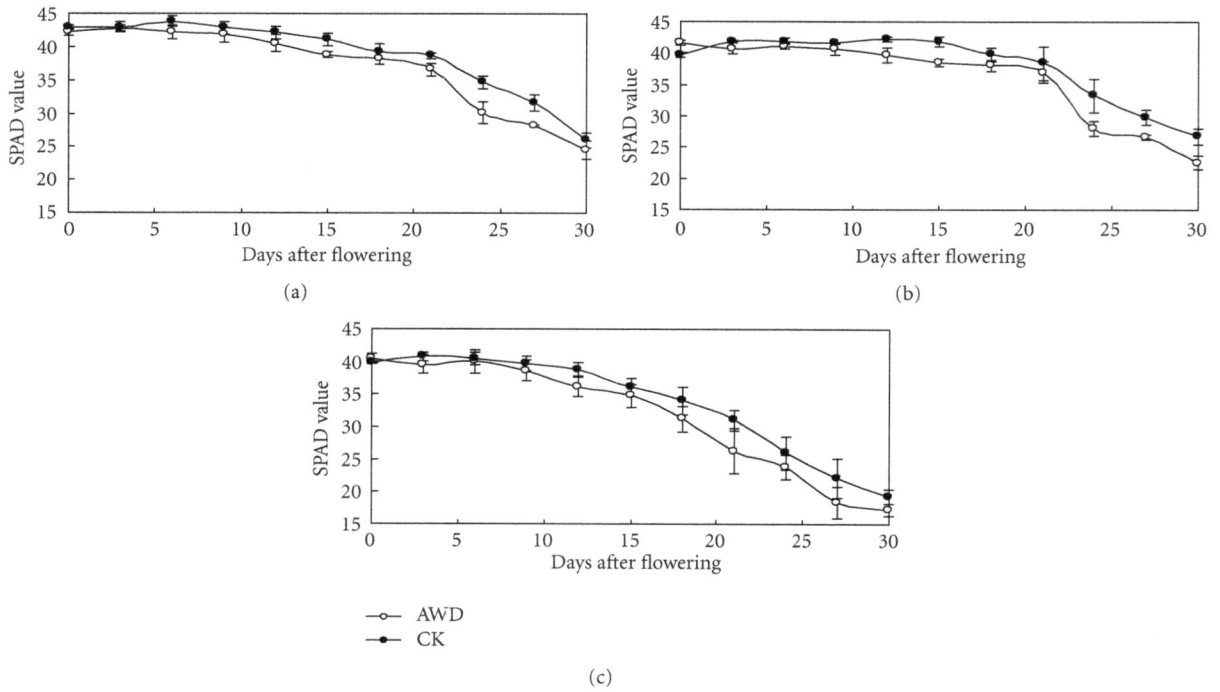

FIGURE 3: SPAD values after flowering under AWD and CF in 2010 DS at the IRRI farm. Three varieties IR72 (a), IR82372H (b), and SL-8H (c) were used in the experiment at high N level.

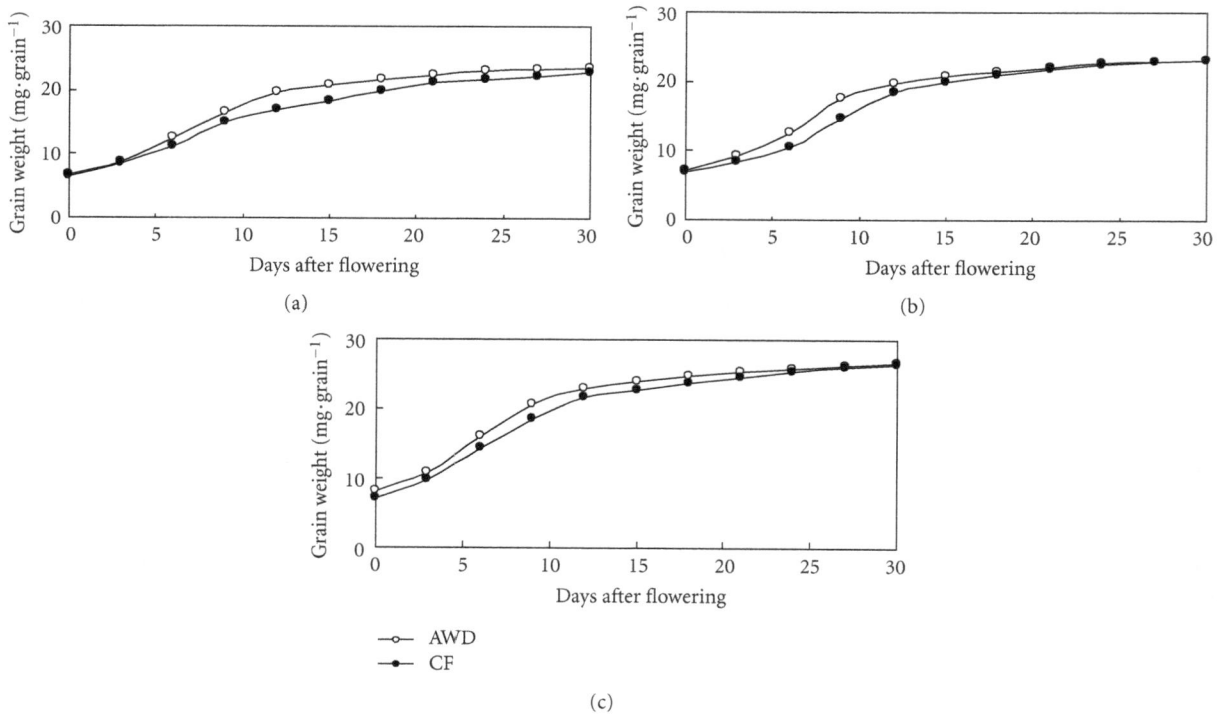

(a)

(b)

AWD
CF

(c)

FIGURE 4: Grain weight after flowering under AWD and CF in 2010 DS. Three varieties IR72 (a), IR82372H (b), and SL-8H (c) were used in the experiment at high N level.

(a)

(b)

(c)

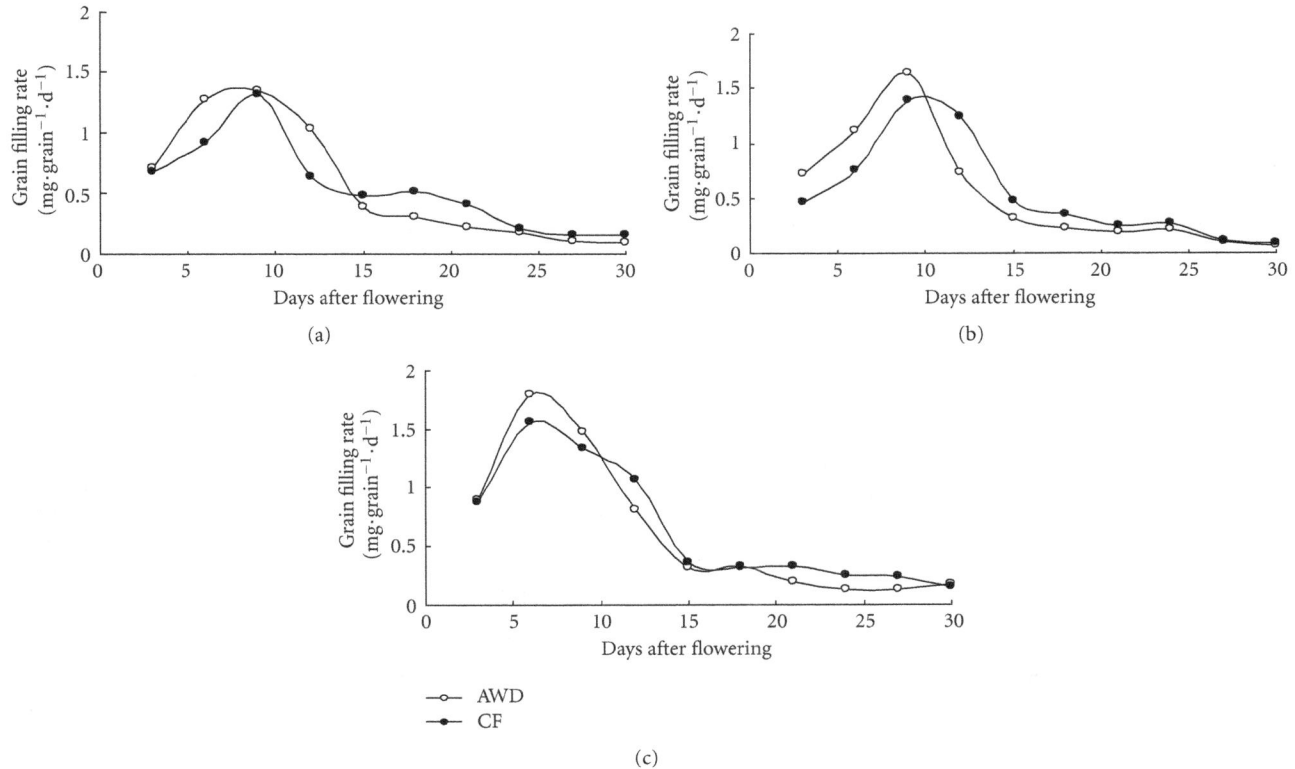

FIGURE 5: Grain filling rate after flowering under AWD and CF in 2010 DS. Three varieties IR72 (a), IR82372H (b), and SL-8H (c) were used in the experiment at high N level.

than in the WS. Average HI was 49.9% in 2009 DS and 44.8% in 2010 DS.

SPAD values were significantly different between AWD and CF in three varieties at the grain filling stage in 2010 DS (Figure 3). The SPAD values of AWD at the first flowering were slightly higher than those of CF. SPAD values rapidly decreased on the 21st day after flowering in IR72 and IR82372H, while values decreased on the 15th day after flowering in SL-8H due to leaf senescence. However, on the 30th day after flowering the SPAD values were obviously lower than those of the CF. Among the three varieties, SPAD values of IR72 and IR8237H were significantly higher than those of SL-8H, particularly on the 30th day after flowering. Therefore, the SPAD value of AWD was higher in the early grain filling stage and lower in the later stage, but CF kept stable leaf senescence and high SPAD value in the later grain filling stage.

Grain filling rates ($mg \cdot grain^{-1} \cdot day^{-1}$) were significantly different between AWD and CF in all the three varieties in grain filling stage (Figures 4 and 5). Grain filling rate of AWD was high at the early grain filling stage and low at the late grain filling stage, but under CF condition grain filling rate was still high at the late grain filling stage. The maximum grain filling rate occurred on the 9th day after flowering in both IR72 and IR8237H and on the 6th day in SL-8H. The maximum grain filling values between AWD and CF were $1.35 \, mg \cdot grain^{-1} \cdot day^{-1}$ and

$1.31 \, mg \cdot grain^{-1} \cdot day^{-1}$ for IR72, $1.64 \, mg \cdot grain^{-1} \cdot day^{-1}$ and $1.39 \, mg \cdot grain^{-1} \cdot day^{-1}$ for IR82372H, and $1.80 \, mg \cdot grain^{-1} \cdot day^{-1}$ and $1.56 \, mg \cdot grain^{-1} \cdot day^{-1}$ for SL-8H, respectively. The average grain filling rate was $0.77 \, mg \cdot grain^{-1} \cdot day^{-1}$ in IR72, $0.77 \, mg \cdot grain^{-1} \cdot day^{-1}$ in IR82372H, and $0.89 \, mg \cdot grain^{-1} \cdot day^{-1}$ in SL-8H. Cereal grains accumulate carbohydrates, proteins, and fatty acids via different pathways during their development [29]. Grain filling plays an important role in grain weight, which is an essential determinant of grain yield in cereal crops, and is characterized by its duration and rate [25]. AWD treatment increased the grain filling rate and shortened grain filling period. Active grain filling period was shortened by 2.1 days and grain filling rate increased by 0.15 mg per day per grain compared with CF. The SPAD value of AWD was higher in the early grain filling stage and lower in the later stage, but CF kept stable leaf senescence and high SPAD value in the later grain filling stage. AWD reduced water input by 25% in tropical area in 2010 DS but decreased grain yield by 5% due to accelerated leaf senescence. High SPAD value and grain filling rate in the later grain filling stage were partially responsible for the high yield of flooded water management. Water deficiency lead to hormonal change, which enhanced whole plant senescence and accelerated grain filling [22].

In conclusion, interaction effects among variety, water management, and N rate were not significant under tropical condition. Grain yield was not significantly different between

AWD and CF in all the seasons through saving water input. Using high nitrogen fertilization and high yield varieties were the two ways to improve water productivity in this study; severe water stress during late grain filling stage accelerated grain filling rate, shortened the grain filling period, and enhanced whole plant senescence, thus reducing grain weight. The study indicated that proper water management greatly contributed to grain yield in the late grain filling stage, and it was critical for safe AWD technology.

Acknowledgments

National Basic Research Program of China (Project no. 2009CB118603) is acknowledged for funding the doctoral studies of the first author, and the Ministry of Science and Technology in China (Project nos. 2011BAD16B14 and 2012BAD04B00) is acknowledged for their financial support. Alex and Eddie are acknowledged for the excellent crop management.

References

[1] D. C. Maclean, B. Dawe, and G. P. Hettel, *Rice Almanac*, International Rice Research Institute, Los Banos, Philippines, 3rd edition, 2002.

[2] D. Normile, "Reinventing rice to feed the world," *Science*, vol. 321, no. 5887, pp. 330–333, 2008.

[3] V. K. Arora, "Application of a rice growth and water balance model in an irrigated semi-arid subtropical environment," *Agricultural Water Management*, vol. 83, no. 1-2, pp. 51–57, 2006.

[4] T. P. Tuong and B. A. M. Bouman, "Rice production in water-scarce environments," in *Water Productivity in Agriculture: Limits and Opportunities for Improvement*, J. W. Kijne, R. Barker, and D. Molden, Eds., vol. 1 of *The Comprehensive Assessment of Water Management in Agriculture Series*, pp. 13–42, CABI, Wallingford, UK, 2002.

[5] L. C. Guerra, S. I. Bhuiyan, T. P. Tuong, and P. Barker, "Producing more rice with less water from irrigated systems," SWIM 5.IWMI/IRRI, Colombo, Sri Lanka, 1998.

[6] R. Barker, D. Dawe, T. P. Tuong, S. I. Bhuiyan, and L. C. Guerra, "The outlook for water resources in the year 2020: challenges for research on water management in rice production," in *Proceeding of the 19th Session of the International Rice Commission*, Cairo, Egypt, 1998, Assessment and orientation towards the 21st century.

[7] B. A. M. Bouman and T. P. Tuong, "Field water management to save water and increase its productivity in irrigated lowland rice," *Agricultural Water Management*, vol. 49, no. 1, pp. 11–30, 2001.

[8] T. P. Tuong, B. A. M. Bouman, M. Martian, and More rice, "less water-integrated approaches for increasing water productivity in irrigated rice-based systems in Asia," in *Proceedings of the 4th International Crop Science Congress*, Brisbane, Australia, 2004.

[9] H. T. Nguyen, K. S. Fischer, and S. Fukai, "Physiological responses to various water saving systems in rice," *Field Crops Research*, vol. 112, no. 2-3, pp. 189–198, 2009.

[10] T. P. Tuong and S. I. Bhuiyan, "Increasing water-use efficiency in rice production: farm-level perspectives," *Agricultural Water Management*, vol. 40, no. 1, pp. 117–122, 1999.

[11] Y. Li, "Water saving irrigation in China," *Irrigation and Drainage*, vol. 55, no. 3, pp. 327–336, 2006.

[12] P. Belder, B. A. M. Bouman, R. Cabangon et al., "Effect of water-saving irrigation on rice yield and water use in typical lowland conditions in Asia," *Agricultural Water Management*, vol. 65, no. 3, pp. 193–210, 2004.

[13] P. Moya, L. Hong, D. Dawe, and C. Chen, "The impact of on-farm water saving irrigation techniques on rice productivity and profitability in Zhanghe irrigation system, Hubei, China," *Paddy Water Environment*, vol. 2, no. 4, pp. 207–215, 2004.

[14] T. R. Sinclar, C. B. Tanner, and J. M. Bennett, "Water use efficiency in crop production," *Bioscience*, vol. 34, no. 1, pp. 36–40, 1984.

[15] R. A. Fischer, "Optimizing the use of water and nitrogen through breeding of crops," *Plant and Soil*, vol. 58, no. 1–3, pp. 249–278, 1981.

[16] C. B. Tanner and T. R. Sinclair, "Efficient water use in crop production: research or research?" in *Limitations to Efficient Water Use in Crop Production*, H. M. Taylor, J. R. Wayne, and S. T. Thomas, Eds., pp. 1–27, American Society of Agronomy, Madison, Wis, USA, 1983.

[17] J. L. Hatfield, T. J. Sauer, and J. H. Prueger, "Managing soils to achieve greater water use efficiency: a review," *Agronomy Journal*, vol. 93, no. 2, pp. 271–280, 2001.

[18] S. Peng, F. V. Garcia, R. C. Laza, and K. G. Cassman, "Adjustment for specific leaf weight improves chlorophyll meter's estimate of rice leaf nitrogen concentration," *Agronomy Journal*, vol. 85, no. 5, pp. 987–990, 1993.

[19] S. Peng, M. R. Laza, F. V. Garcia, and K. G. Cassman, "Chlorophyll meter estimates leaf area-based nitrogen concentration of rice," *Communications in Soil Science and Plant Analysis*, vol. 26, no. 5-6, pp. 927–935, 1995.

[20] V. Buchanan-Wollaston, "The molecular biology of leaf senescence," *Journal of Experimental Botany*, vol. 48, no. 307, pp. 181–199, 1997.

[21] N. Ori, M. T. Juarez, D. Jackson, J. Yamaguchi, G. M. Banowetz, and S. Hake, "Leaf senescence is delayed in tobacco plants expressing the maize homeobox gene knotted1 under the control of a senescence-activated promoter," *Plant Cell*, vol. 11, no. 6, pp. 1073–1080, 1999.

[22] J. Yang, J. Zhang, Z. Wang, Q. Zhu, and L. Liu, "Abscisic acid and cytokinins in the root exudates and leaves and their relationship to senescence and remobilization of carbon reserves in rice subjected to water stress during grain filling," *Planta*, vol. 215, no. 4, pp. 645–652, 2002.

[23] E. S. Ober and T. L. Setter, "Timing of kernel development in water-stressed maize: water potentials and abscisic acid concentrations," *Annals of Botany*, vol. 66, no. 6, pp. 665–672, 1990.

[24] J. Yang, J. Zhang, Z. Wang, Q. Zhu, and W. Wang, "Hormonal changes in the grains of rice subjected to water stress during grain filling," *Plant Physiology*, vol. 127, no. 1, pp. 315–323, 2001.

[25] W. Yang, S. Peng, M. L. Dionisio-Sese, R. C. Laza, and R. M. Visperas, "Grain filling duration, a crucial determinant of genotypic variation of grain yield in field-grown tropical irrigated rice," *Field Crops Research*, vol. 105, no. 3, pp. 221–227, 2008.

[26] D. B. Jones, M. L. Peterson, and S. Geng, "Association between grain filling rate and duration and yield components in rice," *Crop Science*, vol. 19, no. 5, pp. 641–644, 1978.

[27] S. A. S. Institute, SAS Version 9.1.2 2002-2003. SAS Institute, Inc., Cary, NC, 2003.

[28] R. J. Cabangon, E. G. Castillo, and T. P. Tuong, "Chlorophyll meter-based nitrogen management of rice grown under alternate wetting and drying irrigation," *Field Crops Research*, vol. 121, no. 1, pp. 136–146, 2011.

[29] T. Zhu, P. Budworth, W. Chen et al., "Transcriptional control of nutrient partitioning during rice grain filling," *Plant Biotechnology Journal*, vol. 1, pp. 59–70, 2003.

7

An Efficient In Vitro Propagation Protocol of Cocoyam [*Xanthosoma sagittifolium* (L) Schott]

Anne E. Sama,[1] Harrison G. Hughes,[1] Mohamed S. Abbas,[2] and Mohamed A. Shahba[1]

[1] *Department of Horticulture & Landscape Architecture, Fort Collins, CO 80523-1173, USA*
[2] *Department of Natural Resources, Institute of African Research and Studies, Cairo University, Giza 12613, Egypt*

Correspondence should be addressed to Mohamed A. Shahba, shahbam@lamar.colostate.edu

Academic Editor: Daoxin Xie

Sprouted corm sections of "South Dade" white cocoyam were potted and maintained in a greenhouse for 8 weeks. Shoot tips of 3–5 mm comprising the apical meristem with 4–6 leaf primordial, and approximately 0.5 mm of corm tissue at the base. These explants were treated to be used into the culture medium. A modified Gamborg's B5 mineral salts supplemented with $0.05\,\mu M$ 1-naphthaleneacetic acid (NAA) were used throughout the study. Thidiazuron (TDZ) solution containing 0.01% dimethyl sulfoxide (DMSO) was used. Erlenmeyer flasks and test tubes were used for growing cultures. The effect of different media substrate, thidiazuron, and the interaction between TDZ and Benzylaminopurine (BAP) on cocoyam culture were tested. Results indicated that cocoyam can be successfully micropropagated in vitro through various procedures. All concentrations tested (5–$20\,\mu M$ BAP and 1–$4\,\mu M$ TDZ) produced more axillary shoots per shoot tip than the control without cytokinins. Greater proliferation rates were obtained through the use of $20\,\mu M$ BAP and $2\,\mu M$ TDZ, respectively, 12 weeks from initiation. Shoots produced with BAP were larger and more normal in appearance than those produced with TDZ, which were small, compressed, and stunted. The use of stationary liquid media is recommended for economic reasons.

1. Introduction

Cocoyam [*Xanthosoma sagittifolium* (L) Schott] is an herbaceous, monocotyledonous crop that belongs to *Araceae* family. The stem is a starch-rich underground structure, the corm, from which offshoots called cormels develop. Flowering is rare, but when it occurs, the inflorescence consists of a cylindrical spadix of flowers enclosed in a 12–15 cm spathe [1]. It is a staple food in the tropics and subtropics and one of the six most important root and tuber crops worldwide [2]. The corm, cormels, and leaves of cocoyam are an important source of carbohydrates for human nutrition, animal feed [3–5], and of cash income for farmers [6]. Africa produces about 75% of the world production which is about 0.45 million tons [7]. Cocoyam production requires high labor and water. Also, its breeding is difficult in addition to its sensitivity to diseases and pests [8].

Cocoyam usually propagates vegetatively from tuber fragments, which increase pathogens distribution. Vegetatively propagated commercial varieties are highly susceptible to the cocoyam root rot disease caused by *Pythium*

myriotylum [9], and Dasheen mosaic virus that is found in the leaves, corm, and cormels [10]. Trials have been made using conventional procedures to rapidly increase cocoyam-planting material. Micropropagation is an efficient method to mass propagate good-quality materials that substantially improves production. It involves the use of defined growth media supplemented with appropriate growth regulators that enable morphogenesis to occur from naturally growing plant parts. This helps in producing a large number of plants from a single individual in short time and in limited space [11]. Previous studies have shown that shoot multiplication, somatic embryogenesis, and tuberization could be induced in shoot tips of cocoyam cultured in vitro on Murashige and Skoog medium [12] supplemented with various combinations of indol butyric acid (IBA), 1-naphthalene acetic acid (NAA), 2,4-dichlorophenoxyacetic acid (2,4-D), Benzylaminopurine (BAP), and kinetin [13].

The biochemical aspects of induction of in vitro organogenesis have been investigated in a number of plants including carrot [14], pea [15], summer squash [16, 17], winter squash [18], soybean [19], taro [20], watermelon

[21], groundnut [22], asparagus [23], black pepper [24], canola [25], cotton [26], date palm [27], lentil [28], common bean [29], sunflower [30], rice [31], and banana [32]. In spite of its importance in many countries, cocoyam has received very little research attention and is considered insufficiently studied crop [33]. According to Goenaga and Chardon [34], the yield potential of cocoyam is seldom realized, mainly because of a lack of knowledge concerning diseases, proper management practices, and physiological determinants that may limit plant growth and development. In this respect, this study will hopefully contribute to a sustainable cocoyam production. Although it was proposed that large numbers of cocoyam could be produced in vitro, the techniques were not adequately standardized for routine micropropagation. Therefore, the objective of this work is to verify and improve micropropagation of cocoyam via axillary shoot proliferation.

2. Materials and Methods

2.1. Source of Explants. Cocoyam "South Dade" white plants were obtained from the Tropical Fruit Company, Homestead, FL as sprouted corm sections. Each of these sections was potted in polyethylene pots ($\cong 100 \, cm^2$) in a mix of peat, perlite, and vermiculite ($1:1:0.5$ by volume). These plants were maintained in a greenhouse under natural photoperiods. Temperature was maintained at $23 \pm 2°C$. Plants were watered as needed with tap water and fertilized with liquid fertilizer containing $N:P:K$ at $20:10:20$ by volume twice a week. After 8 weeks of planting, sprouts were collected, trimmed to about 5 cm, and washed under running tap water for 30–60 minutes. These were further excised to finally obtain shoot tips of 3–5 mm comprising the apical meristem with 4–6 leaf primordial, and approximately 0.5 mm of corm tissue at the base. These explants were disinfected in a laminar flow hood before transferred into the culture medium.

2.2. Basal Medium (BM). A modified Gamborg's B5 mineral salts [35] supplemented with $0.05 \, \mu M$ 1-naphthaleneacetic acid (NAA) were used throughout the study. The modified component of B5 microsalts was $MnSO_4 \cdot 4H_2O$ at $10 \, mg \, L^{-1}$. Organics consisted of myo-inositol ($100 \, mg \, L^{-1}$), thiamine HCl ($10 \, mg \, L^{-1}$), nicotinic acid ($1 \, mg \, L^{-1}$), and pyridoxine HCl ($10 \, mg \, L^{-1}$). Sucrose was provided at $30 \, g \, L^{-1}$ as a source of carbon and energy. Whenever a semisolid medium was desirable, agar (Sigma agar, type A) was added at a concentration of 0.4%. The pH of the medium was adjusted to 5.7 ± 0.02. Thidiazuron (TDZ) solution containing 0.01% dimethyl sulfoxide (DMSO) was used. Erlenmeyer flasks (125 mL) and test tubes ($25 \times 150 \, mm$) were used for growing cultures. Aliquots of 25 mL and 15 mL were dispensed into the flasks and test tubes, respectively. Flasks were stoppered with nonabsorbent cotton plugs, and then covered with aluminium foil. Test tubes were covered with polypropylene closures, Kaput caps (Bellco Glass, Inc., NJ, USA). The media-containing vessels were then autoclaved for 18 minutes at $121°C$.

2.3. Explant Establishment and Multiplication. To test the effect of different media substrate, solid or liquid with the most efficient shaking pattern, explants were initiated on three media supplemented with either $5.0 \, \mu M$ Benzylaminopurine (BAP), $20.0 \, \mu M$ BAP, or $2.0 \, \mu M$ TDZ. Each medium was either solidified with 0.4% agar or maintained in the liquid state. Liquid media were either continuously shaken on a rotary shaker (Model New Brunswick Scientific, Edison, N. J.) at 80 rpm, held stationary but with the suspension of the explants in the medium, or held stationary with the tissue supported on a filter paper (Whatman no. 1) bridge. Treatments were replicated 10 times, and the whole experiment was repeated twice. Cultures were monitored biweekly and rated from 1 to 4 for survival frequency and shoot elongation, where 1 = creamy or dead cultures with no apparent growth, 2 = growth initiation and appearance of green coloration, 3 = increase in growth, green coloration, and leaf differentiation, and 4 = development of healthy green leaves.

To test TDZ effect on multiplication, the BM was supplemented with TDZ at levels of 1.0, 2.0, 4.0, and $8.0 \, \mu M$ as well as with $5 \, \mu M$ BAP which served as the control. Test tubes of stationary liquid media, without any form of support, were used in all cases. The treatments were replicated 10 times. Explants of 3–5 or 6–10 mm were used. Shoot length, base diameter, and number of axillary shoots as well as roots formed per culture were monitored biweekly for six weeks.

To test the effect of the interaction between TDZ and BAP on cocoyam culture, the explants were cultured on agitated liquid media using six treatments of BAP at 0.0, 10.0, and $20.0 \, \mu M$ factorially combined with TDZ levels of 0.0 and $2.0 \, \mu M$ and supplemented with $0.05 \, \mu M$ NAA. The cultures were replicated 20 times and were maintained in their various initiation media for six weeks. They were monitored biweekly for shoot length, base diameter, axillary shoots, and adventitious root formation. At the end of the initiation phase, shoots were trimmed of any axillary shoots to ensure uniformity and transferred into two media for proliferation. These multiplication media consisted of BM supplemented with either $20.0 \, \mu M$ BAP or $2.0 \, \mu M$ TDZ. Cultures were monitored for eight weeks, principally for the formation of axillary shoots and adventitious roots in addition to shoot length and base diameter. A second subculture was made into fresh media with microshoots serving as explants. Culture was either maintained in their respective treatments on shakers, or subcultured into semisolid media in test tubes. The latter cultures were derived from the $2.0 \, \mu M$ TDZ treatment only. Cultures in the semisolid media were treated in two different ways. They were either maintained in the same media or were subcultured into one that was hormone-free. After six weeks, microshoots from the semisolid media were subsequently subcultured into BM and hormone-free media contained in flasks and test tubes. Cultures were incubated for four weeks, and data were collected on shoot and root formation at two-week intervals. Using another method, cultures were initiated on stationary liquid media in test tubes, and the source material was six week old tissue culture-regenerated plants grown in the greenhouse. The

TABLE 1: Analysis of variance with mean squares and treatment significance of media substrate and growth regulator treatments effect on relative growth of cocoyam shoot tips after 4 weeks of initiation.

Source	DF	Mean squares	P value*
Substrate (S)	3	3.44	0.019
Growth regulators (R)	2	12.32	<0.0001
S × R	6	2.24	0.014
Rep	9	15.6	0.181

*Significant at $P \leq 0.05$.

test media consisted of BM supplemented with BAP at 5.0 and 10.0 μM in factorial combinations with 0.0, 1.0, 2.0, and 4.0 μM TDZ. Also, cultures were initiated on stationary liquid media in Erlenmeyer flasks. This medium consisted of BM supplemented with 5.0 μM BAP. BAP levels used were 0.0, 5.0, 10.0, and 20.0 μM. These levels were combined with TDZ at 0.0, and 2.0 μM in factorial design.

All cultures were incubated in growth chambers maintained at 25 ± 3°C under continuous illumination. These culture conditions were the same for each of the different stages, with slight differences associated with location within the growth room. Light intensities were measured with an LI-185 Quantum/Radiometer/Photometer (Lambda Instruments Corp., Lincoln, Nebraska).

2.4. Data Analysis. Experiments were laid out as a complete block design. All data were subjected to an analysis of variance using unequal replications where contamination was observed. Treatment means were separated by Tukey's Multiple Range Test at a 5% level of significance (SAS Institute, 2006).

3. Results

3.1. Media Substrate Effect on Cocoyam Culture. Generally, the disinfection procedures resulted in low contamination ranging from none in solid media to only 7.0% in agitated and stationary media. All explants survived as evidenced from their enlargement and manifested by an elongation of the shoot tip and a swelling of the base by the second week of culture. At the same time, most cultures changed from the initial creamy color to green coloration. Analysis of variance indicated a significant difference among different medium substrates and among different growth regulator treatments during the initiation of cocoyam tissue cultures (Table 1). Initiation on liquid culture, either on shaker or held stationary, was better than filter bridges or solid medium (Figure 1). Evaluations indicated that 2.0 μM TDZ and 5.0 μM BAP are important in the initiation of cocoyam tissue cultures. Comparisons indicated that 2.0 μM TDZ was significantly better with 5.0 μM BAP intermediate than 20.0 μM BAP which inhibited shoot elongation (Figure 2).

3.2. Media Substrate Effect on Cocoyam Multiplication. Analysis of variance indicated a significant difference in the number of axillary shoots per shoot tip between stationary liquid

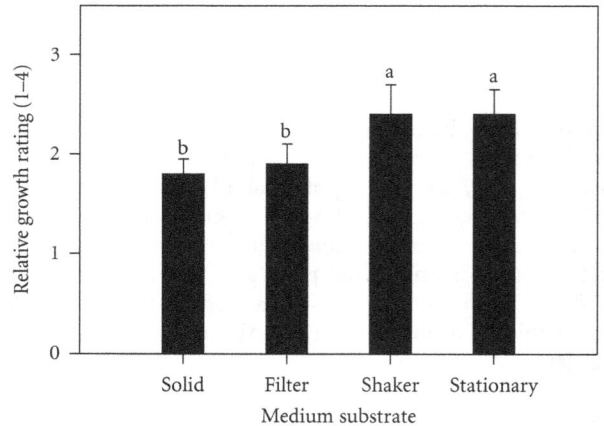

FIGURE 1: Effect of substrates (solid, liquid filter, liquid shaker and liquid stationary) on relative growth of cocoyam shoot tips cultured on growth regulator treatments after 4 weeks of initiation. Relative growth was rated on a scale of 1–4, with 4 be the highest growth. Columns labeled with the same letter are not significantly different at $P = 0.05$ using Multiple Range Test. Vertical bars at the top represent standard errors.

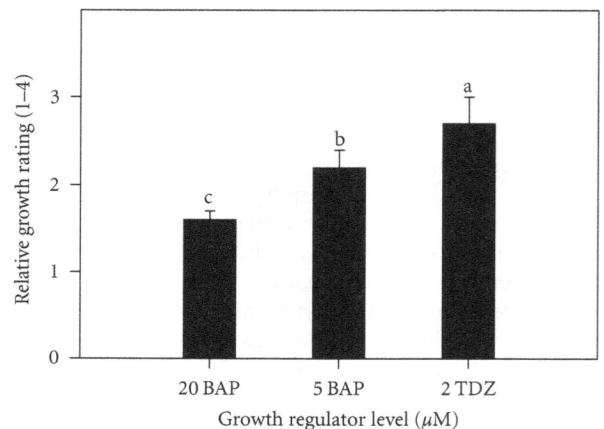

FIGURE 2: Effect of growth regulator levels on relative growth of cocoyam shoot tips cultured on medium substrates after 4 weeks of initiation. Relative growth was rated on a scale of 1–4, with 4 be the highest growth. Columns labeled with the same letter are not significantly different at $P = 0.05$ using Multiple Range Test. Vertical bars at the top represent standard errors.

and agitated liquid media (Figure 3), among the growth regulators treatments (Figure 4) and their interaction. The highest average number of shoots per shoot tip (9.1) was obtained with 1.0 μM TDZ treatment as compared to 5.0 μM BAP (1.9) and 20.0 μM BAP (0.8).

3.3. Thidiazuron Influence on Shoot-Tip Initiation. Analysis of variance indicated no significant difference in number of cocoyam shoots by the second, fourth, or even sixth week of culture, among different TDZ concentrations (1.0, 2.0, 4.0, and 8.0 μM) as compared with 5.0 μM BAP as a control treatment. Initial size of the explants affected its growth and

TABLE 2: Analysis of variance with mean squares and treatment significance of the interaction effect of TDZ and BAP on cocoyam shoot and root proliferation during multiplication after 4 and 6 weeks.

Duration		Four weeks			Six weeks	
Source	DF	Mean squares	P value	DF	Mean squares	P value*
Shoot proliferation:						
Growth regulators	5	1212.3	<0.0001	5	2212.6	<0.0001
Rep	20	2222.2	0.14	20	3256.0	0.44
Root proliferation:						
Growth regulators	5	955.5	<0.0001	5	1462.0	<0.0001
Rep	20	1120.8	0.26	20	1230.0	0.28
Shoot length:						
Growth regulators	5	3200.5	<0.0001	5	4355.0	<0.0001
Rep	20	6612.0	0.65	20	7655.0	0.57

* Significant at $P \leq 0.05$.

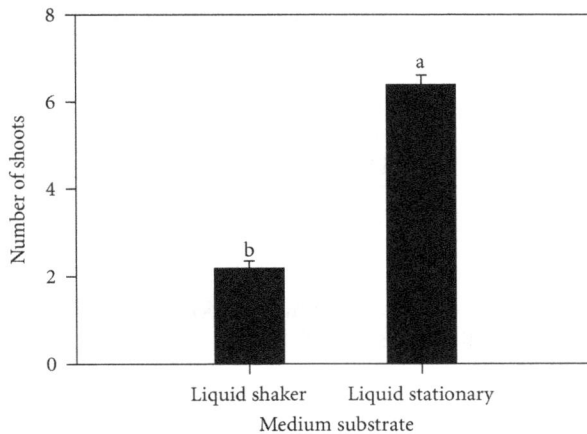

FIGURE 3: Effect of substrates (liquid shaker and liquid stationary) on number of shoots multiplied on different growth regulator levels after 6 weeks of subculture. Columns labeled with the same letter are not significantly different at $P = 0.05$ using Multiple Range Test. Vertical bars at the top represent standard errors.

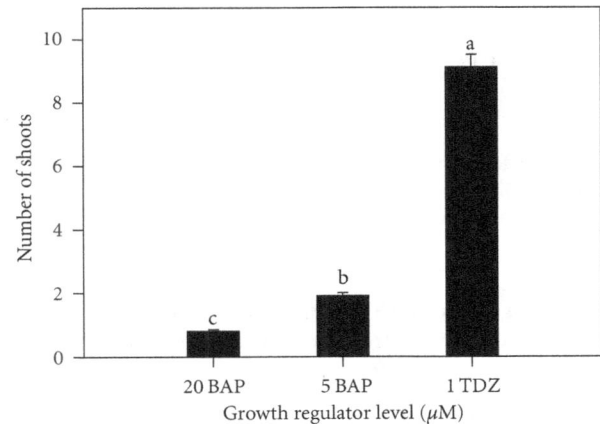

FIGURE 4: Effect of growth regulator levels on the numbers of cocoyam shoots multiplied on liquid substrates after 6 weeks of subculture. Columns labeled with the same letter are not significantly different at $P = 0.05$ using Multiple Range Test. Vertical bars at the top represent standard errors.

development. The larger explants (6–10 mm) established and developed faster that smaller ones (3–5 mm).

3.4. Interaction Effect of TDZ and BAP on Cocoyam Culture on Agitated Liquid Media.

Results indicated that the rate of shoot proliferation was slow under all six treatments during the six weeks of culture initiation with no significant difference among treatments. In multiplication media, the proliferation rate was low under all treatments for the first two weeks, but increased by the fourth and sixth weeks (Figure 6). Analysis of variance indicated that 20.0 μM BAP had a better effect on roots number and shoot length while 2.0 μM TDZ had a better effect on shoots number after 6 weeks. The effect of both TDZ and BAP was similar after 4 weeks of culture. After six weeks of culture, 20.0 μM BAP achieved an average shoots number of 9.7, an average roots number of 6.0, and an average shoot length of 75.0 mm while 2.0 μM TDZ achieved an average shoots number of 13.1, an

average roots number of 1.7, and an average shoot length of 65.1 mm.

Analysis of variance indicated a significant effect of different growth regulators treatments in the initiation culture during the multiplication phase (Table 2). The average number of shoots per culture ranged from 3.0 in the cytokinin-free control to 10.9 in culture with 20.0 μM BAP as well as the culture with 20.0 μM BAP plus 2.0 μM TDZ after 4 weeks of culture (Figure 5(a)). After 6 weeks, it ranged from 4.3 in the control to 16.2 in culture with 2.0 μM TDZ. Roots number and shoots length were similarly affected by the growth regulators treatments. The control treatment had the best effect on roots number (7.7) followed by 10.0 μM BAP (5.8) and 20.0 μM BAP (5.0) (Figure 5(a)). Also, 10.0 μM BAP achieved the highest average shoot length (79.3 mm) (Figure 5(b)).

After subculturing twice, shoot proliferation and root formation were negatively correlated ($r = -0.65$ and -0.57 after 4 and 6 weeks resp.). Mean number of axillary

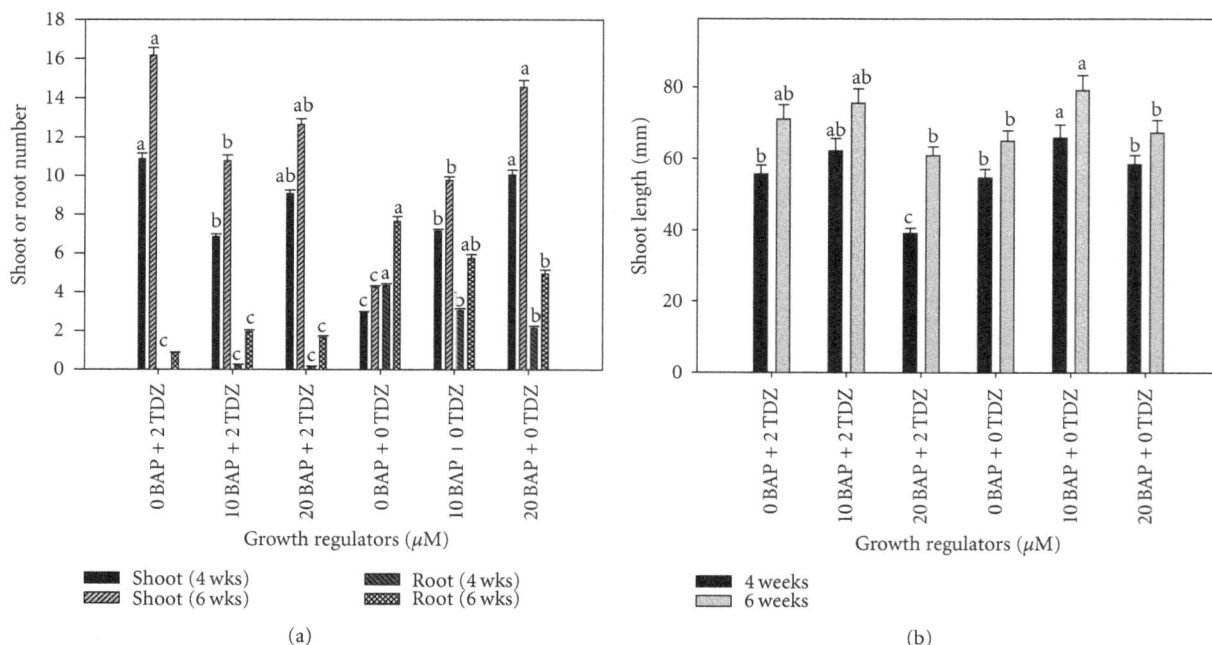

FIGURE 5: Effect of initiation growth regulator treatments on cocoyam shoot proliferation, root formation (a) and shoot length (b) during multiplication. Columns labeled with the same letter are not significantly different at $P = 0.05$ using Multiple Range Test for treatment effect comparison at 4 and 6 weeks. Vertical bars at the top represent standard errors.

shoots ranged from 18.5 with $20.0\,\mu M$ BAP to 26.7 with the cytokinin-free control after 4 weeks and from 19.7 with $20.0\,\mu M$ BAP to 28.7 with the control treatment after six weeks (Figure 6). The superiority of BAP to TDZ on root formation was evident. Shoots proliferated on BAP developed roots after the second week in culture. The poor rooting ability in TDZ as compared to the previous multiplication phase indicated its repressive effect on rooting. Root number ranged from 2.3 and 3.7 for $10.0\,\mu M$ BAP to 4.7 and 6.0 for $20.0\,\mu M$ (Figure 6).

3.5. Interaction Effect of TDZ and BAP on Cocoyam Culture on Stationary Liquid Media in Test Tubes. After six weeks of initiation on stationary liquid media, a few shoots developed and there was no significant treatment effect. The average number of new shoots ranged from 0.1 for the joint effect of $10.0\,\mu M$ BAP and $1.0\,\mu M$ TDZ to 0.7 for the cytokinin-free control. The highest average root number was produced with the control (9.6), with 100% rooting followed by 53% with $5.0\,\mu M$ BAP and 6.7% with $10.0\,\mu M$ BAP.

Analysis of variance indicated a significant difference among treatments after 4 weeks in the multiplication media. The average number of shoots per culture was 0.5 in the cytokinin-free control and was 3.9 in a media with $10.0\,\mu M$ BAP while it was 6.7 with $5.0\,\mu M$ BAP (Figure 7). At six weeks, the pattern of shoot production changed when the greatest number of shoots was produced in $10.0\,\mu M$ BAP combined with $1.0\,\mu M$ TDZ. Generally, there was an approximate doubling of shoot numbers from four to six weeks in all combinations of TDZ with BAP (Figure 7). Growth regulators levels significantly affected rooting, and

there was no significant difference in roots numbers between week four and week six. TDZ completely suppressed root formation (Figure 7).

3.6. Interaction Effect of TDZ and BAP on Cocoyam Culture on Stationary Liquid Media in Erlenmeyer Flasks. Explants subcultured on Stationary Liquid Media in Erlenmeyer Flasks proliferated heavily. After 4 weeks of subculture, an average of 36.6 shoots per culture were formed in media containing $20.0\,\mu M$ BAP combined with $2.0\,\mu M$ TDZ. Shoot proliferation increased two weeks later in the same order, where an average of 44.5 shoots was produced with same previously mentioned combination (Figure 8).

An average of 11 roots was formed per culture after 4 weeks of subculture, and 13.3 roots were formed after 6 weeks in the absence of both BAP and TDZ. In cultures containing only $5.0\,\mu M$ BAP, the average number of roots was 9.1 after 4 weeks and 12.2 after 6 weeks. The presence of TDZ in the culture was completely suppressive to root formation (Figure 8).

4. Discussion

Liquid culture, either on shaker or held stationary, was more efficient during initiation than filter bridges or solid medium. Evaluations indicated that $2.0\,\mu M$ TDZ was significantly better with $5.0\,\mu M$ BAP intermediate than $20.0\,\mu M$ BAP which inhibited shoot elongation. Shoots proliferated on BAP developed roots after the second week in culture. The poor rooting ability in TDZ as compared to the previous multiplication phase indicated its repressive effect

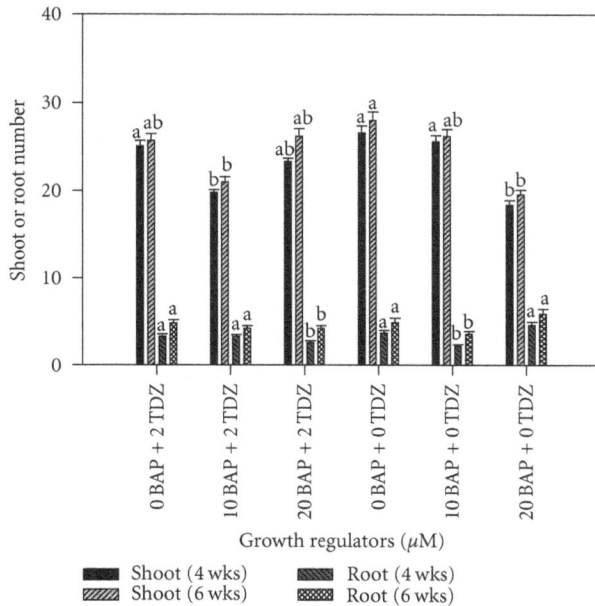

FIGURE 6: Effect of initiation growth regulator treatments on cocoyam shoot and root proliferation, after subculturing twice. Columns labeled with the same letter are not significantly different at $P = 0.05$ using Multiple Range Test for treatment effect comparison at 4 and 6 weeks. Vertical bars at the top represent standard errors.

FIGURE 7: Interaction effect of TDZ and BAP on cocoyam axillary shoot multiplication and root formation on stationary liquid media in test tubes. Columns labeled with the same letter are not significantly different at $P = 0.05$ using Multiple Range Test for treatment effect comparison at 4 and 6 weeks. Vertical bars at the top represent standard errors.

on rooting. These findings agree with earlier findings of Murashige [36]. Acheampong and Henshaw [37] observed that agitated liquid media initiated the development of protocorm-like bodies, which differentiated into plantlets upon transfer into stationary liquid media. The state of the nutrient medium apparently played a significant role in determining the pattern of organogenesis in cocoyam. Jackson et al. [38] noticed the poor growth of taro cultured on agar medium. TDZ completely suppressed root formation. Explants subcultured on Stationary Liquid Media in Erlenmeyer Flasks proliferated heavily. These findings are similar to those reported on grape by Sudarsono and Goldy [39] but contrary to that of Gray and Benton [40] who also studied grape. TDZ produced more compressed shoots, while BAP cultures produced shoots that more easily differentiated into well-defined plantlets with eventual formation of extensive root systems. In muscadine grape, rooting was impossible in the presence of BAP with higher levels affecting subsequent rooting when transferred on media without BAP [41]. Some other reports indicated that TDZ repressed root formation [40] as well. Initiation with 20 μM BAP represses shoots growth, and thus proliferation. When these repressed tissues are further maintained in the same medium, their proliferation is restricted. Lee and Wetzstein [41] observed high mortality of muscadine grape shoots at 20 μM BAP and higher levels. In a study to develop a rapid and efficient shoot regeneration system suitable for the transformation of lentil using TDZ, it was found that MS medium supplemented with 0.25 mg/L, TDZ produced the highest frequency of shoot formation from cotyledonary nodes in both genotypes

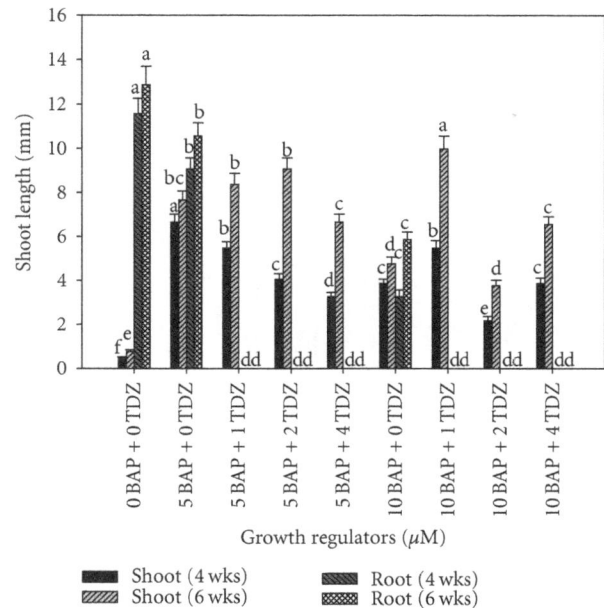

[28]. Induction medium supplemented with 5 mgl-L BAP and 20 or 40 mg/L adenine sulphate (AS) resulted in a higher average of shoots formation when common bean was cultured using MS medium [29]. Alam and Khaleque [22] cultured groundnuts explants on MS medium with different concentration of 2,4-D, BAP, and NAA. 2,4-D at 2 mg/L was found more suitable for good callus induction. MS medium supplemented with different concentrations of BAP produced small shoot bud at different subculture and maximum number of shoot bud differentiation was observed from 2.5 mg/L BAP concentration. They concluded that 2,4-D was the best for callus induction, and BAP was found more suitable for organogenesis compared to NAA. Also, taro plants were regenerated via somatic embryogenesis and organogenesis on Murashige and Skoog (MS) medium [12] with a two-step protocol utilized combinations of 2,4-dichlorophenoxyacetic acid (2,4-D), thidiazuron (TDZ), indole-3-acetic acid (IAA), and 6-benzylaminopurine (BAP) [20]. In this study, TDZ had a tendency to enhance the initial BAP effects. The poor rooting ability in TDZ as compared to multiplication phase indicates that as long as the cultures are on TDZ, the greater its repressive effect on rooting.

The 2.0 μM TDZ medium increased shoot proliferation from four to six weeks of culture. Shoot proliferation in growth regulator-free medium probably reached a maximum at four weeks. It could be better to reculture into a fresh medium at four weeks to optimize proliferation. Continued proliferation of shoots in a medium without growth regulators may be the result of the cumulative effect of growth regulators in previous media. A comparison of

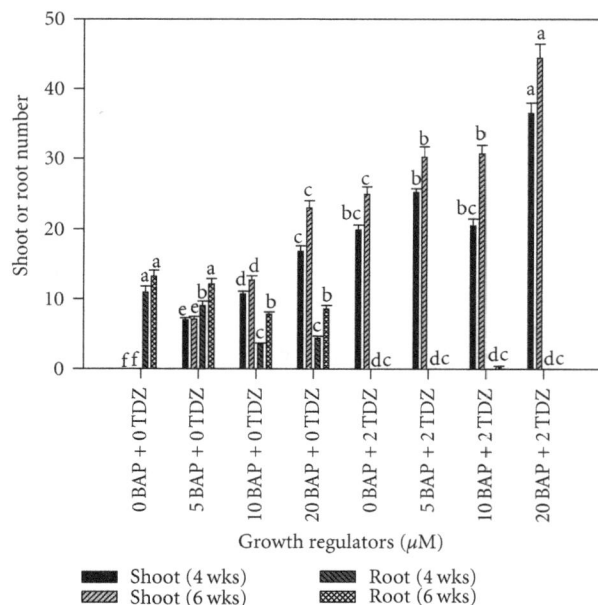

FIGURE 8: Interaction effect of TDZ and BAP on cocoyam shoot and root proliferation after reculturing in stationary liquid media in Erlenmeyer flasks. Columns labeled with the same letter are not significantly different at $P = 0.05$ using Multiple Range Test for treatment effect comparison at 4 and 6 weeks. Vertical bars at the top represent standard errors.

root formation in $2.0\,\mu M$ TDZ and growth regulators-free multiplication media substantiates the inhibition of rhizogenesis by TDZ, while all growth regulators-free cultures rooted. Kerns and Meyer [42] reported similar results for Acer freemanii (autumn blaze maple) cultures. Although the mode of TDZ action in organogenesis is not yet determined, it is possible that root induction is initiated in its presence, but development is delayed and expressed in its absence. Shoots were found to be proliferated in either solid or liquid media, with a tendency of getting larger numbers in solid media. These findings are in contrast to a previous report by Ng and Hahn [43] who did not find cocoyam plantlet formation in agitated liquid cultures.

All TDZ and BAP combinations resulted in an approximate doubling of shoot numbers from four to six weeks. Growth regulators levels significantly affected rooting. Virtually no roots were formed in treatments containing TDZ emphasizing its suppressive effect on rooting. Nyochembeng and Garton [4] found that thidiazuron was more favorable for callus production from petioles than shoot tips. The callus mass from petioles was significantly greater than from shoot tips. Dicamba at $1.36\,\mu M$ produced significantly more callus than other concentrations in the absence of TDZ, while $0.45\,\mu M$ stimulated mainly roots at the base of the petiole explants and later plantlets in both petioles and shoot tips. Thidiazuron had a promotive effect on callus proliferation only at higher dicamba concentrations (4.52 and $13.5\,\mu M$) but delayed callus formation over the entire explants and further proliferation for six weeks compared to dicamba

alone. When callus derived from different treatments was subcultured onto media containing dicamba ($1.36\,\mu M$), proliferation was enhanced and more than 80% became organized into shoots, roots, or a mixture of both. Callus derived from TDZ media initiated shoot organs (bud clumps and single shoot buds) first. Rapid callus initiation and multiplication in cocoyam appear to require potent auxins, that is, dicamba. In the presence of dicamba, darkness was not required for callus initiation. The potency of dicamba has been observed in other tropical monocots, for example, ginger [44] and banana [45]. Subculture of friable and rapidly growing shoot tip callus into B5 basal medium containing dicamba with or without kinetin and 2,4-D with kinetin, followed by agitation, demonstrated that: (i) cocoyam callus can form suspension cultures especially in media containing dicamba alone, and (ii) the organization of callus into bud clumps was greatly enhanced by agitation of suspension cultures of callus tissue, suggesting that the pattern of morphogenesis in cocoyam callus may also be influenced by physical factors of the medium such as aeration and medium matrix. Asokan et al. [46] observed a 3-fold increase in shoot length of X. caracu in liquid shaken media compared to solid media. The differentiation pattern was affected by concentration and culture method utilized. For example, the presence of auxins especially at $1.35\,\mu M$ (dicamba or 2,4-D) inhibited shoot formation whereas kinetin ($0.46\,\mu M$) stimulated shoot formation from initiated bud clumps. Several reports have mentioned the development of protocorms in shoot tip callus cultures [47–50] and directly on agitated axillary, apical, and adventitious bud cultures of cocoyam [37]. Reports on other plants indicated controversial results. Culture on elongation medium supplemented with GA3 was 55% more effective with respect to overall shoot production than that on medium without GA3 for Seedling-derived cotyledon explants of summer squash commercial cultivars True French, Ma'yan, and Goldy [16]. Various concentrations of 2,4-D and NAA were used alone or in combination. 2,4-D at 3–21 μM concentrations in the culture media produced 100% callus induction from soybean-germinated seeds. Roots and shoots were obtained using BAP and Kinetin containing culture media. $5\,\mu M$ BAP was the most effective for that purpose [19]. In a study to develop an efficient method for shoot regeneration of canola and to compare the regeneration capacity of different explants on MS medium with several combinations of plant growth regulators, it was found that the highest shoot regeneration took place when explants were cultivated on medium, containing 1.0 mg/L NAA, 8.0 mg/L BAP and 3.0 mg/L ABA. Also, vitrification of regenerants was promoted by increasing the auxin NAA or cytokinin BAP, and ABA in the nutrient medium [25]. To determine the best combinations of plant growth regulators and other conditions in order to achieve organogenesis and multiplication directly from shoot tips of date palm without callus formation so as to avoid any possibility of undesirable genetic variability, Khierallah, and Bader [27] found that MS-modified medium supplemented with 2.0 mg/L isopentenyladenine (2ip), 1.0 mg/L benzyl adenine (BA), 1.0 mg/L naphthaleneacetic acid (NAA), and 1.0 mg/L naphthoxyacetic acid

(NOA) was the best for bud formation. The maximum morphogenic callus induction rate was observed in summer squash on MS medium supplemented with 2.5 mg/L 2,4-D. The highest percentage of shoot regeneration and highest mean number of shoots per culture were obtained with 0.5 mg/L thidiazuron. Regenerated shoots were rooted in MS medium supplemented with 1.0 mg/L IBA [17]. A variety of explants of winter squash were cultured using media containing different concentrations of 6-benzylaminopurine (BA). Plant regeneration was optimal when the proximal parts of cotyledons from 4-day-old seedlings were cultured on induction medium composed of MS medium with 1 mg/L BA. Adventitious shoots were subcultured on elongation medium composed of MS medium with 0.1 mg/L BA, and the elongated shoots were successfully rooted on MS medium without growth regulators for 2 weeks [18]. Kanmegne and Omokolo [51] mentioned that organogenesis in cocoyam, when induced in the presence of a growth regulator, is preceded by an increase in soluble peroxidase activity, followed by a drop after the appearance of organs. The increase in enzyme activity can be due to the presence of the growth regulator than to organogenesis. This indicates that total peroxidase activity is not a proper marker for the orientation of morphogenesis in cocoyam. In general, during morphogenesis, cellular proteins differ in their function and in their timing and extent of expression during the process [52–54]. It has been shown in a number of plant systems that, during organogenesis, functionally related proteins seem to be encoded by groups of coordinately expressed genes and that plant growth regulators are key moderators [55–58]. Further biochemical analyses, for example, basic isoperoxidases, polyphenoloxidase, and phenol composition are needed for a better understanding of the mechanism underlying differentiation in X. sagittifolium.

In conclusion, cocoyam has been successfully micropropagated in vitro through various procedures: the use of either BAP, TDZ or both; agitated and stationary liquid media at the initiation stage; agitated, stationary or semisolid media at the multiplication stage; stationary liquid and semisolid media at the elongation, but increased substantially during the multiplication phases. All concentrations tested (5–20 μM BAP and 1–4 μM TDZ) produced significantly more axillary shoots per shoot tip than the control without cytokinins. Greater proliferation rates were obtained through the use of 20 μM BAP and 2 μM TDZ, respectively, 12 weeks from initiation. Shoots produced with BAP were larger and more normal in appearance than those produced with TDZ, which were small, compressed, and stunted. Based on our results, the use of stationary liquid media is recommended because of economic reasons.

Abbreviations

MS: Murashige and Skoog (1962) medium
TDZ: Thidiazuron
BAP: Benzylaminopurine
BM: Basal medium
NAA: 1-naphthaleneacetic acid
AS: Adenine sulphate.

References

[1] J. W. Purseglove, *Tropical Crops: Monocotyledons*, Longmans, London, UK, 1985.

[2] I. C. Onwueme and W. B. Charles, "Cultivation of cocoyam," in *Tropical Root and Tuber Crops. Production, Perspectives and Future Prospects*, pp. 139–161, FAO Plant Production and Protection Paper 126, Rome, Italy, 1994.

[3] D. O. Ndoumou, G. N. Tsala, G. Kanmegne, and A. P. Balange, "In vitro induction of multiple shoots, plant generation and tuberization from shoot tips of cocoyam," *Comptes Rendus de l'Académie des Sciences*, vol. 318, pp. 773–778, 1995.

[4] L. M. Nyochembeng and S. Garton, "Plant regeneration from cocoyam callus derived from shoot tips and petioles," *Plant Cell, Tissue and Organ Culture*, vol. 53, no. 2, pp. 127–134, 1998.

[5] S. Sefa-Dedeh and E. K. Agyir-Sackey, "Chemical composition and the effect of processing on oxalate content of cocoyam *Xanthosoma sagittifolium* and *Colocasia esculenta* cormels," *Food Chemistry*, vol. 85, no. 4, pp. 479–487, 2004.

[6] J. T. Tambong, X. Ndzana, J. G. Wutoh, and R. Dadson, "Variability and germplasm loss in the Cameroon national collection of cocoyam (Xanthosoma sagittifolium Schott (L.)," *Plant Genetic Resources Newletters*, vol. 112, pp. 49–54, 1997.

[7] FAO, Food and agriculture organization statistical database: world production of fruits and vegetables, 2006, http://www.ers.usda.gov/publications/vgs/tables/world.pdf.

[8] I. C. Onwueme, *The Tropical Tuber Crops: Yams, Cassava, Sweet Potato, Cocoyams*, John Wiley & Sons, Chichester, UK, 1978.

[9] R. P. Pacumbaba, J. G. Wutoh, A. E. Sama, J. T. Tambong, and L. M. Nyochembeng, "Isolation and pathogenicity of rhizosphere fungi of cocoyam in relation to the cocoyam root rot disease," *Journal of Phytopathology*, vol. 135, pp. 265–273, 1992.

[10] J. Chen and M. J. Adams, "Molecular characterisation of an isolate of Dasheen mosaic virus from *Zantedeschia aethiopica* in China and comparisons in the genus *Potyvirus*," *Archives of Virology*, vol. 146, no. 9, pp. 1821–1829, 2001.

[11] P. C. Debergh and P. E. Read, "Micropropagation," in *Micropropagation, Technology and Application*, P. C. Debergh and R. H. Zimmerman, Eds., pp. 1–13, Kluwer Academic Publishers, Dodrecht, The Netherlands, 1991.

[12] T. Murashige and F. Skoog, "A revised medium for rapid growth and bioassays with tobacco tissue culture," *Plant Physiology*, vol. 15, pp. 473–497, 1962.

[13] N. D. Omokolo, N. G. Tsala, G. Kanmegne, and A. P. Balange, "Production of multiple shoots, callus, plant regeneration and tuberization in *Xanthosoma* sagittifolium cultured in vitro," *Comptes Rendus de l'Académie des Sciences*, vol. 318, pp. 773–778, 1995.

[14] J. H. Choi and Z. R. Sung, "Two-dimensional gel analysis of carrot somatic embryonic proteins," *Plant Molecular Biology Reporter*, vol. 2, no. 3, pp. 19–25, 1984.

[15] S. Stirn and H. J. Jacobsen, "Marker proteins for embryogenic differentiation patterns in pea callus," *Plant Cell Reports*, vol. 6, no. 1, pp. 50–54, 1987.

[16] G. Ananthakrishnan, X. Xia, C. Elman et al., "Shoot production in squash (*Cucurbita pepo*) by in vitro organogenesis," *Plant Cell Reports*, vol. 21, no. 8, pp. 739–746, 2003.

[17] S. P. Pal, I. Alam, M. Anisuzzaman, K. K. Sarker, S. A. Sharmin, and M. F. Alam, "Indirect organogenesis in summer squash (*Cucurbita pepo* L.)," *Turkish Journal of Agriculture and Forestry*, vol. 31, no. 1, pp. 63–70, 2007.

[18] Y. K. Lee, W. Chung, and H. Ezura, "Efficient plant regeneration via organogenesis in winter squash (Cucurbita maxima Duch.)," *Plant Science*, vol. 164, no. 3, pp. 413–418, 2003.

[19] E. Y. Joyner, L. S. Boykin, and M. A. Lodhi, "Callus induction and organogenesis in soybean [Glycine max (L.) Merr.] cv. Pyramid from mature cotyledons and embryos," *The Open Plant Science Journal*, vol. 4, pp. 18–21, 2010.

[20] V. M. Verma and J. J. Cho, "Plantlet development through somatic embryogenesis and organogenesis in plant cell cultures of Colocasia esculenta (L.) schott," *Asia-Pacific Journal of Molecular Biology and Biotechnology*, vol. 18, no. 1, pp. 167–170, 2010.

[21] M. G. Z. Krug, L. C. L. Stipp, A. P. M. Rodriguez, and B. M. J. Mendes, "In vitro organogenesis in watermelon cotyledons," *Pesquisa Agropecuaria Brasileira*, vol. 40, no. 9, pp. 861–865, 2005.

[22] A. K. M. M. Alam and M. A. Khaleque, "In vitro response of different explants on callus development and plant regeneration in groundnut (Arachis hypogeae L.)," *International Journal of Experimental Agriculture*, vol. 1, pp. 1–4, 2010.

[23] B. Sarabi and K. Almasi, "Indirect organogenesis is useful for propagation of Iranian edible wild asparagus (Asparagus officinalis L.)," *Asian Journal of Agricultural Sciences*, vol. 2, pp. 47–50, 2010.

[24] R. Sujatha, L. C. Babu, and P. A. Nazeem, "Histology of organogenesis from callus cultures of black pepper (Piper nigrum L.)," *Journal of Tropical Agriculture*, vol. 41, pp. 16–19, 2010.

[25] G. B. Kamal, K. G. Illich, and A. Asadollah, "Effects of genotype, explant type and nutrient medium components on canola (Brassica napus L.) shoot *in vitro* organogenesis," *African Journal of Biotechnology*, vol. 6, no. 7, pp. 861–867, 2007.

[26] I. I. Ozyigit, "Phenolic changes during *in vitro* organogenesis of cotton (Gossypium hirsutum L.) shoot tips," *African Journal of Biotechnology*, vol. 7, no. 8, pp. 1145–1150, 2008.

[27] H. S. M. Khierallah and S. M. Bader, "Micropropagation of date palm (Phoenix dactylifera L.) var. Maktoom through direct organogenesis," *Acta Horticulturae*, vol. 736, pp. 213–224, 2007.

[28] K. M. Khawar, C. Sancak, S. Uranbey, and S. Özcan, "Effect of Thidiazuron on shoot regeneration from different explants of lentil (Lens culinaris Medik.) via organogenesis," *Turkish Journal of Botany*, vol. 28, no. 4, pp. 421–426, 2004.

[29] A. M. Gatica Arias, J. M. Valverde, P. R. Fonseca, and M. V. Melara, "*In vitro* plant regeneration system for common bean (Phaseolus vulgaris): effect of N6-benzylaminopurine and adenine sulphate," *Electronic Journal of Biotechnology*, vol. 13, no. 1, pp. 1–8, 2010.

[30] M. L. Mayor, G. Nestares, R. Zorzoli, and L. A. Picardi, "Analysis for combining ability in sunflower organogenesis-related traits," *Australian Journal of Agricultural Research*, vol. 57, no. 10, pp. 1123–1129, 2006.

[31] Y. R. An, X. G. Li, H. Y. Su, and X. S. Zhang, "Pistil induction by hormones from callus of Oryza sativa in vitro," *Plant Cell Reports*, vol. 23, no. 7, pp. 448–452, 2004.

[32] N. Banerjee and E. de Langhe, "A tissue culture technique for rapid clonal propagation and storage under minimal growth conditions of Musa (Banana and plantain)," *Plant Cell Reports*, vol. 4, no. 6, pp. 351–354, 1985.

[33] K. Z. Watanabe, "Challenges in biotechnology for abiotic stress tolerance on root and tubers," *JIRCAS Working Reports*, pp. 75–83, 2002.

[34] R. Goenaga and U. Chardon, "Growth, yield and nutrient uptake of taro grown under upland conditions," *Journal of Plant Nutrition*, vol. 18, no. 5, pp. 1037–1048, 1995.

[35] O. L. Gamborg, R. A. Miller, and K. Ojima, "Nutrient requirements of suspension cultures of soybean root cells," *Experimental Cell Research*, vol. 50, no. 1, pp. 151–158, 1968.

[36] T. Murashige, "Plant propagation through tissue cultures," *Annual Review of Plant Physiology*, vol. 25, pp. 135–166, 1974.

[37] E. Acheampong and G. G. Henshaw, "*In vitro* methods for cocoyam improvement," in *Tropical Root Crops. Production and Uses in Africa. Proceedings of the 2nd Triennial Symposium of the International Society for Tropical Root*, E. R. Terry, E. V. Doku, O. B. Arene, and N. M. Mahungu, Eds., pp. 165–168, Africa Branch, Douala, Cameroon International Development Research Center (IDRC), Ottawa, Canada, 1984, IDRC- 221.

[38] G. V. H. Jackson, E. A. Ball, and J. Arditti, "Tissue culture of taro, Colocasia esculenta (L.) Schott," *Journal of Horticultural Science*, vol. 52, pp. 373–382, 1977.

[39] Sudarsono and R. G. Goldy, "Effect of some growth regulators on in vitro culture of three Vitis rotudifolia cultivars," *HortScience*, vol. 23, p. 757, 1988.

[40] D. J. Gray and C. M. Benton, "*In vitro* micropropagation and plant establishment of muscadine grape cultivars (Vitis rotundifolia)," *Plant Cell, Tissue and Organ Culture*, vol. 27, no. 1, pp. 7–14, 1991.

[41] N. Lee and H. Y. Wetzstein, "In vitro propagation of muscadine grape by axillary shoot proliferation," *Journal of the American Society for Horticultural Science*, vol. 115, pp. 324–329, 1990.

[42] H. R. Kerns and M. M. Meyer Jr., "Tissue culture propagation of Acer x freemanii using thidiazuron to stimulate shoot tip proliferation," *HortScience*, vol. 21, no. 5, pp. 1209–1210, 1986.

[43] S. Y. Ng and S. K. Hahn, "Applications of tissue culture to tuber crops at IITA," in *Biotechnology in International Agricultural Research. Proceedings of the Inter-Center Seminar on International Agricultural Research Centers (IARCs) and Biotechnology, Manila, Philippines*, pp. 29–40, 1985.

[44] A. Kackar, S. R. Bhat, K. P. S. Chandel, and S. K. Malik, "Plant regeneration via somatic embryogenesis in ginger," *Plant Cell, Tissue and Organ Culture*, vol. 32, no. 3, pp. 289–292, 1993.

[45] F. J. Novak, R. Afza, M. Van Duren, M. Perea-Dallos, G. V. Conger, and T. Xiaolang, "Somatic embryogenesis and plant regeneration in suspension cultures of dessert (AA and AAA) and cooking (ABB) bananas (Musa spp)," *Bio/Technology*, vol. 7, pp. 154–159, 1989.

[46] M. P. Asokan, S. K. O'Hair, and R. E. Litz, "Rapid multiplication of Xanthosoma caracu by in vitro shoot tip culture," *HortScience*, vol. 19, pp. 885–886, 1984.

[47] M. M. Abo El-Nil and F. W. Zettler, "Callus initiation and organ differentiation from shoot tip cultures of Colocasia esculenta," *Plant Science Letters*, vol. 6, no. 6, pp. 401–408, 1976.

[48] L. P. Nyman, C. J. Gonzales, and J. Arditti, "Reversible structural changes associated with callus formation and plantlet development from aseptically cultured shoots of taro (Colocasia esculenta var antiquorum)," *Annals of Botany*, vol. 51, no. 3, pp. 279–286, 1983.

[49] T. Q. Nguyen and V. U. Nguyen, "Aroid propagation by tissue culture: shoot tip culture and propagation of Xanthosoma violaceum," *HortScience*, vol. 22, pp. 671–672, 1987.

[50] S. Sabapathy and H. Nair, "*In vitro* propagation of taro, with spermine, arginine and ornithine. II. Plantlet regeneration via callus," *Plant Cell Reports*, vol. 14, no. 8, pp. 520–524, 1995.

[51] G. Kanmegne and N. D. Omokolo, "Changes in phenol content and peroxidase activity during in vitro organogenesis in *Xanthosoma sagittifolium* L," *Plant Growth Regulation*, vol. 40, no. 1, pp. 53–57, 2003.

[52] B. Arnholdt-Schmitt, "Rapid changes in amplification and methylation pattern of genomic DNA in cultured carrot root explants (*Dacus carota* L.)," *Theoretical and Applied Genetics*, vol. 85, pp. 792–800, 1993.

[53] K. Cui, L. Ji, G. Xing, L. Jianlong, L. Wang, and Y. Wang, "Effect of hydrogen peroxide on synthesis of proteins during somatic embryogenesis in *Lycium barbarum*," *Plant Cell, Tissue and Organ Culture*, vol. 68, no. 2, pp. 187–193, 2002.

[54] L. E. Kay and D. V. Basile, "Specific peroxidase isoenzymes are correlated with organogenesis," *Plant Physiology*, vol. 84, pp. 99–105, 1987.

[55] B. Deumling and L. Clermont, "Changes in DNA content and chromosomal size during cell culture and plant regeneration of *Scilla siberica*: selective chromatin diminution in response to environmental conditions," *Chromosoma*, vol. 97, no. 6, pp. 439–448, 1989.

[56] H. Klee and M. Estelle, "Molecular genetic approaches to plant hormone biology," *Annual Review of Plant Physiology and Plant Molecular Biology*, vol. 42, no. 1, pp. 529–551, 1991.

[57] B. Parthier, "Hormone-induced alterations in plant gene expression," *Biochemie und Physiologie der Pflanzen*, vol. 185, pp. 289–314, 1989.

[58] K. Skriver and J. Mundy, "Gene expression in response to abscisic acid and osmotic stress," *Plant Cell*, vol. 2, no. 6, pp. 503–512, 1990.

8

Socioeconomic Importance of the Banana Tree (*Musa Spp.*) in the Guinean Highland Savannah Agroforests

Pierre Marie Mapongmetsem,[1] Bernard Aloys Nkongmeneck,[2] and Hamide Gubbuk[3]

[1] *Department of Biological Sciences, Faculty of Sciences, University of Ngaoundere, P.O. Box 454, Ngaoundere, Cameroon*
[2] *Department of Plant Biology, Faculty of Sciences, University of Yaounde I, P.O. Box 818, Yaounde, Cameroon*
[3] *Department of Horticulture, Faculty of Agriculture, Akdeniz University, 07059 Antalya, Turkey*

Correspondence should be addressed to Pierre Marie Mapongmetsem, piermapong@yahoo.fr

Academic Editor: Vergel C. Concibido

Home gardens are defined as less complex agroforests which look like and function as natural forest ecosystems but are integrated into agricultural management systems located around houses. Investigations were carried out in 187 households. The aim of the study was to identify the different types of banana home gardens existing in the periurban zone of Ngaoundere town. The results showed that the majority of home gardens in the area were very young (less than 15 years old) and very small in size (less than 1 ha). Eleven types of home gardens were found in the periurban area of Ngaoundere town. The different home garden types showed important variations in all their structural characteristics. Two local species of banana are cultivated in the systems, *Musa sinensis* and *Musa paradisiaca*. The total banana production is 3.57 tons per year. The total quantity of banana consumed in the periurban zone was 3.54 tons (93.5%) whereas 1.01 tons were sold in local or urban markets. The main banana producers belonged to home gardens 2, 4, 7, and 9. The quantity of banana offered to relatives was more than what the farmers received from others. Farmers, rely on agroforests because the flow of their products helps them consolidate friendship and conserve biodiversity at the same time.

1. Introduction

Agroforestry systems aim to optimize the benefits from biological interaction created where trees and shrubs sometimes are deliberately combined with crops as well as animals [1, 2]. They promote many forms of diversity in the agro ecosystem such as creation of habitats for wild life and beneficial organisms, offering greater diversity of products and lowering the need for external inputs [3–5]. Agroforestry systems have huge potential in meeting the challenges of food security, due to increasing population and degraded environment [6–8]. For a variety of agroforestry systems found in Cameroon, home gardens are among the most favoured land use systems as they enhance the farming family's nutritional and income status considerably [9–11]. Diversification of their production systems has become a keystone to the sustainability of those families. Fruit tree species play an important role in the socioeconomy conditions of the farmers. Bananas are among the most preferred fruit

tree species growing in this traditional system but their production remains insufficient for the population. Also, the available species are local; therefore, it is important to evaluate the existing banana production as well as the socioeconomic and ecological characteristics that result from these interactions. As has been pointed out, policies that promote the linkage between domestication and commercialization of nonwood forest products are one of the important areas for further work [12]. In this regard, there is also a need for better integration of food needs and other products with those of the subsistence farmer [13]. Very few works have been done on this problem at the national scale. The first step is the development of better knowledge of the potential utilization of the various species, products, and of the constraints associated with banana production.

The global aim is to evaluate the structure and the functioning of the GHS agroforests.

The specific objectives are to identify the different types of home gardens existing in the area and evaluate the flow of banana (production, consumption, gift, and commercialization).

This information will serve as baseline to develop appropriate management techniques. These techniques could help farmers introduce and grow new species of banana in order to get maximum benefits from their production system.

2. Materials and Methods

2.1. Study Site. The Guinean Highland Savannah (GHS) is located between latitude 7°2′36 N and longitude 13°34′72 E. The climate is guinean type with one active dry season (October–March) and a rainy season covering the remaining of the year. The yearly average total precipitation is 1315.6 mm with a yearly total mean evaporation of 1902.95 mm. The distribution of the rainfall is monomodal. Two main winds blow in the region notably the monsoon during the rainy season from the South and the Harmattan from the North responsible for the drought [11, 14]. The soil of the area is rich in ferruginous compounds derived from granites, granodiorites, and of gneiss after rejuvenation and is composed of red ferralitic developed on old basalts [15]. The vegetation is mainly composed of shrubby and/or woody savannahs with consistent predominance of *Daniellia oliveri* and *Lophira lanceolata* [16]. Nowadays, the density of these species has significantly decreased under the influence of human activities [17].

2.2. Methodological Approaches. This study was undertaken in the periurban zone of Ngaoundere notably in Gangassaou, Dene, Sabongari, Wack, Tagboum, Sassa Mbersi, Biskewal and Daran. A total of 187 households distributed in four ethnolinguistic groups (Fulbe, Gbaya, Mboum and Dii) were interviewed and a systemic approach was used. Participatory and reiterative ethnobotanical interviews were conducted through questionnaires containing open, closed and oriented questions. In each household a detailed survey of the composition and management practices were made. The main rubrics of the questionnaire dealt with the main characteristics of farmers (farmer age, gender, family size, beliefs, matrimonial status, strategies used) and agroforests (age, area, inputs), main crops cultivated and/or protected species; constraints.

On the basis of the typology elaborated, the second phase of the work was carried out with farmers verse with technological innovations and who agreed to collaborate with the authors. Only 110 of them accepted to work with the team. Investigations focused only on the banana production. Evaluation in each household included the quantity of banana produced, consumed, commercialised, offered or received from other farmers, the price, incomes generated, and so forth. Farmers involved in the study were train to manage the record book. Home gardens were visited twice a month and discussion held with the garden head.

2.3. Data Collection and Analysis. Structural and functional parameters were computed and descriptive statistics were used. For classification of the 110 home gardens, a hierarchical cluster analysis was applied using age, size as main variables. Correlation and regression analyses were used to determine relation between banana production parameters on annual basis. The statistic programs used were Statistica and Statgraphic plus 6.

3. Results

3.1. Farmer' Socioeconomic Characteristics

3.1.1. Beliefs, Gender and Land Acquisition. Investigations showed that 80% of the respondents are Muslims. 94% of the respondents were men against 6% of women. Regarding the tradition of the area, the land is acquired by donation or inheritance from parents. In each village, the Chief "Djaoro" distributes land. But due to the urbanisation and the extension of the town toward the neighbourhoods, land transactions have been reported at Daran, Biskewal, and Dang. About 55.70% of gardens have been obtained from the chief and 45.63% inherited from parents. However, the farmer can lose his property as soon as he/she leaves the village. In addition, according to the beliefs of Muslims, only men can receive the inheritance from the parents; however, in the other groups like Gbaya and Dii, men as well as women can get it. In these groups, women have their farm different from the one of their husband. In the Foulbe, women are not allowed to work in the farm. They have small plots where they grow some vegetables (*Hibiscus sabdariffa, Hibiscus exculenthus, Solanum spp., Brassica oleracea, Sesamum indicum, Capsicum fructescens*, etc.).

3.1.2. Farmer Ages. The percentage of the farmers' ages varied significantly (0.0419 < 0.05) from 9.20% who were below 30 years to 28.80% in the group of farmers with more than 60 years old (Table 1). The majority of the farmers interviewed (60.58%) were over forty years old. Farmers in the age group under 30 years were less represented.

Despite the general trend, disparities exist between the villages. In Sabongari, no peasant was older than 60 years, whereas in Tagboum, none were less than 30 years in age. However, in Dene, farmers with ages between 41 and 50 years did not exist.

3.1.3. Matrimonial Status and Scholarship. Almost 62.76% of the farmers were monogamous against the 27.58% who were polygamous (Table 1). Bachelors and widows occupied 6.56 and 5.67%, respectively. About 72% of farmers with age less than 51 years old were monogamous. Despite the fact that the majority of the population were Muslim, young farmers have chosen monogamy. In Sassa Mbersi, there were no bachelors and no widows. There is a significant difference (0.000 < 0.001) among the farmers in terms of marital status.

Concerning the intellectual level of the population, the analysis shows that more than 50% of farmers are illiterate. Among those who have been in school, 30% attained the primary school whereas 19.46% reached the second cycle of secondary school and only 9.12% went up to the level of upper

TABLE 1: Farmer percentages according to their socioeconomic characteristics (age, marital status, family size) in nine villages of the Guinean Highland Savannahs. BA; Bachelor, MO; Monogamy, PO; Polygamy, WI; Widow.

Villages	Age (years)					Marital status				Family size				
	<30	30–40	41–50	51–60	>60	BA	MO	PO	WI	<5	6–10	11–15	16–20	21–30
Gangassaao	7,9	27.7	28.9	18.4	21.0	7.9	60.5	31.6	0	13.1	52.6	15.8	15.8	2.6
Sabongarii	7.7	23.1	23.1	7.7	28	7.7	46.1	46.1	0	15.4	61.5	15.4	0	7.7
Dene	7.7	46.1	0	23.1	23.1	30.8	30.8	20.1	15.4	46.1	38.5	23.1	7.7	0
Tagboum	0	20	28	24	26.7	4	76	16	4	20	72	4	4	0
Sassa Mbersi	7.7	23.1	23.1	7.7	28	0	84.6	15.4	0	7.7	84.6	7.7	0	0
Wack	3.7	25.9	14.8	18.5	37	3.7	70.4	22.2	3.7	14.8	59.2	18.5	7.4	0
Mbe	5.7	31.0	18.9	22.5	20.1	5	60	15	20	15	40	30	5	10
Biskewal	35	5.8	15	16.7	28.5	8.8	61.2	23.8	6.33	21	59	14.5	5.5	0
Daran	7.5	25.5	0	24	43	25.5	70	3.5	1	14.5	72	7.5	6	0
Mean	9.2	25.5	16.9	17.0	28.8	6.56	62.8	27.6	5.7	16.6	55.6	5.6	5.6	2.3

sixth. The highest percentage of literacy was in Mbe. The possible explanation is that Mbe is the Administrative centre of the Mbe Subdivision. In addition, there exist many educational centres. For the other villages where illiteracy is high, it is due to the fact that children have to spend their time looking after cattle instead of going to school. The Dii people are known to be good herdsmen in the Adamawa region.

3.1.4. Family Size. Among the farmers interviewed, 55.56% live with 5 to 10 persons in the homegarden whereas 2.26% live with 21 to 30 persons. The number of persons living in home gardens is significant (0.000 < 0.001). Similar result has been obtained in East Asia [18] by Mapongmetsem et al., in 2000. This general trend with high number of family members (21–30) is peculiar at the village level (Table 1). Illustration is given by Gangassaou, Sabongari, and Mbe villages. However Sassa Mbersi is singularised by the absence of families with more than 15 persons. Despite the high number of people under the responsibility of the head of the family, there is no significant correlation between the family size and the active member who can help the head of the family during farming works ($r = 0.081; 1.564 > 0.05$). Chiefs of family in their majority (46.6%) have 2 to 3 active persons.

For farming activities, 92% of the farmers do not receive help from visiting family members who spend some of their times in the household. Small fractions among them (8%) receive help from 1 to 5 persons. Fortunately, collective help exists in each village of the region locally known as "surge" to solve the problem of labour. There is no significant correlation ($r = 0.2; 0.8913 > 0.05$) between farmer age and family size nor between active members considered ($r = 0.046; 1,234 > 0.05$).

3.2. Agroforest's Characteristics

3.2.1. Home Gardens Age. The distribution of home gardens per village and age shows that 35.3% of home gardens are less than 15 years old. Home gardens of 15 to 30 years are ranked second (28.45%), those of more than 45 years, third (19.19%), and lastly those of 31–45 years (16.97%) (Table 2).

This result is comparable with that obtained in East Asia [18]. In some ethnolinguistic groups like the Dii, home gardens farming has been recently introduced.

Home gardens more than 30 years old are absent in Sabongari as compared to those which are 15–30 years old. The situation is due to the fact that this village has been created recently between 1970 and 1980 with the massive migration of the Dii population from the Savannah to near the road [19, 20]. There is no significant correlation between the farmer age and that of home gardens ($r = 0.4; 0.823 > 0.05$).

3.2.2. Home Gardens Area. Three landholding sizes were identified: small (<1 ha), medium (1-2 ha), and large (>2 ha). Similar results have been reported in Kerala [21]. Majority (80.37%) of the Periurban agroforests are concentrated in areas less than one hectare (<1 ha) against 8.68% for those in which the surface area is equal to or more than 2.1 hectares (Table 2). In the rural zone, the same trend is observed. The explanation is given by the traditional law which limits the area per inhabitant [22]. Exceptionally farmer can go beyond 0.25 ha. In Sabongari, 15.4% of home gardens with more than 3 ha are found (Table 2). Some of them are located at the beginning whereas others are situated at the end of the villages along the main road.

This information is very important in the elaboration of the typology of home gardens of the region which could serve as baseline to address efficiently the various constraints enumerated by farmers in the region.

3.2.3. Differentiation Multistrata Type. The garden's ages as well as their surface areas are the two important criteria to distinguish the home gardens. On the basis of a cluster analysis using the nearest neighbour method and the squared euclidean, the 187 selected households were categorised into 11 types (Figure 1). These different types of agroforests are conceivedand managed by farmers, over the Guinean Highland Savannahs (GHSs).

3.2.4. Structural Characteristics. Every house has fruit trees growing, although not all are productive. Less than a quarter

TABLE 2: Distribution percentages of home gardens according to age and area in villages.

Villages	AGE (years)				AREA (ha)			
	<15	15–30	31–45	>45	<1	1–2	2.1–3	>3
Gangassao	21	31.60	18.40	28.90	86.8	10.5	2.7	0
Sabongari	15.4	84.60	0	0	61.5	15.4	7.7	14.4
Dene	53.8	23.10	15.4	7.7	76.9	8.5	11.5	3.1
Tagboum	36.0	16.00	24.00	24.00	80	4	12	4
Sassa Mbersi	61.5	15.4	7.7	7.7	84.6	15.4	0	0
Wack	14.8	18.5	33.33	33.33	85.2	11.1	0	3.70
Mbe	45	10	20	25	85	15	0	0
Biskewal	56	4	30	10	76.5	5.5	10	8
Daran	48	24	17	21	87	10.5	2.5	0
Mean	39.06	25.24	18.43	17.51	80.39	10.70	5.16	3.69

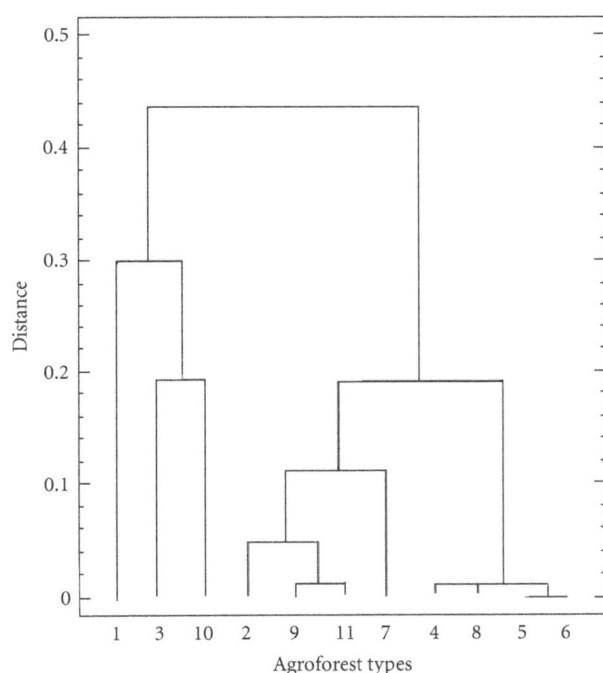

FIGURE 1: Hierarchical classification showing 11 types of agroforests existing in the tropical Guinean Highland savannahs region.

of all fruit trees recorded actually bear fruit. Those without fruits were newly planted and are expected to start giving fruit within the next few years. Once these trees start producing, the potential value of the crop per household and for the whole village will be substantially higher.

The Guinean Highland Savannah agroforests harbour a substantial diversity of tree species. The number of species in the various traditional system types ranges from 23 to 62 species. The maximum number of layers found generally in home gardens is 5 [23]. Five stratas are effectively present in Types 5, 9, 10, and 11 which are old whereas only four of them exist in the rest. The layer number three is absent in Type 7.

Concerning the age criteria, home garden Types 1, 2, and 3 are young (less than 15 years) whereas 9, 10, and 11 are

the oldest (more than 45 years old). The rest of types are intermediate.

For the area, home garden Types 1, 3, 6, and 9 are very small (less than 1 ha) whereas Types 5, 8, and 11 are the largest (more than 2 ha). Types 2, 4, 7, and 10 are intermediate (1-2 ha). The most predominant agroforestry home garden types in the GHS are 1 (26.174%), 3 (22.148%), 6 (16,107%), and 9 (20.142%) types. Similar characteristics are reported in tropical Asia mainly in India [21].

Here in after are described the main traits of the eleven types of homesteads found in the GHS.

Type 1. Gardens belonging to the Type 1 are the most represented (26.17%). It is found in all the villages of the region. In Sassa Mbersi and Dene, 15.38% of the owners do not have other farms. The majority of farmers belonging to this group are between 30 and 50 years old and 72% of them are monogamous. They practice cattle farming (12%) whereas 46% of them rear goats and sheep. A total of 16 species are present among which are 9 crops and 7 fruit trees.

Type 2. It represents 5.36% of the study sample. They are found in Gangassao, Dene, Sasa-Mbersi, Wack, and Mbe. Young (<30) and old farmers (>60 years) are absent. Of these 50% are cattle farming farmers. A total of 19 species are found among which 11 are crops.

Type 3. This type represents 22.81% of the total sample and is found in all the villages. It is more abundant in Sabongari. Cattle percentage is 24% against 40% for goats and sheep. The total number of plant species found is 26 distributed into 13 crops and 13 fruit trees.

Type 4. It is less represented (1.34%) and exists only in Sabongari. Their owners have not been to school. Their family size is between 5 and 10 persons. A total of 62 plant species are cultivated. The majority of these species are represented by 46 fruit trees. This type tends to become a forest. The three types of species that are trees of the past, trees of future, and trees of the present described by Limier (1978) are well represented.

Type 5. It is concentrated in Wack and Sabongari where it is very abundant (23.07%). All the farmers have more than 30 years old. They rear goats and cattle. The number of plant species is 47 among which are 34 fruit trees like in Type 4.

Type 6. This type represents 16.81% of the sample. It is found in 5 villages among which Gangassaou, Daran, Biskewal, and Wack (29.62%). Some of its owners (8.33%) do not have another type of farms. About 23 plant species are cultivated there among which 10 crops.

Type 7. This type representing only 2.01% of the sample is found in Gangassaou, Dene, and Wack. Its owners in majority (66.66%) are more than 60 years old. This result suggests that the three villages are ageing. Among these farmers, 33.33% do not have another type of farms. In addition they are monogamous. They rear only chickens. The number of plant species is 36 with 15 crops and 21 fruit trees.

Type 8. A few farmers (1.34%) are in practice. Farming exists only in Gangassaou and Dene. They are illiterate and rear cattle. Plant species cultivated are composed of 13 crops and 4 fruit trees.

Type 9. It is one of the most represented types (20.13%) and rank third after Types 1 and 3. It is found in all the villages of the region. Their owners rear various types of animals of whom 13% rear cattle and 30% rear goats and/or sheep. The total number of species found is 25 among which 9 crops and 15 fruit trees.

Type 10. It is less represented in the region (1.34%) and is found in Gangassaou and Mbe. Their owners are monogamous and their family size ranges from 16 to 20 persons. They are literate. Among them, 50% rear cattle. They grow 11 crops and 15 fruit trees.

Type 11. This type is found only in wack village. It is very old (60 years old) and very large (8 ha). Bush fallowing and rotation are practiced. Some important wild tree species are protected in this type. Among these species are *Borassus Aethiopum, Vitex Spp., Parkia Biglobosa, and vitellaria paradoxa.* these wild edible plants are among the most preferred species, "the top 14" in the area [24, 25]. The number of plant species is high and mainly based on fruit trees.

3.3. Functional Characteristics. The eleven types differ in their functional characteristics. Various crops are grown in tropical multilayer agroforestry systems. Due to the nutrition habits and choice of the population, the crops are grouped into main categories such as cereals, tubers, leguminous foods, vegetables, fruits, wood, and medicines. Fruits, tubers, and cereals are the most important in the region (Figure 2). The quasitotality of the production in home gardens is consumed. However, the surplus of products could be sold in order to buy products which are not produced by the farmers. The incomes generated from the commercialization of the various products help to buy soap, meat, salt, oil, clothes, and so forth. The management of the system varies

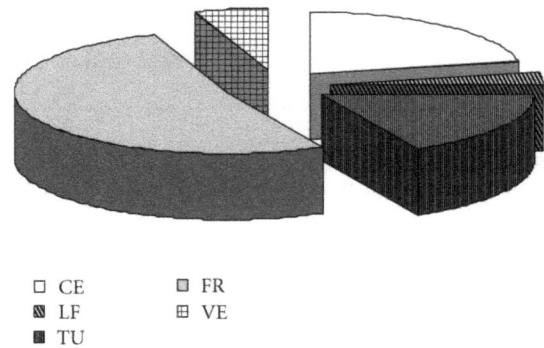

☐ CE ▧ FR
▨ LF ⊞ VE
▪ TU

FIGURE 2: Diversity of food products in home gardens: cereals (CE), tubers (TU), leguminous food (LF), fruits (FR), and vegetables (VE)

according to the socioeconomic strategy of the head of the family. Two main choices that are commonly practised in the region are subsistence and marketing strategies. The fruits are the most important products sold directly by the farmers along the road or in the local and/or regional markets. The most common fruit tree species represented in the system are *Persea americana* (87.75%), *Citrus limon* (75.79%), *Mangifera indica* (73.56%), *Musa spp.* (73.52%), *Anacardium occidentale* (68.58%), and *Carica papaya* (69.07%). Among these species, bananas occupy an important rank and contribute equally to income generation in the family while contributing to the daily diet of the family. It appears necessary to estimate its production in the context of the GHS of Cameroon.

The total amount of food produced in gardens evaluated is equal to 23.195 tons among which fruits (11.99 t), cereals (5.07 t), and tubers (4.01 t) contribute the most (Figure 2). These foods are mostly consumed. According to the farmer's choices and strategies, the excess products can be commercialized in order to buy other products and/or to pay school fees of children.

The contribution of the fruits on the gardens production is consistent. Bananas are among the main fruits consumed in the region but only a few species grow in the Guinean highland savannahs. The quantity produced is not enough to supply the population. In addition, only local species are grown by farmers. This situation induces farmers to import considerable quantities of banana and plantain from the humid lowland forest of Cameroon. Thus, it is necessary to appreciate its production in this tradition system in order to estimate the approximate quantity for importation as well as to develop strategies to introduce new banana species. The different types of gardens can be classified along a gradient from small to big banana producers.

3.4. Banana in the System

3.4.1. Description of Banana Tree. Two local species of banana are cultivated in the region. They are *M. sinensis* var *petit nain* and *M. paradisiaca* locally known as kouni and kodon in Fulfulde, respectively. The tree of the first species is short (2.5 m to 3 m). *M. paradisiaca* is very tall (6 to 8.5 m).

The bunch size of each species as well as the number of hands and fingers varies according to the management techniques. Also, fruits of *Musa paradisiaca* are hard like what is found in plantain. Maybe, it is why they call it locally "kodon" which means plantain. Banana is consumed and commercialized. However, the size remains small as compared to the improved species and varieties grown in the Humid Lowland forest. When ripped, bananas are eaten and cooked when unripe depending on the ethnolinguistic habits of the group.

3.4.2. Management of Banana Tree and Uses. There is no special management of the banana tree in this system concerning the fertilisation. Farmers do not used chemicals as it is usual in the multistorey agroforestry systems of the area. Organic matter is the main input used by the farmers. The soil fertility is maintained through dust, animal dungs, and litter from trees toward decomposition. The propagation of the tree is by suckers.

Very often, the farmers remove old and dried leaves of the banana tree and put them under the tree. They cut young leaves for domestic uses (conservation of food or meat, etc.). Different parts of the tree are also used in traditional medicine. Yellow leaves associated with other plants (*Bidens pilosa, Psydium guajava, Harungana madagascariensis, Chrisanthellum americanum*, etc.) are decocted to cure typhoid fever whereas those associated with egg and fried treat diarrhea and amoeba. Bananas have been integrated in this traditional system as a vital component of the local diet and economy. In areas where erosion is common, farmers use *Musa sinensis* to stabilize the soils (to control erosion).

3.4.3. Yield of Banana and Significant Relationship between Farmers. The banana production varies from farmer to farmer depending on the importance given to the product. The total mean yield of banana in the periurban home gardens is estimated at 3.57 tons. Year^{-1}. Three categories of banana producers exist in the area. The first group for which the quantity produced is more than 300 kg, is belonging to agroforest Types 2 (598 kg), 4 (707 kg), 7 (546 kg), and 9 (770 kg) types. The second group comprises those who produce between 100 and 300 kg that are represented by gardens of Types 1, 5, 6, and 10 whereas the last category regroups those of 3 and 11 types which produce less than 100 kg of banana per year (Figure 3). Significant differences exist between the different types of homesteads found in the area.

The total quantity of banana consumed in the periurban zone is 3.5439 tons year^{-1} (93.5%). The quantity consumed can derive from the farmer's own production or from gifts by relatives. Farmers who do not grow banana receive them free from their neighbours. It is the case of those belonging to the eleventh type of home gardens (Figure 3). Part of the periurban production is sent to the urban zone as gift or sold for food.

Tropical agroforest plays an important social role beyond the economic and ecological aspects by promoting the family integration through sharing farming, managing, harvesting

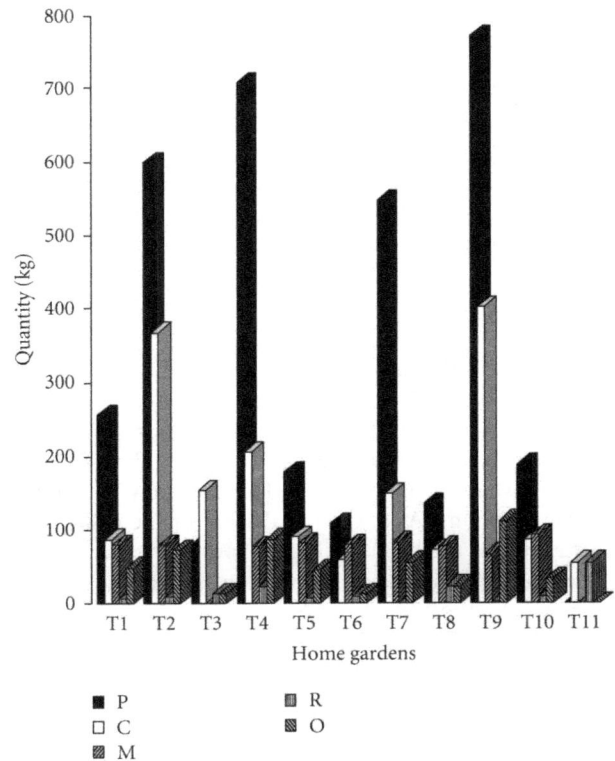

FIGURE 3: Banana annual flow in 11 agroforests (Type 1, 2, ..., 11): production (P), consumption (C), marketing (M), reception (R), and offers (O).

activities, and products with other members of the communities. In this regard, the total quantity of banana offered to relatives is 0.55 tons against 0.112 tons received. Among the household evaluated, gardens belonging to Type 11 do not cultivate banana. Nevertheless, it can be mentioned that, in rare cases, the two species are found in the same plot. In general, there is a tendency for the supply to exceed demand in all the garden types except for Type 11 where the quantity of banana supply is equal to demand (Figure 3). There is a significant correlation between banana production and consumption ($r = 0.834$; $0.00013 < 0.001$). Generally, the banana produced in agroforests is consumed (93.5%). Only a few quantities (6.5%) are commercialized.

3.5. Commercialisation of Banana. The species are commercialized either ripe or unripe. It is sold in bunches, hands, and/or fingers. When the fruit reaches the consumption maturity state, the cost of one is 10 francs CFA. The bunch costs 600–1200 francs CFA depending on its size. Due to the fact that plantains are difficult to grow in the area, *Musa paradisiaca* is more preferred by the local population. The total quantity sold in local markets is 1.01 t. Very often, "buyam and sellam" buy banana and plantain from the humid lowland forest and distribute them all over the two agroecological zones (Guinean highland savannahs, Sudanosahelian). In some cases, these products are transferred to neighbouring countries markets (Chad, Central African Republic). Despite the fact that each farmer belongs to a particular type and susceptible to sell agricultural products, the quantity sold varies among the farmers. Farmers from the agroforests Types 2, 4, 7, and 9 are the main sellers. The mean annual income generated is 1689.14$. The income generated helps the farmers to pay school fees of children, food that they do not produce, oil, salt, meat, and animals, and manufactured products.

4. Discussion

The majority of the farmers interviewed were Muslims. The high percentage of the above group can be explained by the double domination of the Adamawa population by Ousman Dan Fodio and Lamido of Rey-Bouba [19, 20]. Existence of the Christians stems from the fact that they had contact with the Germans. Also, the installation of the colonial administration in the Wack's village favoured the contact during this period. The rest of ethnolinguistic groups were still dominated by the Lamido of Ngaoundere [19]. This result suggests that home gardens are owned by men. Women who are head of family are rare. They are represented by widows or bachelors. Widows inherited farm from their late husbands whereas bachelors can obtain them from their parents either while they are alive or after the death. Among the Dii ethnolinguistic group, men as well as women can inherit the "properties" from their parents.

Their education level is low. The representation of the women is low. Predominance of monogamy in the periurban zone of Ngaoundere is in disagreement with the polygamy found Muslims areas [25]. The majority of agroforests found in the zone are very young and their area is less than one hectare. This can be due to the recent installation of the population. It can be also due to the mobility of the population. Similar results were reported in the Sodano-Guinean savannahs [26]. Eleven home garden types exist in the Sudano-Guinean savannah of Cameroon. They showed important variations in all their structural characteristics.

The results of this work indicate that a considerable array of plant species are actively cultivated in this traditional agroforestry system in GHS on residential plots and that they have a significant cash and home consumption value. The production of the system is usually consumed; only small quantities are sold. In the South African rural villages, similar trends have been observed [27]. The preponderant use of garden products for domestic consumption indicates that from the point of view of food security, it is a valuable option for small-scale farmers [28]. These findings corroborate those of other studies also indicating the importance of fruits to rural communities [22, 27, 29, 30].

The contribution of the banana production to the agroforest's yield is consistent. Among the 11 types found, the main banana producers belong to home garden Types 2, 4, 7, and 9. However, the banana production in the GHS agroforests is lower than that obtained in agroforests in the humid lowland forest zone of Cameroon [29]. The major constraints for banana are climate and soils. The quantity

produced remains insufficient to satisfy local demand. It has been beneficial for the population to introduce new varieties in the area. The study demonstrates the important socioeconomical role played by tropical home gardens. The diversity of species found in the Guinean Highland Savannahs' agroforests may serve as another example for the suggestion of Janzen (1998) by Khaya et al. [31] for which "A gardenification of the tropics is capable of preserving large lumps of wild biodiversity". The flow of products in the system is important in consolidating friendship relation among the farmers.

Thus, the growing of crops for home consumption represents considerable cash saving. This work complements others demonstrating the multifaceted nature of rural livelihoods. Coupled with the extraction of resources from adjacent communal lands it indicates the importance of small-scale activities and secondary resources in contributing to food security, household well-being, and the informal economy.

Besides being a potentially interesting source of income, banana also contributes to the structural diversification of the agroforests, which is important for creating habitat for local fauna and flora. Farmers rely on agroforests for their food, income, creation, and consolidation of relationships, treating diseases and reducing spending money.

Acknowledgments

The authors owe a debt of gratitude to the Centre of Interface, Research and Applications for the Sustainable Development in Africa (CIRADA) which funded the present work. They are equally grateful to anonymous reviewers who helped to improve the quality of the manuscript with their comments and the useful comments from Nguepi Raymond on the revised draft of this paper.

References

[1] O. S. Abdoellah, "Home gardens in Java and their future development," in *Tropical Home Gardens, UNU*, K. Landauer and M. Brazil, Eds., pp. 69–79, Banding Brazil, 1990.

[2] C. Devendra and D. Thomas, "Crop-animal interactions in mixed farming systems in Asia," *Agricultural Systems*, vol. 71, no. 1-2, pp. 27–40, 2002.

[3] R. S. Malik and P. S. Shehrawat, "Constraints in adoption of Poplar (*Populus deltoides*) based agroforestry in haryara state, India," in *Proceedings of the 1st World Agroforestry Congress*, Book of Abstracts, p. 69, Orlando, Fla, USA, 2004.

[4] J. C. Okafor and E. C. M. Fernandes, "Compound farms of south eastern Nigeria: a predominant agroforestry home garden system with crops and small livestock," *Agroforestry Systems*, vol. 5, no. 2, pp. 153–168, 1987.

[5] B. N. Okigbo, "Home gardens in tropical Africa," in *Tropical Home Gardens*, K. Landauer and M. Brazil, Eds., pp. 21–40, Banding Brazil, 1990.

[6] M. Bentes-Gama and J. R. V. Gama, "Tropical homegardens in riverine communities of Amazonian estuary, Marajo island, Brazil," in *Proceedings of the 1st World Agroforestry Congress*, Book of Abstracts, p. 109, Orlando, Fla, USA, 2004.

[7] H. D. de Foresta and G. Michon, "Etablissement et gestion des agroforêts paysannes en Indonosie. Quelques enseignements pour l'Afrique forestière," in *Alimentation en Forêt Tropicale: Bioculturelles et Perspectives de Développement*, C. M. Hladick, A. Hladick, H. Pagezy, F. O. Linares, J. A. G. Koppert, and A. Froment, Eds., vol. 2, pp. 1081–1099, 1996.

[8] R. R. Thaman, "Mixed home gardens in the pacific islands: present status and future prospects," in *Tropical Home Gardens, UNU*, K. Landauer and M. Brazil, Eds., pp. 41–65, Banding Brazil, 1990.

[9] F. Tetio-Kagho, C. M. Tankou, and Y. J. Pinta, "Contribution des jardins de case dans la diversité génétique en milieu traditionnel camerounais," in *Actes du Colloque sur la Conservation et L'utilisation des Ressources Phytogénétiques du 23–25 Mars*, pp. 281–296, 1994.

[10] M. Tchatat, H. Puig, and T. Manga, "Les jardins de case des provinces du Centres et du Sud du Cameroun : description et utilisation d'un système agroforestier traditionnel," *Journal D'agriculture Traditionnelle et de Botanique Appliquées*, vol. 37, no. 2, pp. 165–182, 1995.

[11] P. M. Mapongmetsem, B. A. Nkongmeneck, D. A. Yves, A. Salbai, and M. Haoua, "Jardins de case et sécurité alimentaires dans les savanes soudano-guinéennes du Cameroun," p. 15, Com. Mega-Tchad. Nanterre, France, 2002.

[12] D. C. P. A. Dewees and S. J. Scherr, "Policies and markets for non—timber forest products," EPTD Discussion Paper 16, IFPRI, Washington, DC, USA, 1996.

[13] R. R. B. Leakey, P. Greenwell, and M. N. Hall, "Domestication of indigenous fruit trees in West and Central Africa," in *Proceedings of the 3rd International Workshop on the Improvement of Safou and other Non-Conventional Oil Crops*, J. Kengue, C. Kepseu, and G. J. Kayem, Eds., pp. 73–92, Actes, Yaounde, Cameroon, 2000.

[14] P. M. Mapongmetsem, *Phénologie et apports au sol des substances biogènes par la litière des fruitiers sauvages des savanes soudano-guinéennes (Adamaoua, Cameroun)*, thèse de Doctorat, Université de Yaoundé I, Cameroun, 2005.

[15] S. Yonkeu, *Végétation des pâturages de l'Adamaoua (Cameroun): écologie et potentialités pastorales*, thèse de Doctorat, Université de Rennes 1, France, 1993.

[16] R. Letouzey and P. Lechevalier, Eds., *Etude Phytogéograpique du Cameroun*, 1968.

[17] P. M. Mapongmetsem, D. Y. Alexandre, A. Ibrahima, and T. F. N. Fohouo, "Situation et dynamique des jardins de case dans les savanes soudano-guinéennes du Cameroun," in *Biosciences Proceedings*, C. M. Mbofung and X. F. Etoa, Eds., vol. 7, pp. 403–413, Ngaoundéré, Cameroun, 2000.

[18] P. M. Mapongmetsem, I. Diedhio, N. Layaïda, and D. Y. Alexandre, "Situation des jardins de case en Asie du Sud Est: cas de la province de Kandal (Cambodge)," in *Biosciences Proceedings*, C. M. Mbofung and X. F. Etoa, Eds., vol. 7, pp. 396–402, Ngaoundéré, Cameroun, 2000.

[19] J. C. Müller, ""Merci à vous les blancs, de nous avoir libérés" Le cas des Dii de l'Adamaoua (Nord-Cameroun)," *Miroirs du Colonialisme*, pp. 59–72, 1997.

[20] H. Bakoet, *Patrmoine végétal du peuple Dii de l'Adamaoua au Nord-Cameroun : tradition et mutations. Mémoire de Maîtrise*, Faculté des Arts, Lettres et Sciences Humaines.Université de Ngaoundéré, Cameroun, 1999.

[21] A. Peyre, A. Guidal, K. F. Wiersum, and F. Bongers, "Dynamics of homegarden structure and function in Kerala, India," *Agroforestry Systems*, vol. 66, no. 2, pp. 101–113, 2006.

[22] P. M. Mapongmetsem, Y. Hamawa, M. C. Niwah, M. Froumsia, C. F. Kossebe, and L. Zigro, "Gestion et conservation de la biodiversité dans les agroforêts de la zone soudano-guinéenne," Yaoundé, Cameroun, Com. AETFAT, 2008.

[23] G. Michon and H. de Foresta, "Agroforest: an original agroforestry model from small holder farmers for environmental conservation and sustainable development," in *Proceedings of the UNESCO*, 1996.

[24] C. Tchiegang-Megueni, P. M. Mapongmetsem, C. H. Akagou Zedong, and C. Kapseu, "An ethnobotanical study of indigenous fruit trees in Northern Cameroon," *Forests Trees and Livelihoods*, vol. 11, no. 2, pp. 149–158, 2001.

[25] P. M. Mapongmetsem, "Représentations et gestion paysanne des jardins de case agroforestiers dans la zone périurbaine de Ngaoundéré (Adamaoua, Cameroun)," *Cameroon Journal of Ethnobotany*, vol. 1, no. 1, pp. 92–102, 2005.

[26] P. M. Mapongmetsem, "Valorisation de la biodiversité dans les agroforêts tropicales," *Procédés Biologiques et Alimentaires*, vol. 3, no. 1, pp. 82–103, 2006.

[27] C. High and C. M. Shackleton, "The comparative value of wild and domestic plants in home gardens of a South African rural village," *Agroforestry Systems*, vol. 48, no. 2, pp. 141–156, 2000.

[28] R. P. Miller and P. K. R. Nair, "Indigenous agroforestry systems in Amazonia: from prehistory to today," *Agroforestry Systems*, vol. 66, no. 2, pp. 151–164, 2006.

[29] D. J. Sonwa, B. A. Nkongmeneck, S. F. Weise, M. Tchatat, A. A. Adesina, and M. J.J. Janssens, "Diversity of plants in cocoa agroforests in the humid forest zone of Southern Cameroon," *Biodiversity and Conservation*, vol. 16, no. 8, pp. 2385–2400, 2007.

[30] E. C. M. Fernandes, A. O'kting'Ati, and J. Maghembe, "The Chagga home gardens: a multi-storeyed agro-forestry cropping system on Mt. Kilimanjaro, Northern Tanzania," *Agroforestry Systems*, vol. 2, pp. 2385–2400, 1984.

[31] M. Kaya, L. Kammesheidt, and H. J. Weidelt, "The forest garden system of Saparua island, Central Maluku, Indonesia, and its role in maintaining tree species diversity," *Agroforestry Systems*, vol. 54, no. 3, pp. 225–234, 2002.

9

Planting *Jatropha curcas* on Constrained Land: Emission and Effects from Land Use Change

M. S. Firdaus and M. H. A. Husni

Department of Land Management, Faculty of Agriculture, Universiti Putra Malaysia, Selangor, 43400 Serdang, Malaysia

Correspondence should be addressed to M. H. A. Husni, husni@agri.upm.edu.my

Academic Editors: A. S. Hursthouse and G. O. Thomas

A study was carried out to assess carbon emission and carbon loss caused from land use change (LUC) of converting a wasteland into a *Jatropha curcas* plantation. The study was conducted for 12 months at a newly established *Jatropha curcas* plantation in Port Dickson, Malaysia. Assessments of soil carbon dioxide (CO_2) flux, changes of soil total carbon and plant biomass loss and growth were made on the wasteland and on the established plantation to determine the effects of land preparation (i.e., tilling) and removal of the wasteland's native vegetation. Overall soil CO_2 flux showed no significant difference ($P < 0.05$) between the two plots while no significant changes ($P < 0.05$) on soil total carbon at both plots were detected. It took 1.5 years for the growth of *Jatropha curcas* to recover the biomass carbon stock lost during land conversion. As far as the present study is concerned, converting wasteland to *Jatropha curcas* showed no adverse effects on the loss of carbon from soil and biomass and did not exacerbate soil respiration.

1. Introduction

The onset of the Industrial Revolution has seen an increase of 110 ppmv carbon dioxide (CO_2) in our atmosphere resulting in the current atmospheric CO_2 concentration to be more than 390 ppmv [1]. Burning of fossil fuel had been the largest source of CO_2 emission followed by agriculture and land use change where, in 2005, the two activities made up 48% and 31% of the total global CO_2 emission, respectively [2].

Nonconventional fuel such as biodiesel from *Jatropha curcas* had been largely explored to be the alternative to fossil fuel in reducing CO_2 emission. The advantage of using *Jatropha curcas* compared to other biofuel crops is that it is nonedible; therefore, it does not create a conflict of utilizing food for fuel. It is also the third highest oil producing crop in terms of oil yield per hectare after oil palm and coconut where it could produce 2.236 L oil ha^{-1} under optimum field condition [3]. Being a perennial crop, it does not require frequent removal of biomass and soil tillage and can continue producing seeds of which the oil is extracted from before needing to be replanted after 25 years. Apart from that, it is a hardy plant which requires minimum irrigation and

fertilization making it possible to be cultivated on marginal soil and on dry parts of the world [4].

Biodiesel from *Jatropha curcas* was claimed to emit less greenhouse gases in particular CO_2 compared to diesel from fossil fuel. A number of life cycle analysis (LCA) studies were carried out on the production and combustion of biodiesel from *Jatropha curcas* and most studies concluded that there is a reduction in emission of greenhouse gases when compared to conventional diesel. Life cycle analysis studies by Kritana and Gheewala [5], Dehue and Hettinga [6], and Ndong et al. [7] for instance showed 77%, 68%, and 72% of greenhouse gas reduction, respectively.

Nevertheless, most of the LCA studies that are currently available had not put much emphasis on the emissions from land use change (LUC) of establishing a *Jatropha curcas* plantation despite of LUC being the second largest contributor of the total global greenhouse gas emission. None of the reviewed LCAs had used actual primary data on the effects on land use change particularly on changes of carbon stock and CO_2 emission from soil. Most of the LCA studies either used standard factors published by the IPCC [8], secondary data from other published studies, or simply made assumptions without concrete scientific basis. Past LCA studies had emphasized on CO_2 emission caused by energy consumption during production and transportation of *Jatropha curcas* biodiesel and emission from its combustion with less focus being placed on gathering data on the effects of LUC.

This study was, therefore, carried out to make field assessment on the emissions or perhaps sequestration that arises as a result of converting a wasteland to a *Jatropha curcas* plantation. Subsequently, assessment from this study would be able to fill the gap left out by most of the LCA studies on emissions from LUC. The specific objectives of this study were to (i) quantify carbon stock loss during land conversion and regeneration of new carbon stock through *Jatropha curcas* biomass production, (ii) to compare the soil CO_2 fluxes between a *Jatropha curcas* plantation and a wasteland, and (iii) to quantify changes in soil carbon at both land use types.

2. Materials and Methods

2.1. Site Description. The study was conducted at Tanah Merah Estate at Port Dickson, Negeri Sembilan, Malaysia (2° 39′ 32.51″N, 101° 47′ 07.55″E). The estate is an oil palm plantation with a total area of slightly less than 500 ha. Approximately, 11 km of 550 kV electricity pylon runs through the estate covering an area of 45.68 ha of which could not be planted with oil palm due to height restriction. The area under the pylon had since turned into wasteland which is defined as a barren uncultivated land covered with wild shrubs. The wasteland underneath the pylon was predominantly covered by *Lantana camara, Paspalum sp., Digitaria sp., Axonopus sp.,* and *Argyreia sp.*

In 2008, the company that owned the oil palm plantation had decided to convert the wasteland into a *Jatropha curcas* plantation since its height which was kept at 2 m would not obstruct the cables of the pylon. Land preparation activities which included removing the native vegetation using a bulldozer and two repetitions of soil plowing was carried out in November 2008. The first planting of *Jatropha curcas* saplings began in January 2009.

Average monthly temperature at the site was 30°C with an annual rainfall of 2200 mm [9]. The soil at the area is a Typic Paleudult with a sandy clay to sandy clay loam texture [10].

2.2. Experimental Setup. The study began in August 2009 until July 2010. Two study plots were established at the site. The plot that was planted with *Jatropha curcas* was designated as plot P while the wasteland was designated as plot S. The size of both plots is approximately one hectare locating next to each other with both plots having an approximately 7% slope.

At plot P, five 3 × 3 m quadrats were randomly placed. Each quadrat in plot P comprised of four *Jatropha curcas* trees. Soil sampling and soil flux measurements were conducted within the five quadrats at plot P. Stem diameter of each tree in every quadrat ($n = 20$) was measured monthly at plot P.

At plot S, five 3 × 3 m quadrats were also randomly placed. Within these quadrats, however, only soil CO_2 flux measurements were made. Another five 3 × 3 m quadrats were randomly placed at plot S for the removal of native vegetation at the wasteland during the initial stage of the study.

2.3. Biomass Determination

2.3.1. Biomass at Plot P. Destructive sampling of *Jatropha curcas* trees of different stem diameter and ages was made to determine the allometry relationship between increments of stem diameter with increments of its biomass dry weight in accordance with the methods of Basuki et al. [11]. The destructive sampling was carried out in a *Jatropha curcas* plot at Universiti Putra Malaysia, Serdang, Malaysia. It was assumed that *Jatropha curcas* trees at Universiti Putra Malaysia would be representative to the trees at Port Dickson as soil at both sites are of the same soil series, and both sites were also planted with the same local "Sengkarang" accession.

Fifteen trees were randomly chosen and were sawn off as close as possible to the ground. The aboveground sections of the trees were then separated to its main stems, branches, and leaves. All the parts were directly weighed in the field for the determination of their fresh weight. The belowground sections of the trees (i.e., roots) were excavated using a backhoe, and soil particles attached to the roots were washed with water before the roots were weighed for determination of fresh weight.

Three replications of subsamples were made on the root, stem, and branch for moisture content determination where the samples were oven dried at 60°C to a constant weight. Dry matter of the whole tree was then calculated by subtracting the moisture content from the fresh weight of the felled trees.

Equations based on the allometry relationship were then generated by plotting the natural log transformed aboveground and belowground biomass dry weight against its respective transformed stem diameters. The plots were then fitted to a linear regression, and the linear models generated were used to estimate the above and belowground biomass dry weight

$$\ln(\text{BDM}) = c + a\ln(\text{SD}), \qquad (1)$$

where BDM is biomass dry matter in kg, SD is stem diameter in cm, c is the intercept, and a is the slope coefficient of the regression.

At plot P, stem diameter of 20 *Jatropha curcas* trees was measured every month using a digital caliper (Mitutoyo, Japan). Two measurements were made on each tree running latitudinal and longitudinally, and the average of the two measurements was recorded as its stem diameter. The stem diameter measurements were then used to calculate the current month's above and belowground biomass by using the allometric equations generated. The average biomass for the sampled 20 trees was then extrapolated to a hectare scale based on the assumed planting density of 1600 trees ha^{-1} [12].

TABLE 1: Mean ± standard error of stem diameter, dry weight of the aboveground and belowground section of *Jatropha curcas,* and its moisture content at different ages.

Age (month)	Stem diameter (cm)	DWAG (kg)	Moisture (%)	DWBG (kg)	Moisture (%)
46	14.6 ± 1.0	7.72 ± 0.25	65.8 ± 0.5	3.17 ± 0.07	67.1 ± 0.2
32	12.5 ± 0.9	5.88 ± 0.58	68.4 ± 0.8	2.27 ± 0.05	64.7*
24	9.3 ± 0.4	2.23 ± 0.21	70.7 ± 0.1	1.03 ± 0.37	67.2 ± 2.1
10	7.6 ± 0.0	0.45 ± 0.02	73.0 ± 0.6	0.47 ± 0.02	58.8*
<6	1.6 ± 0.2	0.02 ± 0.01	n.a	0.01 ± 0.00	n.a

DWAG: dry weight aboveground.
DWBG: dry weight belowground.
n.a: not applicable.
*: no replications were made for moisture content determination.

2.3.2. Biomass at Plot S. Vegetations within the five quadrats randomly placed at plot S was clipped and were directly weighed to determine its fresh weight. Three replications of subsamples from the vegetation at each quadrat were made and oven dried at 60°C to a constant weight for moisture content determination. Biomass dry weight was then estimated by subtracting the fresh weight of the removed biomass by its moisture content. Estimated biomass from the sampled area of 45 m^2 was then extrapolated to an area of one hectare.

2.3.3. Quantification of Litterfall Production. Litterfall was collected monthly to quantify biomass production through litterfall production. Five litter traps were constructed under the canopy of five randomly selected *Jatropha curcas* trees at plot P. The litter traps was made of nylon fishing net covering an area of 4 m^2 under each canopy. The trapped litterfall was removed monthly and transferred into paper bags. The litterfall were then oven dried at 60°C to a constant weight, and the final weights were recorded as its dry matter.

2.4. Determination of Carbon in Biomass. Subsamples of the different parts from the destructive harvesting of *Jatropha curcas* trees and monthly collected litterfall were used in the determination of carbon in biomass. The dried sampled parts of the *Jatropha curcas* trees were sheared to smaller pieces using secateurs before they were ground to sizes of less than 5 mm using a kitchen mill. Litterfall was ground to sizes of less than 2 mm using a cutting mill (IKA, Germany). All samples were then analyzed for carbon content by the combustion method using the CR-412 carbon analyzer (LECO, Mich, USA).

The percentages of carbon in the different parts of the biomass were then used to estimate the mass of carbon in biomass based on biomass dry weight. The estimated biomass per hectare was then converted into mass of carbon in biomass per hectare.

2.5. Soil Total Carbon. Soil sampling was carried out to determine the monthly changes of total soil carbon. One soil sample was sampled from each quadrat at both plot P and plot S at a depth of 0 to 20 cm from the soil surface using a soil auger every month. The soil samples were then air dried ground and sieved through a 2 mm sieve before being sent for total carbon analysis also by using the CR-412 carbon analyzer (LECO, Mich, USA).

2.6. Soil CO_2 Flux. Soil CO_2 flux was measured to determine the amount of carbon lost from soil as CO_2 to the atmosphere at both plots. Measurements were made by using the LI-8100 automatic soil CO_2 flux system (LI-COR Biosciences, Neb, USA). Flux measurements were conducted at both plots where five measurements were made within the assigned quadrats of each plot.

Prior to making flux measurement, a soil collar made of a 10 × 10 cm (d × h) PVC pipe was inserted into the soil for at least an hour to allow the disrupted soil and sheared fine roots to stabilize. Soil CO_2 flux measurements were made between 1000 and 1200 hrs where daily soil CO_2 flux was at its highest rate [13].

The flux rate was calculated by fitting the changes in concentration of CO_2 within the chamber of the flux system to either an exponential or a linear regression which was done by the system's software and given in units of μmol CO_2 m^{-2} s^{-1}.

2.7. Statistical Analyses. Comparison of soil CO_2 fluxes between plots P and S was made by using a one-way t-test. Linear regression analysis was conducted to determine the relationship between two parameters of interest (e.g., biomass versus stem diameter). Analysis of variance and mean separation by Tukey's test were used for the determination of significant differences among means. All data sets were analyzed for outliers by using the Grubb's test [14].

3. Results

3.1. Dry Matter Production and Sequestered Carbon in Biomass. The mean stem diameter, mean dry weight, and mean moisture content of the aboveground and belowground sections of the 15 *Jatropha curcas* sampled are presented in Table 1 categorized by age.

The natural log transformed biomass dry weight of the aboveground and the belowground sections of the trees

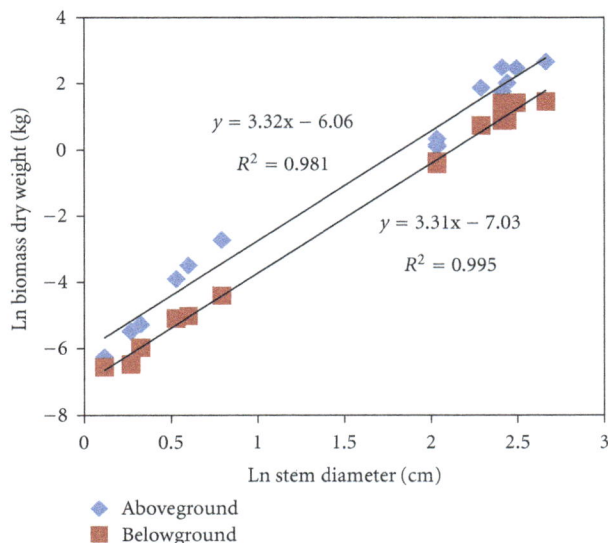

FIGURE 1: Linear regression of the stem diameter and biomass dry weight of the aboveground and belowground sections of *Jatropha curcas*.

TABLE 2: Mean carbon content of the different parts of *Jatropha curcas*.

Plant part	Carbon content (%)
Aboveground	45.60
Belowground	44.86
Leaf	46.46

plotted against its respective log transformed stem diameter are shown in Figure 1. The two plots were then regressed to a linear model.

The two allometric equations generated based on the linear regressions of the aboveground and belowground sections were

$$\ln(\text{AGDW}) = 3.32(\ln \text{SD}) - 6.06, \quad (2)$$

$$\ln(\text{BGDW}) = 3.11(\ln \text{SD}) - 7.03, \quad (3)$$

where AGDW and BGDW are the aboveground and belowground dry weights in kg, respectively, while SD is the stem diameter in cm.

The percentage of carbon in leaves and the aboveground and belowground sections of the composite biomass samples are presented in Table 2.

From the generated allometric equations (1) and (2), estimation of monthly biomass dry weight was made based on measured stem diameter where subsequently mass on carbon in biomass was calculated based of carbon content in biomass (Table 3). Litterfall production and mass of carbon in litterfall was also listed in Table 3. All figures had been extrapolated to a hectare scale.

Total biomass dry weight of the vegetation removed from plot S from five quadrats was 11.45 kg for an area of 45 m^2 (Table 4). By extrapolating it to a hectare scale, the

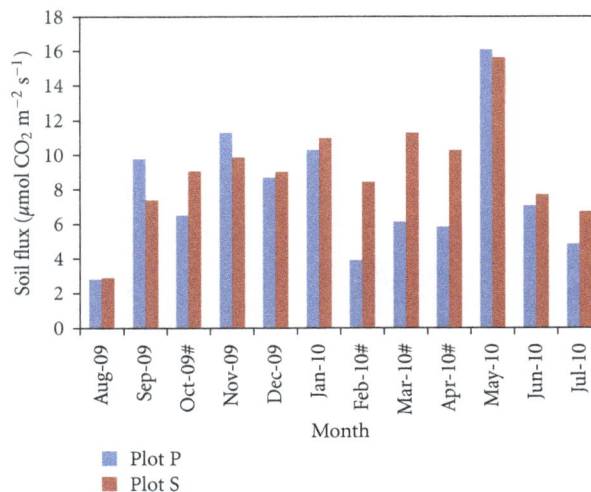

FIGURE 2: Mean soil flux at plot P and plot S. Months with a # sign indicates that there is a significant difference by t-test ($P > 0.05$).

mean biomass dry matter at plot S was estimated to be 2.54 Mg ha^{-1}. An assumption of $1:2.5$ root to shoot ratio typical for tropical shrubs and undergrowth [15] was used to calculate then the belowground biomass of the removed vegetation, and it was estimated to be 1.02 Mg ha^{-1}. Adding up the above- and belowground sections yields 3.56 Mg of biomass dry matter per hectare.

3.2. Soil Carbon. Table 5 shows soil carbon content at both plots sampled from the five quadrats at each plot expressed in percentage of carbon. Both plots showed no significant differences ($P < 0.05$) between months by ANOVA within each plot.

3.3. Soil CO$_2$ Flux. Soil CO$_2$ fluxes from plot P and plot S are presented in Figure 2. Mean soil CO$_2$ flux at plot P ranged from 2.83 μmol m^{-2} s^{-1} to 16.08 μmol m^{-2} s^{-1}. Meanwhile at plot S, soil CO$_2$ flux ranged from 2.91 μmol m^{-2} s^{-1} to 15.59 μmol m^{-2} s^{-1}. The minimum and maximum fluxes at both plots were recorded at August 2009 and May 2010, respectively.

During most months, monthly soil CO$_2$ flux at plot P was not significantly different ($P < 0.05$) by t-test from flux at plot S with the exception during October, February, March and April where fluxes at plot S were significantly higher ($P > 0.05$) than at plot P.

4. Discussion

4.1. Biomass Production and Carbon in Biomass. Carbon content in aboveground and belowground biomass of 45.60% and 44.68%, respectively, was found to be lower than the carbon content observed by CMSCRI [16] of 50.9% and 51.5% for the aboveground and belowground biomass, respectively, of a one-year-old *Jatropha curcas* tree. Mean dry weight of *Jatropha curcas* trees of the present study, however,

TABLE 3: Estimated monthly dry weight of biomass and litterfall of *Jatropha curcas* and mass of carbon stored in each respective part (mean ± standard error).

Month	Year	Biomass dry weight	Carbon in biomass	Litterfall dry weight	Carbon in litterfall
		(Mg ha^{-1})		(Mg ha^{-1})	
August	2009	0.60 ± 0.38	0.27 ± 0.17	0.05 ± 0.02	0.03 ± 0.01
September	2009	0.83 ± 0.49	0.38 ± 0.22	0.03 ± 0.02	0.01 ± 0.01
October	2009	1.02 ± 0.48	0.46 ± 0.22	0.08 ± 0.05	0.04 ± 0.03
November	2009	1.04 ± 0.50	0.47 ± 0.23	0.11 ± 0.08	0.05 ± 0.04
December	2009	1.44 ± 0.69	0.65 ± 0.31	0.18 ± 0.09	0.08 ± 0.04
January	2010	1.64 ± 0.76	0.74 ± 0.34	0.08 ± 0.04	0.04 ± 0.02
February	2010	2.06 ± 0.98	0.93 ± 0.44	0.11 ± 0.03	0.05 ± 0.02
March	2010	2.42 ± 1.07	1.10 ± 0.48	0.05 ± 0.05	0.02 ± 0.02
April	2010	2.99 ± 1.53	1.18 ± 0.58	0.12 ± 0.09	0.06 ± 0.04
May	2010	3.69 ± 1.88	1.46 ± 0.73	0.12 ± 0.04	0.06 ± 0.02
June	2010	4.09 ± 1.82	1.86 ± 0.82	0.16 ± 0.07	0.08 ± 0.03
July	2010	4.68 ± 1.51	2.13 ± 0.68	0.22 ± 0.07	0.10 ± 0.03
Cumulated Total		4.08	1.86	1.29	0.60

TABLE 4: Biomass dry weight of aboveground vegetation removed at plot S from five replications of 9 m^2 quadrats.

Quadrat (3 × 3 m)	Biomass dry weight (kg)
1	2.30
2	1.05
3	3.85
4	1.45
5	2.80
Total	11.45

was observed to be higher than that of CMSCRI where mean biomass dry weight of the present study is 1.64 Mg ha^{-1} compared to 0.49 Mg ha^{-1} of CMSCRI despite being the same age. Recorded biomass dry weight of the present study at 46 months was also found to be higher than that of CSMCRI [15] at 42 months where mean total biomass dry weight of the present study is 10.89 kg tree^{-1} compared to 5.5 kg tree^{-1} of CSMCRI albeit a four months difference between the trees.

The discrepancies in biomass carbon content of the two studies might be due to the different lignin content in the biomass from the two studies [17]. Lignin was not analyzed in the two studies but an analysis by Vaithanomsat and Apiwatanapiwat [18] found lignin content in *Jatropha curcas* stem to be 24.11%. The discrepancies of the different biomass dry weight of the two studies might possibly be due to the agronomic practices of the two plantations and site characteristics of the two studies.

Nevertheless, based on the biomass carbon content and dry weight, an estimated 0.74 Mg C ha^{-1} was sequestered in biomass of a one-year-old *Jatropha curcas* from the present study as opposed to only 0.25 Mg C ha^{-1} of CMSCRI [15]. Meanwhile, carbon sequestration in three-years-old *Jatropha*

curcas of the present study, and that of CMSCRI was 7.84 and 4.40 Mg C ha^{-1}, respectively. The large differences between the two studies might suggest that quantification of biomass production of *Jatropha curcas* have to be made according to specific sites.

Total litterfall production of 1.29 Mg ha^{-1} of the present study somewhat agrees with the result of Abugre et al. [19] who found that litterfall production of *Jatropha curcas* planted at planting distances of 1 m × 1 m, 2 m × 1 m, and 3 m × 1 m to be 2.27, 1.10 and 0.79 Mg ha^{-1}, respectively. According to the same study by Abugre et al. [19], after 120 days of decomposition, between 2.45 and 34.6% carbon is still left from *Jatropha curcas* litterfall. The large difference in the decomposition rate is due to the difference in sunlight exposure on the litterfall [20].

The amount of carbon stock that was removed when converting the wasteland into *Jatropha curcas* was estimated to be 1.78 Mg carbon ha^{-1} assuming the carbon content of the shrubs at plot S to be 50%. This value is lower than estimated value of 3.10 Mg C ha^{-1} when converting tropical grassland to *Jatropha curcas* [6].

Based on *Jatropha curcas* biomass growth of the present study, it only took 1.5 years for *Jatropha curcas* to recover back the initial carbon stock that was lost during the land clearing process. Carbon stock of plot P at 18 months after planting was 1.86 Mg ha^{-1}. As far as the time required for replenishing lost carbon stock from land conversion is concerned, result of the present study is faster than what was concluded by Fargione et al. [21] and Romijn [22] that estimated 20 to 30 years to recover lost carbon from biomass as a result of LUC for biofuel production.

4.2. Soil CO$_2$ Flux. Soil fluxes at the two sites showed no significant differences throughout the observation period apart from during October 2009, February, March, and April 2010

TABLE 5: Changes of soil carbon content at plot P and plot S.

Month	2009						2010					
	Aug	Sep	Oct	Nov	Dec	Jan	Feb	Mar	Apr	May	Jun	Jul
						Carbon (%)						
Plot S	2.12	2.09	1.76	2.59	1.80	2.35	2.67	1.83	2.13	2.05	1.87	1.81
Plot P	1.63	2.20	1.94	1.16	1.35	1.42	1.62	1.56	1.74	1.80	1.66	1.75

where soil flux at plot S was significantly higher than that at plot P. This indicated that soil preparation activities (i.e., the removal of native vegetation and soil tillage) did not cause an increase in CO_2 emission from the soil which contradicts with other previous observations [23–25]. As a matter of fact, four out of the twelve months of observation showed that CO_2 fluxes at plot S were higher than those at plot P.

No changes in total soil carbon content were detected at both plots during the observation period. This again showed that land preparation activity did not have much influence on soil carbon. The results contradicted with the report of Romijn [22] who found a net loss of soil organic carbon as much as $32\,Mg\,ha^{-1}$ on a *Jatropha curcas* plantation converted from a virgin Miombo Woodland. The LCA study by Dehue and Hettinga [6] on the other hand assumed that no carbon buildup occurs in the soil even after 20 years of planting.

5. Conclusion

No significant losses were detected at least during the first year of cultivation on soil carbon and by means of soil CO_2 fluxes. Within less than one and a half year, the initial carbon stock that was removed during land preparation was recovered back by the growth of *Jatropha curcas* trees. It could, therefore, be concluded that converting a wasteland into a *Jatropha curcas* plantation does not show any degrading effects on LUC at least in the case of the present study.

References

[1] NOAA, "Recent Global CO_2," in *Trends in Carbon Dioxide*, 2011.

[2] T. Herzog, "World greenhouse gas emissions in 2005," in *WRI Working Paper*, World Resource Institute, 2009.

[3] P. McDougall, "Jatropha—this particular oil well holds a lot of future promise," in *Bayer CropScience Editorial Service*, Bayer CropScience, 2008.

[4] J. Heller, *Physic nut. Jatropha curcas L. Gatersleben and Rome*, Institute of Plant Genetics and Crop Plant Research/International Plant Genetic Resources Institute, 1996.

[5] P. Kritana and S. Gheewala, "Energy and greenhouse gas implications of biodiesel production from *Jatropha curcas* L.," in *Proceedings of the 2nd Joint International Conference on Sustainable Energy and Environment*, Bangkok, Thailand, 2006.

[6] B. Dehue and W. Hettinga, *GHG Performance Jatropha Biodiesel*, Ecofys bv, Utrecht, The Netherlands, 2008.

[7] R. Ndong, M. Montrejaud-Vignoles, O. Saint Girons et al., "Life cycle assessment of biofuels from *Jatropha curcas* in West Africa: a field study," *GCB Bioenergy*, vol. 1, pp. 197–210, 2009.

[8] S. Eggleston, L. Buendia, K. Miwa, T. Ngara, and K. Tanabe, Eds., *2006 IPCC Guidelines for National Greenhouse Gas Inventories*, IGES, Japan, 2006.

[9] Malaysian Meteorological Department, *Rainfall Data Tanah Merah, Port Dickson (2009-2010)*, Malaysian Meteorological Department, Petaling Jaya, 2011.

[10] Soil Survey Staff, *Keys to Soil Taxonomy*, USDA-National Resource Conservation Service, Washington, DC, USA, 11th edition, 2010.

[11] T. M. Basuki, P. E. van Laake, A. K. Skidmore, and Y. A. Hussin, "Allometric equations for estimating the above-ground biomass in tropical lowland Dipterocarp forests," *Forest Ecology and Management*, vol. 257, no. 8, pp. 1684–1694, 2009.

[12] N. Carels, K. Jean Claude, and D. Michel, "*Jatropha curcas*: a review," in *Advances in Botanical Research*, J.-C. Kader and M. Delseny, Eds., pp. 39–86, Academic Press, 2009.

[13] Y. Luo and X. Zhou, "Temporal and spatial variations in soil respiration," in *Soil Respiration and the Environment Burlington*, pp. 107–131, Academic Press, 2006.

[14] F. Grubbs, "Procedures for detecting outlying observations in samples," *Technometrics*, vol. 11, pp. 1–21, 1969.

[15] CDM-EB, "Estimation of GHG emissions due to clearing, burning and decay of existing vegetation attributable to a CDM A/R project activity," Report EB 50, Annex 22: UNFCCC, 2009.

[16] CSMCRI, "*Data Measurements and Expert Judgement*," P. Ghosh, J. Patolia, and M. Gandhi, Eds., Central Salt & Marine Chemicals Research Institute, Bhavnagar, India, 2007.

[17] K. W. Ragland, D. J. Aerts, and A. J. Baker, "Properties of wood for combustion analysis," *Bioresource Technology*, vol. 37, no. 2, pp. 161–168, 1991.

[18] P. Vaithanomsat and W. Apiwatanapiwat, "Feasibility study on vannilin production from Jatropha curcas stem using steam explosion as a pretreatment," *International Journal of Chemical and Biological Engineering*, vol. 2, pp. 211–214, 2009.

[19] S. Abugre, C. Oti-Boateng, and M. Yeboah, "Litter fall and decomposition trend of *Jatropha curcas* L. leaves mulches under two environmental conditions," *Agriculture and Biology Journal of North America*, vol. 2, pp. 462–470, 2011.

[20] X. Feng, K. M. Hills, A. J. Simpson, J. K. Whalen, and M. J. Simpson, "The role of biodegradation and photo-oxidation in the transformation of terrigenous organic matter," *Organic Geochemistry*, vol. 42, no. 3, pp. 262–274, 2011.

[21] J. Fargione, J. Hill, D. Tilman, S. Polasky, and P. Hawthorne, "Land clearing and the biofuel carbon debt," *Science*, vol. 319, no. 5867, pp. 1235–1238, 2008.

[22] H. A. Romijn, "Land clearing and greenhouse gas emissions from Jatropha biofuels on African Miombo Woodlands," *Energy Policy*, vol. 39, no. 10, pp. 5751–5762, 2011.

[23] U. M. Sainju, J. D. Jabro, and W. B. Stevens, "Soil carbon dioxide emission and carbon content as affected by irrigation, tillage, cropping system, and nitrogen fertilization," *Journal of Environmental Quality*, vol. 37, no. 1, pp. 98–106, 2008.

[24] D. Chatskikh and J. E. Olesen, "Soil tillage enhanced CO_2 and N_2O emissions from loamy sand soil under spring barley," *Soil and Tillage Research*, vol. 97, no. 1, pp. 5–18, 2007.

[25] Y. Nouvellon, D. Epron, A. Kinana et al., "Soil CO_2 effluxes, soil carbon balance, and early tree growth following savannah afforestation in Congo: comparison of two site preparation treatments," *Forest Ecology and Management*, vol. 255, no. 5-6, pp. 1926–1936, 2008.

10

The Interdependence between Rainfall and Temperature: Copula Analyses

Rong-Gang Cong[1] and Mark Brady[1,2]

[1] Centre for Environmental and Climate Research (CEC), Lund University, Lund S-22362, Sweden
[2] AgriFood Economics Centre, Department of Economics, Swedish University of Agricultural Sciences, Lund S-22007, Sweden

Correspondence should be addressed to Rong-Gang Cong, ronggang.cong@cec.lu.se

Academic Editors: G. O. Thomas and C. Varotsos

Rainfall and temperature are important climatic inputs for agricultural production, especially in the context of climate change. However, accurate analysis and simulation of the joint distribution of rainfall and temperature are difficult due to possible interdependence between them. As one possible approach to this problem, five families of copula models are employed to model the interdependence between rainfall and temperature. Scania is a leading agricultural province in Sweden and is affected by a maritime climate. Historical climatic data for Scania is used to demonstrate the modeling process. Heteroscedasticity and autocorrelation of sample data are also considered to eliminate the possibility of observation error. The results indicate that for Scania there are negative correlations between rainfall and temperature for the months from April to July and September. The student copula is found to be most suitable to model the bivariate distribution of rainfall and temperature based on the Akaike information criterion (AIC) and Bayesian information criterion (BIC). Using the student copula, we simulate temperature and rainfall simultaneously. The resulting models can be integrated with research on agricultural production and planning to study the effects of changing climate on crop yields.

1. Introduction

Weather is the key source of uncertainty affecting crop yield especially in the context of climate change [1–3]. For example, Vergara et al. studied the potential impact of catastrophic weather on the crop insurance industry and found that 93% of crop loss was directly related to unfavorable weather [4]. Accurate modeling of multivariate weather distributions would allow farmers to make better decisions for reducing their exposure to weather risk or take advantage of favorable climatic relationships [5]. Among variables relevant to weather, rainfall and temperature are two important factors which have a large effect on crop yield [6–9]. Typically, temperature affects the length of the growing season and rainfall affects plant production (leaf area and the photosynthetic efficiency) [10, 11].

There is a lot of literature studying the effects of temperature and rainfall on crop yield. Erskine and El Ashkar quantified the effect of rainfall on lentil seed yield and found

that rainfall accounted for 79.8% of the variance of seed yield [12]. Lobell et al. studied 12 major Californian crops and found rainfall was able to explain more than 60% of the observed variability in yields for most crops [13]. Cooper et al. found that not only the seasonal rainfall totals and their season-to-season variability were important, but also the "within season" variability had a major effect on crop productivity [14], which implies that monthly data is needed in crop production analysis.

Muchow et al. found that lower temperature increased the length of time that the maize could intercept radiation and hence grow [15]. Lobell and Asner found a roughly 17% relative decrease in both corn and soybean yield in the USA for each degree of increase in growing season temperature [16]. In summary, it is well established that rainfall and temperature are two important climatic factors affecting agricultural production [17–19].

Since temperature and rainfall are critical determinants of crop yield, accurate simulation of temperature and rainfall is important not only for meteorology but also for agricultural economics. However, in reality it is difficult to simulate rainfall and temperature simultaneously due to the interdependence (correlation) between them [20–22]. Spatially, it is generally believed that there exists significant correlation between rainfall and temperature over tropical oceans and land [23]. For example, Aldrian and Dwi Susanto examined the relationship between rainfall and sea surface temperature and found that Indonesian rainfall variability revealed some sensitivity to sea-surface temperature variability in adjacent parts of the Indian and Pacific Oceans [24]. Black also studied the relationship between Indian Ocean sea surface temperature and East Africa rainfall and concluded that strong East African rainfall was associated with warming in the Pacific and Western Indian Oceans and cooling in the Eastern Indian Ocean [25].

Temporally, it is generally believed that the correlation between rainfall and temperature changes between months. For example, Rajeevan et al. examined the temporal relationship between land surface temperature and rainfall [26]. They found that temperature and rainfall were positively correlated during January and May but negatively correlated during July. Using annual data Huang et al. also found a negative correlation between rainfall and temperature in Yellow River basin of China [27].

To take the interdependence between rainfall and temperature into account, multivariate probability simulation is needed. Traditionally multivariate probability density functions, however, are generally limited to the multivariate normal distribution or mixtures of it [28]. A possible method that provides an alternative is the copula method. Copulas are advantageous because they can model joint distributions of random variables with greater flexibility both in terms of marginal distributions and the dependence structure [29]. Copulas have been used in financial economics for quite some time [30–32]. However, there are relatively few applications to agricultural weather simulation.

In respect to temperature and rainfall, AghaKouchak et al. applied two different elliptical copula families, namely, Gaussian and t-copula, to simulate the spatial dependence of rainfall and found that using the t-copula might have significant advantages over the well-known Gaussian copula particularly with respect to extremes [33]. Serinaldi also studied the spatial dependence of rainfall and confirmed that only positive contemporaneous pairs of rainfall observations correctly described the intersite dependence [34]. Laux et al. highlighted the importance of pretreatment of meteorological data in the copula modeling process [35]. Laux et al. used the Clayton copula to construct the bivariate distribution of drought duration and intensity [36]. Similar applications of the Clayton copula can also be found in the studies of Favre et al. and Shiau et al. [37, 38]. Furthermore, they raised the question as to which copula model best fitted the empirical data. The only literature concerning the application of copula simulation to model the interdependence between temperature and rainfall up to now is that of Schölzel and Friederichs [39]. They used a simple statistical model based on the copula approach to describe the phenomenon that cold periods were accompanied by small precipitation amounts.

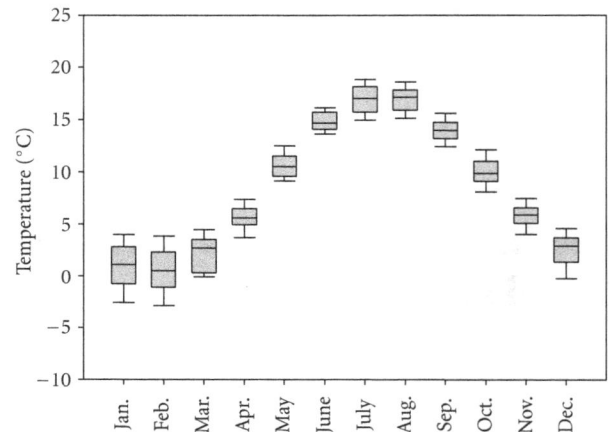

FIGURE 1: Monthly average temperature in Scania, Sweden, from 1961 to 2010. Note: the boundary of the box closest to zero indicates the 25th percentile, a line within the box marks the median, and the boundary of the box farthest from zero indicates the 75th percentile. Whiskers (error bars) above and below the box indicate the 90th and 10th percentiles.

Inspired by Dupuis's study on hydrological random variables [40], the purpose of this paper is to illustrate the pretreatment process of meteorological data, demonstrate the application of different copulas to modeling of joint distributions of rainfall and temperature, select the most suitable copula function according to information criteria, and finally simulate rainfall and temperature simultaneously.

2. Materials and Methods

2.1. Study Area. Scania is Sweden's southernmost province and one of Northern Europe's most fertile farming districts with the main crops being winter wheat, rapeseed, sugar beets, and barley. As Scania is surrounded by water on three sides (the Baltic Sea, the Kattegat Sea, and the Öresund Sound), it has a maritime climate, especially along the south and east coasts. The winters are mild (few days of snow), but the summers are similar to those in the rest of southern Sweden.

2.2. Data Collection and Preliminary Analysis. Monthly temperature and rainfall data for Scania from 1961 to 2010 was obtained from the Swedish Meteorological and Hydrological Institute.

2.2.1. Temperature. Monthly average temperature in Scania shows a clear seasonal cycle from 1961 to 2010 (Figure 1). The average temperature usually reaches its peak in July and its bottom in February. From April to November, the average temperature is always above 0°C. The variability of average temperature in January and February is though relatively large. Some descriptive temperature statistics are listed in Table 1.

TABLE 1: Descriptive statistics for monthly average temperature from 1961 to 2010 (unit: °C).

	Jan.	Feb.	Mar.	Apr.	May	June
Maximum	5.10	5.20	5.80	8.30	13.10	18.00
Minimum	−5.20	−5.20	−2.30	1.90	8.40	12.00
Mean	0.83	0.54	2.21	5.68	10.58	14.78
Standard deviation	2.48	2.39	1.85	1.36	1.20	1.07
Variation coefficient	2.99	4.44	0.84	0.24	0.11	0.07
	July	Aug.	Sep.	Oct.	Nov.	Dec.
Maximum	21.00	21.50	16.90	13.20	8.30	7.10
Minimum	13.90	14.60	11.60	7.60	2.70	−2.80
Mean	16.99	17.04	13.98	10.03	5.80	2.49
Standard deviation	1.59	1.45	1.20	1.33	1.27	1.88
Variation coefficient	0.09	0.09	0.09	0.13	0.22	0.76

TABLE 2: Descriptive statistics for monthly total rainfall from 1961 to 2010 (unit: mm).

	Jan.	Feb.	Mar.	Apr.	May	June
Maximum	70.00	50.00	73.50	87.90	90.60	123.3
Minimum	1.00	5.00	3.30	3.80	6.30	0.1
Mean	35.19	25.14	30.07	32.20	39.74	46.28
Standard deviation	17.07	11.38	16.25	18.86	20.46	26.82
Variation coefficient	0.49	0.45	0.54	0.59	0.51	0.58
	July	Aug.	Sep.	Oct.	Nov.	Dec.
Maximum	147.60	189.90	161.90	106.30	95.00	106.00
Minimum	7.40	5.70	7.30	4.50	17.00	4.80
Mean	51.15	58.33	49.20	45.90	45.73	40.80
Standard deviation	30.76	39.97	31.48	24.26	19.65	19.03
Variation coefficient	0.60	0.69	0.64	0.53	0.43	0.47

2.2.2. Rainfall. Compared with temperature, monthly total rainfall in Scania does not show a clear seasonal cycle from 1961 to 2010. From June to November, the average monthly total rainfall is relatively high (Figure 2). Some descriptive rainfall statistics are listed in Table 2.

2.2.3. The Relationship between Rainfall and Temperature. The physical rationale behind the relationship between rainfall and temperature is that rainfall may affect soil moisture which may in turn affect surface temperature by controlling the partitioning between the sensible and latent heat fluxes [41]. Because the sample data is non-Gaussian distributed and skewed, the Kendall correlation coefficient is employed to calculate the correlation between monthly rainfall and temperature. It is found that there are negative correlations between rainfall and temperature from April to July and in September (at the 10% confidence level) (Table 3).

2.3. Methods. Here we use the copula functions to model the interdependence between the probability distributions of a certain month's temperature and rainfall. Let X and Y be continuous random variables representing temperature and rainfall, with cumulative distribution functions $F_X(x) = \Pr(X \leq x)$ and $G_Y(y) = \Pr(Y \leq y)$, respectively. Following Sklar [42], there is a unique function C such that

$$\Pr(X \leq x, Y \leq y) = C(F(x), G(y)), \quad (1)$$

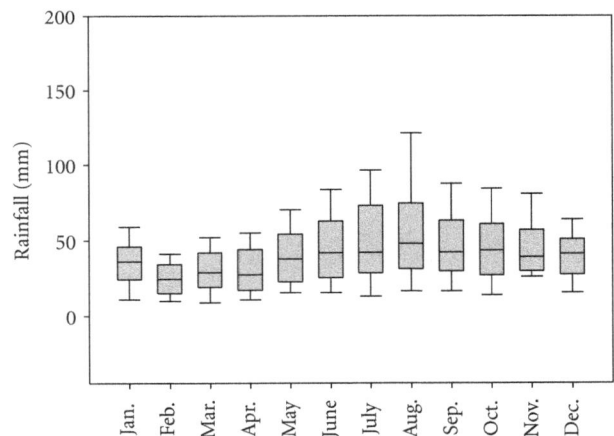

FIGURE 2: Monthly total rainfall in Scania, Sweden, from 1961 to 2010. Note: the boundary of the box closest to zero indicates the 25th percentile, a line within the box marks the median, and the boundary of the box farthest from zero indicates the 75th percentile. Whiskers (error bars) above and below the box indicate the 90th and 10th percentiles.

where $C(u, v) = \Pr(U \leq u, V \leq v)$ is the distribution of the pair $(U, V) = (F(X), G(Y))$ whose margins are uniform on $[0, 1]$. The function C is called a copula. As argued by Joe [43]

TABLE 3: Correlation analysis for monthly temperature and rainfall from 1961 to 2010.

	Jan.	Feb.	Mar.	Apr.	May	June
Kendall correlation coefficients	0.12	0.13	0.07	**−0.27**	**−0.3**	**−0.17**
P value	0.22	0.19	0.49	**0.007**	**0.002**	**0.08**
	July	Aug.	Sep.	Oct.	Nov.	Dec.
Kendall correlation coefficients	**−0.3**	−0.02	**−0.19**	−0.13	−0.02	0.09
P value	**0.002**	0.84	**0.06**	0.19	0.85	0.37

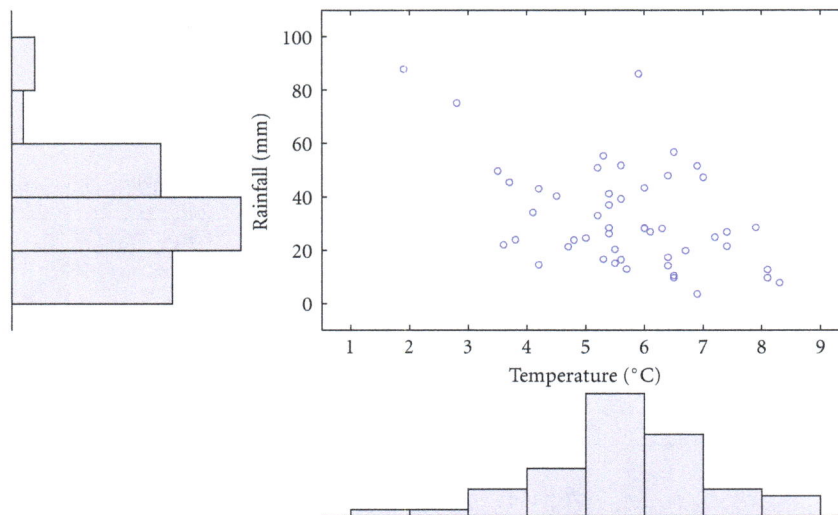

FIGURE 3: Temperature and rainfall in April from 1961 to 2010.

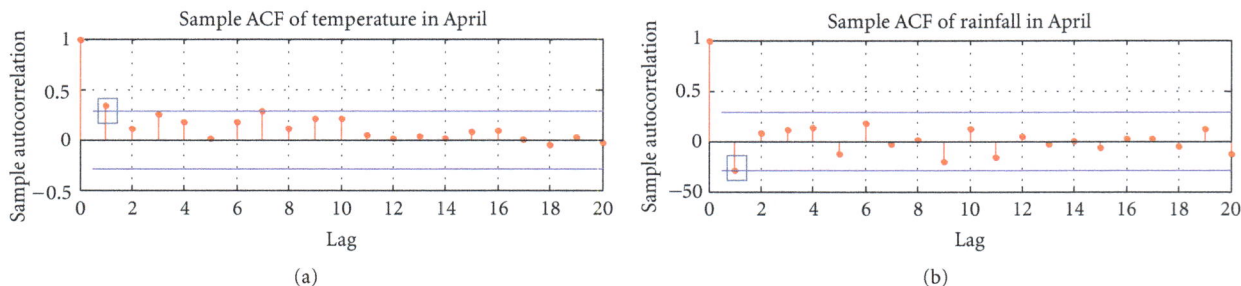

(a)

(b)

FIGURE 4: Sample autocorrelation function (ACF) of temperature and rainfall in April from 1961 to 2010.

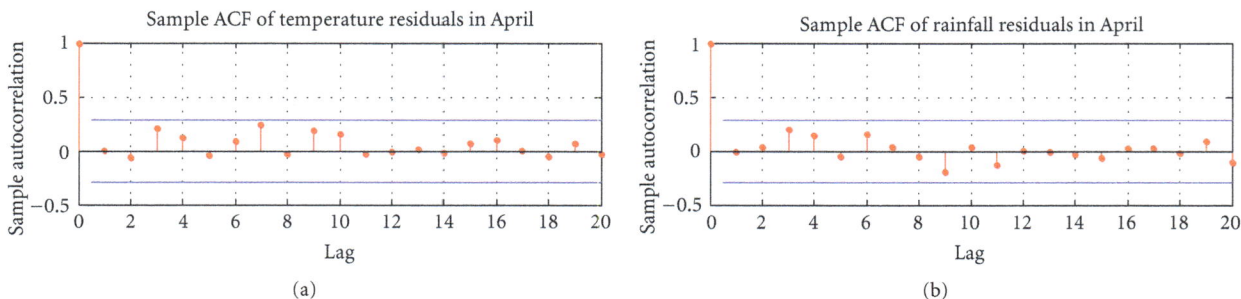

(a)

(b)

FIGURE 5: Sample autocorrelation function (ACF) of AR adjusted temperature and rainfall in April.

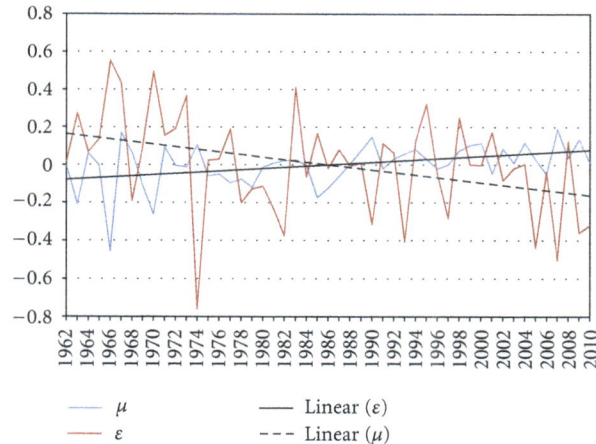

FIGURE 6: Residuals for AR adjusted temperature and rainfall in April.

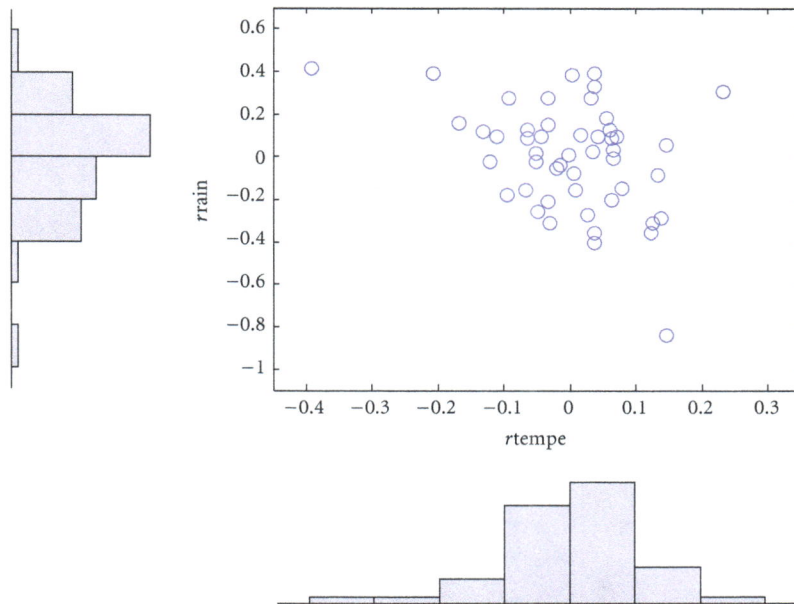

FIGURE 7: Scatters of residuals for trend adjusted temperature and rainfall in April.

and Nelsen [44] among others, C characterizes the dependence in the pair (X, Y). There are many parametric copula families available, which usually have parameters that control the strength of dependence. Among these, five families of commonly used copulas are considered. They are listed in Table 4, along with their parameter ranges. The first three are Archimedean [43] and the last two are metaelliptical [45].

After calculating the parameters of each copula, it is necessary to decide which family is the best representation of the dependence structure between the variables of interest. There are a few techniques to select the best copula. One of them is based on distance measures pertaining to the distributions of the candidate models (copulas) and the empirical distribution of the data [46, 47]. Alternative methods include likelihood ratio tests and approaches related to information criteria [31], such as Akaike [48] and Schwarz's Bayesian [49]

Information Criteria. Information criteria are adopted here because they can describe the tradeoff between bias (accuracy) and variance (complexity) in model construction. The Akaike information criterion (AIC) is a measure of the relative goodness of fit of a statistical model. Its definition is

$$AIC = 2k - 2\ln(L), \tag{2}$$

where k is the number of parameters in the copula and L is the maximized value of the likelihood function for the copula. The Bayesian information criterion (BIC) was developed by Schwarz using Bayesian formalism. Its definition is

$$BIC = -2\ln(L) + k\ln(N), \tag{3}$$

where N is the sample size.

(a)

(b)

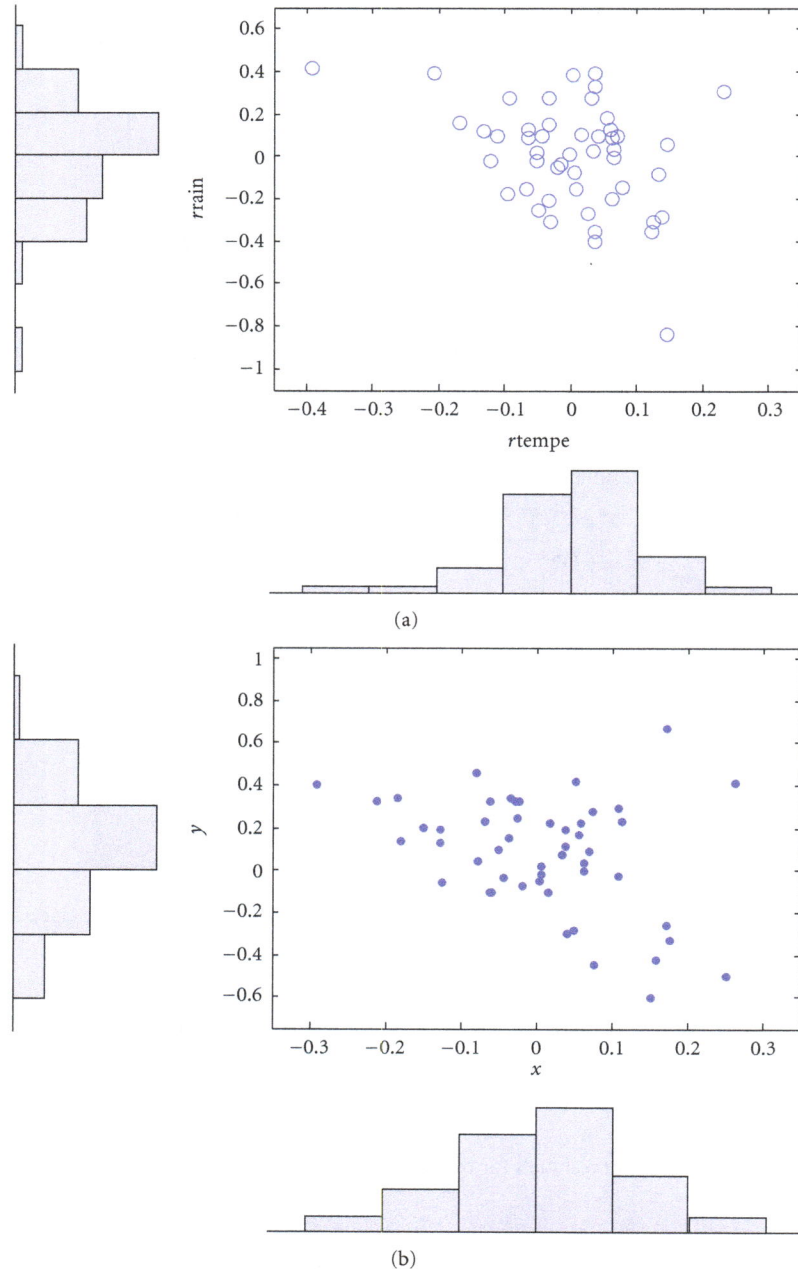

FIGURE 8: Scatter plots of real residuals (a) and student-based copula simulated residuals (b).

3. Results and Discussion

Temperature and rainfall data in April from 1961 to 2010 is employed as an example to demonstrate the modeling process (Figure 3). There is a significant negative relationship (Kendall correlation coefficient is -0.27, P-value = 0.007) between temperature and rainfall in April. Temperature has negative skewness (-0.35) and rainfall has positive skewness (1.07), which may cause a heteroscedasticity problem when fitting the model [50]. Following Kim and Ahn [51], the temperature and rainfall data are log-transformed to remove

this effect. The logarithmic transformation for the data is invertible, which will not affect the fitting results.

Following Benth and Šaltyte-Benth's instructions [52], the time series of temperature and rainfall are tested for auto-correlation using the Q-statistics (Figure 4). Autocorrelation describes the correlation between values of temperature (or rainfall) at different points in time, as a function of the time difference. The presence of autocorrelation increases the variances of residuals and estimated coefficients, which reduces the model's efficiency. The Ljung-Box Q test is a type of statistical test of whether autocorrelations of a time series

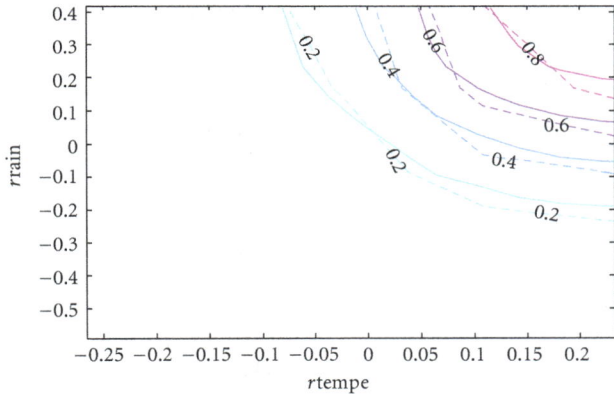

FIGURE 9: Real versus Gauss fitted CDF. Note: The dashed lines are the real CDFs while the solid lines are the simulated CDFs.

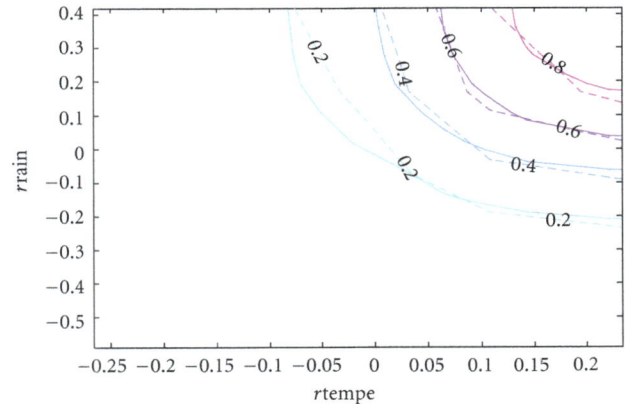

FIGURE 11: Real versus Clayton fitted CDF. Note: The dashed lines are the real CDFs while the solid lines are the simulated CDFs.

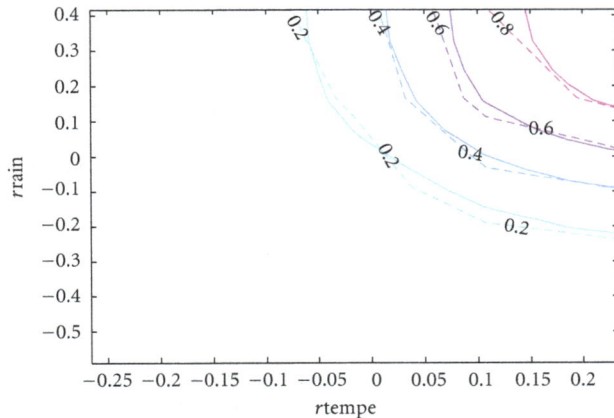

FIGURE 10: Real versus student fitted CDF. Note: The dashed lines are the real CDFs while the solid lines are the simulated CDFs.

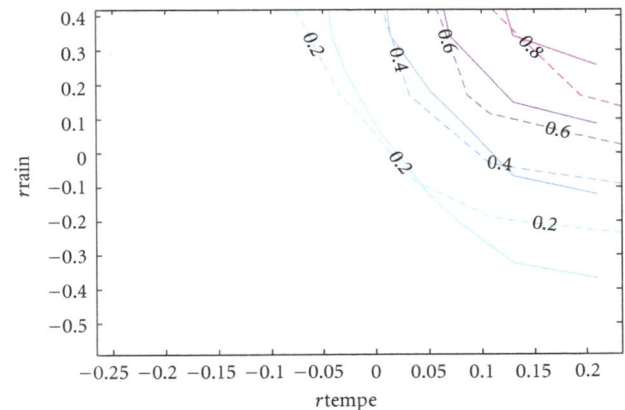

FIGURE 12: Real versus Frank fitted CDF. Note: The dashed lines are the real CDFs while the solid lines are the simulated CDFs.

TABLE 4: Five families of copulas.

Family	$C(u, v)$	Range of θ
Normal	$N_\theta\left(\Phi^{-1}(u), \Phi^{-1}(v)\right)$	$[-1, 1]$
Student	$T_{\theta,\gamma}\left(T_\gamma^{-1}(u), T_\gamma^{-1}(v)\right)$	$[-1, 1]$
Clayton	$\left(u^{-\theta} + v^{-\theta} - 1\right)^{-1/\theta}$	$(0, \infty)$
Frank	$-\theta^{-1} \ln\left\{1 + (e^{-\theta u} - 1)(e^{-\theta v} - 1)/(e^{-\theta} - 1)\right\}$	$(-\infty, \infty)$
Gumbel	$\exp\left\{-[(-\ln u)^\theta + (-\ln v)^\theta]^{1/\theta}\right\}$	$[1, \infty)$

Φ: cumulative distribution function (CDF) of a $N(0,1)$.
N_θ: CDF of a standard bivariate normal distribution with Pearson correlation θ.
T_γ: CDF of a student distribution with γ degrees of freedom.
$T_{\theta,\gamma}$: CDF of a bivariate student distribution with γ degrees of freedom.
Source: [46].

TABLE 5: Results of different copula models for temperature and rainfall in April.

	Normal	Student	Clayton	Frank	Gumbel
θ	−0.34	−0.31	0.001	0.001	1.1
Log likelihood	3.05	4.11	−0.0007	−0.0002	−1.86
AIC	−6.06	−8.15	0.042	0.041	3.75
BIC	−6.02	−8.07	0.081	0.08	3.79

are different from zero [53]. The Q-statistics is defined as follows:

$$Q = N(N + 2) \sum_{a=1}^{h} \frac{\hat{p}_a^2}{N - a}, \tag{4}$$

where $\hat{p}_a^2$ is the sample autocorrelation at lag a, and h is the number of lags being tested. The first-order autocorrelations

are found to be strong both for temperature (Q-stat = 6.32, P value = 0.01) and rainfall (Q-stat = 4.52, P value = 0.03), as shown in Figure 4.

Therefore, an AR(1) model is used to eliminate the autocorrelation in the series as follows:

$$\text{tempe}_t = 0.48 + 0.35 \times \text{tempe}_{t-1} + \varepsilon_t$$

$$(4.7^{**})(2.56^{**}),$$

$$\text{rain}_t = 1.85 - 0.29 \times \text{rain}_{t-1} + \mu_t \tag{5}$$

$$(9.06^{**})(-2.1^{**}).$$

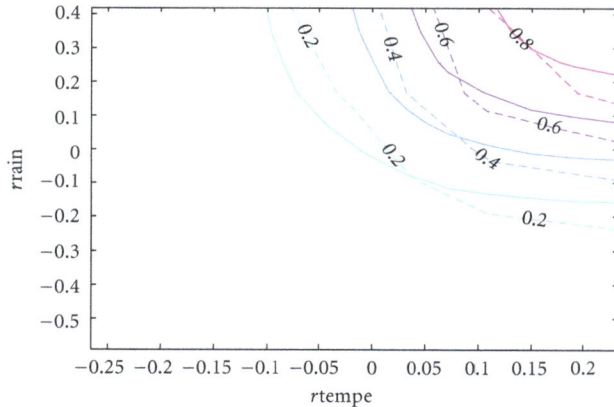

FIGURE 13: Real versus Gumbel fitted CDF. Note: The dashed lines are the real CDFs while the solid lines are the simulated CDFs.

Note that the numbers in the bracket are t-values and ** stands for the statistical significance at the 95% confidence level.

Residuals ε_t and μ_t are tested where only weak autocorrelations are found (Figure 5).

In addition to autocorrelation, time trends are also found in the series of ε_t and μ_t. Based on Manton et al.'s research [54], the time trends should be removed from the series to obtain a stationary process. The functions used to detrend the time series are

$$\varepsilon_t = -0.08 + 0.0032 \times t + \varphi_t$$
$$(-2.65**)(3.04**),$$
$$\mu_t = 0.17 - 0.007 \times t + \gamma_t$$
$$(2.3**)(-2.65**).$$
(6)

We find that temperature has an increasing trend and rainfall has a decreasing trend in April from 1961 to 2010 (Figure 6). The annual rate of increase in temperature in April is $0.0032°C$ and decrease in rainfall is $0.007\,mm$ per year. The trend adjusted data are shown in Figure 7 where $rtempe_t$ and $rrain_t$ are used to represent the corrected values of φ_t and γ_t, respectively.

The residuals for the trend adjusted variables have negative skewness: temperature (-1) and rainfall (-0.7). Based on the inference for the margins (IFM) [55], the parameter estimates and model evaluation indices for each copula for $rtempe_t$ and $rrain_t$ are presented in Table 5.

The log-likelihood ratio is largest and the AIC and BIC are smallest for the student copula, which means that the student copula is the most suitable model.

A comparison of the real and simulated residuals of temperature and rainfall is shown in Figure 8.

Since the purpose of this paper is to develop a copula model of the bivariate distribution of rainfall and temperature that can be used in simulation studies, the accuracy of the resulting model is of utmost importance. Although Table 5 has provided some statistical support for the model

and Figure 8 has given some visual evidence, the contours of the cumulative distribution functions can best show the difference between the real and simulated data.

In Figures 9, 10, 11, 12, and 13, the contours of the cumulative distribution functions (CDFs) for the real and simulated data from the five copula models are plotted to visualize the difference or similarity in the distributions as the case may be. It is found that the student copula model fits the real data best according to the similarity of the contour lines. Consequently the student copula is the best choice of model according to all our criteria.

Based on the estimated parameters, 1,000 draws are made from the Student copula model. The simulated data is then transformed to the original scale and compared with the real data in Figure 14.

4. Conclusions

This paper presents a copula-based methodology for modeling the joint distribution of temperature and rainfall, which are of utmost importance for agricultural production especially in the context of climate change. Copulas have been used extensively in the financial literature, but have not been widely used in weather simulation. The copula approach provides a powerful and flexible method to model multivariate distributions and thus goes beyond joint normality, regression, and mean-variance criterion. Accurate simulation of weather events may help to improve risk management in agricultural planning.

A shortcoming of the copula method is the arbitrariness of the selection of a particular copula. The main purpose of this paper is to present a complete copula modeling framework to model the interdependence of rainfall and temperature. In contrast to Schölzel and Friederichs [39], we compare different copulas and show how to select the optimal copula based on information criteria (AIC and BIC). The advantage of this approach is that it does not require any assumptions and is primarily data driven thus minimizing the subjectivity introduced by the researcher. The model selection criteria indicate that the Student copula produces the best model to simulate the dependence structure between rainfall and temperature in Scania, Sweden.

Although the month of April was chosen as our working example, we have also tested the data for other months with similar results. The study is only based on meteorological data for a single region. The most suitable copula family for rainfall and temperature might change from one region to another due to differences in geographical and geophysical conditions. Our approach however can be applied in studies of other parts of the world to select the most appropriate copula model. A potentially valuable extension of this research is to connect the analysis with crop production planning and agricultural economics. If the relationship among temperature, rainfall, and crop yield can be determined, then it could be used in developing risk reducing strategies for farmers, something which will become increasingly important in the face of climate change. This is the focus of our ongoing research.

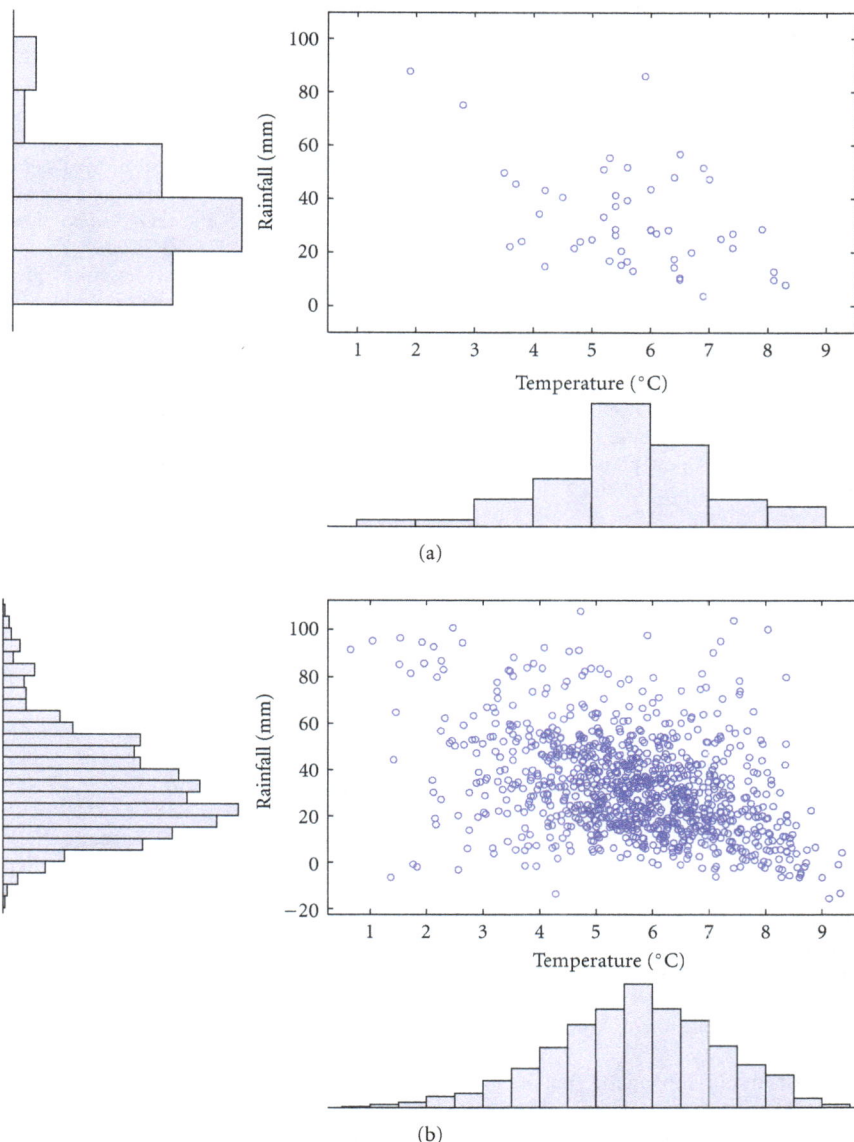

FIGURE 14: Real (a) and student-based copula simulated (b) temperature and rainfall data for Scania in April.

Acknowledgments

The authors gratefully acknowledge the good research environments provided by the Centre for Environmental and Climate Research (CEC), Lund University, and AgriFood Economics Centre. We also thank the editors and anonymous referees for their helpful suggestions on the earlier draft of our paper according to which we improved the content. This research is funded by "Biodiversity and Ecosystem services in a Changing Climate (BECC)" and SAPES. The funders had no role in study design, data collection and analysis, decision to publish, or preparation of the paper. The contents are the responsibility of the authors and do not necessarily reflect the views of the Centre for Environmental and Climate Research (CEC), Lund University, or AgriFood Economics Centre.

References

[1] J. R. Black and S. R. Thompson, "Some evidence on weather-crop-yield interaction," *American Journal of Agricultural Economics*, vol. 60, no. 3, pp. 540–543, 1978.

[2] V. A. Alexandrov and G. Hoogenboom, "The impact of climate variability and change on crop yield in Bulgaria," *Agricultural and Forest Meteorology*, vol. 104, no. 4, pp. 315–327, 2000.

[3] O. Chloupek, P. Hrstkova, and P. Schweigert, "Yield and its stability, crop diversity, adaptability and response to climate change, weather and fertilisation over 75 years in the Czech Republic in comparison to some European countries," *Field Crops Research*, vol. 85, no. 2-3, pp. 167–190, 2004.

[4] O. Vergara, G. Zuba, T. Doggett, and J. Seaquist, "Modeling the potential impact of catastrophic weather on crop insurance industry portfolio losses," *American Journal of Agricultural Economics*, vol. 90, no. 5, pp. 1256–1262, 2008.

[5] J. W. Jones, J. W. Hansen, F. S. Royce, and C. D. Messina, "Potential benefits of climate forecasting to agriculture," *Agriculture, Ecosystems and Environment*, vol. 82, no. 1–3, pp. 169–184, 2000.

[6] E. C. A. Runge, "Effects of rainfall and temperature interactions during the growing season on corn yield," *Agronomy Journal*, vol. 60, no. 5, pp. 503–507, 1968.

[7] P. E. Abbate, J. L. Dardanelli, M. G. Cantarero, M. Maturano, R. J. M. Melchiori, and E. E. Suero, "Climatic and water availability effects on water-use efficiency in wheat," *Crop Science*, vol. 44, no. 2, pp. 474–483, 2004.

[8] D. F. Calderini, L. G. Abeledo, R. Savin, and G. A. Slafer, "Effect of temperature and carpel size during pre-anthesis on potential grain weight in wheat," *Journal of Agricultural Science*, vol. 132, no. 4, pp. 453–459, 1999.

[9] M. Medori, L. Michelini, I. Nogues, F. Loreto, and C. Calfapietra, "The impact of root temperature on photosynthesis and isoprene emission in three different plant species," *The Scientific World Journal*, vol. 2012, Article ID 525827, 10 pages, 2012.

[10] P. Cantelaube and J. M. Terres, "Seasonal weather forecasts for crop yield modelling in Europe," *Tellus A*, vol. 57, no. 3, pp. 476–487, 2005.

[11] J. E. Olesen and M. Bindi, "Consequences of climate change for European agricultural productivity, land use and policy," *European Journal of Agronomy*, vol. 16, no. 4, pp. 239–262, 2002.

[12] W. Erskine and F. El Ashkar, "Rainfall and temperature effects on lentil (Lens culinaris) seed yield in Mediterranean environments," *Journal of Agricultural Science*, vol. 121, no. 3, pp. 347–354, 1993.

[13] D. B. Lobell, K. N. Cahill, and C. B. Field, "Historical effects of temperature and precipitation on California crop yields," *Climatic Change*, vol. 81, no. 2, pp. 187–203, 2007.

[14] P. J. M. Cooper, J. Dimes, K. P. C. Rao, B. Shapiro, B. Shiferaw, and S. Twomlow, "Coping better with current climatic variability in the rain-fed farming systems of sub-Saharan Africa: an essential first step in adapting to future climate change?" *Agriculture, Ecosystems and Environment*, vol. 126, no. 1-2, pp. 24–35, 2008.

[15] R. C. Muchow, T. R. Sinclair, and J. M. Bennett, "Temperature and solar-radiation effects on potential maize yeld across locations," *Agronomy Journal*, vol. 82, no. 2, pp. 338–343, 1990.

[16] D. B. Lobell and G. P. Asner, "Climate and management contributions to recent trends in U.S. Agricultural yields," *Science*, vol. 299, no. 5609, p. 1032, 2003.

[17] D. B. Lobell and C. B. Field, "Global scale climate-crop yield relationships and the impacts of recent warming," *Environmental Research Letters*, vol. 2, no. 1, Article ID 014002, 2007.

[18] R. K. Kaufmann and S. E. Snell, "A biophysical model of corn yield: integrating climatic and social Determinants," *American Journal of Agricultural Economics*, vol. 79, no. 1, pp. 178–190, 1997.

[19] S. J. Riha, D. S. Wilks, and P. Simoens, "Impact of temperature and precipitation variability on crop model predictions," *Climatic Change*, vol. 32, no. 3, pp. 293–311, 1996.

[20] J. Shukla and B. M. Misra, "Relationships between sea surface temperature and wind speed over the central Arabian Sea, and monsoon rainfall over India," *Monthly Weather Review*, vol. 105, pp. 998–1002, 1977.

[21] A. F. Moise, R. A. Colman, and J. R. Brown, "Behind uncertainties in projections of Australian tropical climate: analysis of 19 CMIP3 models," *Journal of Geophysical Research*, vol. 117, Article ID D10103, 2012.

[22] M. Tanarhte, P. Hadjinicolaou, and J. Lelieveld, "Intercomparison of temperature and precipitation data sets based on observations in the Mediterranean and the Middle East," *Journal of Geophysical Research D*, vol. 117, no. 12, Article ID D12102, 2012.

[23] R. F. Alder, G. Gu, J. J. Wang, G. J. Huffman, S. Curtis, and D. Bolvin, "Relationships between global precipitation and surface temperature on interannual and longer timescales (1979–2006)," *Journal of Geophysical Research D*, vol. 113, no. 22, Article ID D22104, 2008.

[24] E. Aldrian and R. Dwi Susanto, "Identification of three dominant rainfall regions within Indonesia and their relationship to sea surface temperature," *International Journal of Climatology*, vol. 23, no. 12, pp. 1435–1452, 2003.

[25] E. Black, "The relationship between Indian Ocean sea-surface temperature and East African rainfall," *Philosophical Transactions of the Royal Society A*, vol. 363, no. 1826, pp. 43–47, 2005.

[26] M. Rajeevan, D. S. Pai, and V. Thapliyal, "Spatial and temporal relationships between global land surface air temperature anomalies and indian summer monsoon rainfall," *Meteorology and Atmospheric Physics*, vol. 66, no. 3-4, pp. 157–171, 1998.

[27] Y. Huang, J. Cai, H. Yin, and M. Cai, "Correlation of precipitation to temperature variation in the Huanghe River (Yellow River) basin during 1957–2006," *Journal of Hydrology*, vol. 372, no. 1–4, pp. 1–8, 2009.

[28] D. S. Wilks, *Statistical Methods in the Atmospheric Sciences*, Academic Press, 2nd edition, 2005.

[29] A. AghaKouchak, A. Bárdossy, and E. Habib, "Conditional simulation of remotely sensed rainfall data using a non-Gaussian v-transformed copula," *Advances in Water Resources*, vol. 33, no. 6, pp. 624–634, 2010.

[30] Y. Malevergne and D. Sornette, "Testing the Gaussian copula hypothesis for financial assets dependences," *Quantitative Finance*, vol. 3, no. 4, pp. 231–250, 2003.

[31] A. J. Patton, "Copula-based models for financial time series," in *Handbook of Financial Time Series*, pp. 767–785, Springer, Berlin, Germany, 2009.

[32] C. Genest, M. Gendron, and M. Bourdeau-Brien, "The advent of copulas in finance," *European Journal of Finance*, vol. 15, no. 7-8, pp. 609–618, 2009.

[33] A. AghaKouchak, A. Bárdossy, and E. Habib, "Copula-based uncertainty modelling: application to multisensor precipitation estimates," *Hydrological Processes*, vol. 24, no. 15, pp. 2111–2124, 2010.

[34] F. Serinaldi, "Analysis of inter-gauge dependence by Kendall's τK, upper tail dependence coefficient, and 2-copulas with application to rainfall fields," *Stochastic Environmental Research and Risk Assessment*, vol. 22, no. 6, pp. 671–688, 2008.

[35] P. Laux, S. Vogl, W. Qiu, H. R. Knoche, and H. Kunstmann, "Copula-based statistical refinement of precipitation in RCM simulations over complex terrain," *Hydrology and Earth System Sciences*, vol. 15, no. 7, pp. 2401–2419, 2011.

[36] P. Laux, S. Wagner, A. Wagner, J. Jacobeit, A. Bárdossy, and H. Kunstmann, "Modelling daily precipitation features in the Volta Basin of West Africa," *International Journal of Climatology*, vol. 29, no. 7, pp. 937–954, 2009.

[37] A. C. Favre, S. E. Adlouni, L. Perreault, N. Thiémonge, and B. Bobée, "Multivariate hydrological frequency analysis using copulas," *Water Resources Research*, vol. 40, no. 1, pp. 1–12, 2004.

[38] J. T. Shiau, S. Feng, and S. Nadarajah, "Assessment of hydrological droughts for the Yellow River, China, using copulas," *Hydrological Processes*, vol. 21, no. 16, pp. 2157–2163, 2007.

[39] C. Schölzel and P. Friederichs, "Multivariate non-normally distributed random variables in climate research—introduction to the copula approach," *Nonlinear Processes in Geophysics*, vol. 15, no. 5, pp. 761–772, 2008.

[40] D. J. Dupuis, "Using copulas in hydrology: benefits, cautions, and issues," *Journal of Hydrologic Engineering*, vol. 12, no. 4, pp. 381–393, 2007.

[41] J. Huang and H. M. Van Den Dool, "Monthly precipitation-temperature relations and temperature prediction over the United States," *Journal of Climate*, vol. 6, no. 6, pp. 1111–1132, 1993.

[42] A. Sklar, "Random variables, joint distribution functions, and copulas," *Kybernetika*, vol. 9, no. 6, pp. 449–460, 1973.

[43] H. Joe, *Multivariate Models and Dependence Concepts*, Chapman and Hall, London, UK, 1997.

[44] R. Nelsen, *An Introduction to Copulas*, Springer, New York, NY, USA, 1999.

[45] H. B. Fang, K. T. Fang, and S. Kotz, "The meta-elliptical distributions with given marginals," *Journal of Multivariate Analysis*, vol. 82, no. 1, pp. 1–16, 2002.

[46] V. Gregoire, C. Genest, and M. Gendron, "Using copulas to model price dependence in energy markets," *Energy Risk*, vol. 5, pp. 58–64, 2008.

[47] E. Kole, K. Koedijk, and M. Verbeek, "Selecting copulas for risk management," *Journal of Banking and Finance*, vol. 31, no. 8, pp. 2405–2423, 2007.

[48] H. Akaike, "A new look at the statistical model identification," *IEEE Transactions on Automatic Control*, vol. AC-19, no. 6, pp. 716–723, 1974.

[49] G. E. Schwarz, "Estimating the dimension of a model," *Annals of Statistics*, vol. 6, no. 2, pp. 461–464.

[50] S. T. Katircioglu, "Research Methods In Banking And Finance," http://www.emu.edu.tr/salihk/courses/bnfn504/chp-11.pdf.

[51] T. W. Kim and H. Ahn, "Spatial rainfall model using a pattern classifier for estimating missing daily rainfall data," *Stochastic Environmental Research and Risk Assessment*, vol. 23, no. 3, pp. 367–376, 2009.

[52] F. E. Benth and J. Šaltyte-Benth, "Stochastic modelling of temperature variations with a view towards weather derivatives," *Applied Mathematical Finance*, vol. 12, no. 1, pp. 53–85, 2005.

[53] G. M. Ljung and G. E. P. Box, "On a measure of lack of fit in time series models," *Biometrika*, vol. 65, no. 2, pp. 297–303, 1978.

[54] M. J. Manton, P. M. Della-Marta, M. R. Haylock et al., "Trends in extreme daily rainfall and temperature in southeast Asia and the south Pacific: 1961–1998," *International Journal of Climatology*, vol. 21, no. 3, pp. 269–284, 2001.

[55] H. Joe and J. J. Xu, *The Estimation Method of Inference Function for Margins for Multivariate Models*, Department of Statistics, University of British Columbia, 1996.

11

Synthetic *Brassica napus* L.: Development and Studies on Morphological Characters, Yield Attributes, and Yield

M. A. Malek,[1,2] **M. R. Ismail,**[2] **M. Y. Rafii,**[2] **and M. Rahman**[3]

[1] *Plant Breeding Division, Bangladesh Institute of Nuclear Agriculture, Mymensingh 2202, Bangladesh*
[2] *Institute of Tropical Agriculture, Universiti Putra Malaysia, 43400 Serdang, Malaysia*
[3] *Department of Plant Science, North Dakota State University, Fargo, ND 58108-6050, USA*

Correspondence should be addressed to M. A. Malek, mamalekbina@yahoo.com and M. R. Ismail, razi@putra.upm.edu.my

Academic Editors: J.-F. Hausman and S. Thewes

Brassica napus was synthesized by hybridization between its diploid progenitor species *B. rapa* and *B. oleracea* followed by chromosome doubling. Cross with *B. rapa* as a female parent was only successful. Among three colchicine treatments (0.10, 0.15, and 0.20%), 0.15% gave the highest success (86%) of chromosome doubling in the hybrids (AC; $2n = 19$). Synthetic *B. napus* (AACC, $2n = 38$) was identified with bigger petals, fertile pollens and seed setting. Synthetic *B. napus* had increased growth over parents and exhibited wider ranges with higher coefficients of variations than parents for morphological and yield contributing characters, and yield per plant. Siliqua length as well as beak length in synthetic *B. napus* was longer than those of the parents. Number of seeds per siliqua, 1000-seed weight and seed yield per plant in synthetic *B. napus* were higher than those of the parents. Although flowering time in synthetic *B. napus* was earlier than both parents, however the days to maturity was little higher over early maturing *B. rapa* parent. The synthesized *B. napus* has great potential to produce higher seed yield. Further screening and evaluation is needed for selection of desirable genotypes having improved yield contributing characters and higher seed yield.

1. Introduction

Allopolyploids are widely spread in the plant kingdom. Their success might be explained by positive interactions between homoeologous genes on their different genomes, similar to the positive interactions between different alleles of one gene causing heterosis in heterozygous diploid genotypes [1]. Amphidiploid species are a form of polyploids that have evolved from interspecific hybridization between two or more diploid species, either through the fusion of unreduced gametes or through interspecific hybridization followed by spontaneous chromosome doubling. Many wild species as well as major field crops like wheat, oat, soybean, cotton, and rapeseed are the result of spontaneous interspecific hybridization, showing the high potential of allopolyploid species.

Allotetraploid *B. napus* (AACC, $2n = 38$) has evolved from a natural cross between *B. rapa* (AA, $2n = 20$)

and *B. oleracea* (CC, $2n = 18$) along the Mediterranean coast with uncertain evolutionary origin time approximate ranging from 0.12 to 1.37 million years ago [2, 3]. The short domestication history and traditional breeding schedule of *B. napus* has led to a narrow genetic range in the population. As a whole, although the allopolyploid species has been rapidly and widely cultivated globally as an oilseed due to the advantages of high yield and wide adaptation, rapeseed breeding and heterosis utilization have undergone genetic bottlenecks due to exhaustion of the genetic variation [4, 5]. Artificial synthesis of the naturally occurring amphidiploid *B. napus* by hybridization between its progenitors followed by chromosome doubling provides a means to increase the usable genetic variability [6]. Artificial *B. napus* was also synthesized earlier by Schranz and Osborn [7], Albertin et al. [8], and Gaeta et al. [9]. The present investigation was, therefore, aimed for development of synthetic *B. napus* from the hybrids of its two progenitor species and to study the C_2

(second colchiploid generation) synthetic *B. napus* compared to its parents regarding some morphological characters, yield attributes, and yield.

2. Materials and Methods

The experiments were conducted during November to February each of 2005-2006, 2006-2007, and 2007-2008 at the Bangladesh Institute of Nuclear Agriculture, Mymensingh, Bangladesh.

2.1. Plant Materials. Binasarisha-6 of *B. rapa* var. "yellow sarson" and Alboglabra-1 of *B. oleracea* var. "alboglabra" were used as parental genotypes for the development of interspecific hybrids. Interspecific hybrids were induced to double chromosome number for the development of synthetic *B. napus*. Synthetic *B. napus* was compared with parental genotypes.

2.2. Crossing and Collection of Hybrid Seeds. Flower buds of each of the female parents, expected to be opened in the next morning, were selected for emasculation. The emasculated buds were immediately pollinated with fresh pollen grains collected from the male parent. Pollinated flowers were covered with thin brown paper bags. The siliqua bearing F_1 seeds were collected after proper maturation. The hybrid (F_1) seeds were threshed, dried, and stored for the next season to grow the F_1 hybrids.

2.3. Chromosome Count of F_1 Hybrid and Pollen Fertility Study. Root tips from the germinating seeds were fixed in acetic alcohol (1 : 3) after pretreatment in saturated aqueous monobromonaphthalene solution for 2.5 hours. The tips were hydrolyzed in 10% HCl for 12 minutes at 60°C and then stained with 1% acetocarmine. Individual chromosome was counted with microscope at 100 times magnification. Acetocarmine (1%) was used for pollen fertility study. Intensely stained and normal shaped pollen grains were scored as fertile while the unstained and collapsed pollen grains were scored as sterile according to Sheidai et al. [10]. The ratio of stained pollen to the total was expressed as percentage of pollen fertility.

2.4. Colchicine Application and Development of C_1 Synthetic B. napus. Cotton plug method was followed to double chromosome number in the hybrids. Three concentrations (0.10%, 0.15%, and 0.20%) of colchicine were applied. Colchicine treatments on hybrids were applied at five to six leaves stages. Hybrids grown in pots were placed under shade and the twigs of each hybrid were removed. Two leaf axils of each hybrid plant were selected for treatment. A small cotton wool ball was placed on each of the selected leaf axils. The cotton wool balls were then soaked with colchicine at six hours intervals with 10-microlitre solution following the modified version of Gland [11]. Duration of treatment was maintained for 24 hours. The chromosome-doubled shoots developed from the hybrid plants were named as C_1 (first colchiploid generation).

2.5. Growing of C_2 Synthetic B. napus with Parents and Collection of Data. C_2 seeds collected from the C_1 plants having higher percentages of pollen fertility and siliqua setting along with higher number of seeds per siliqua were used for growing C_2 *B. napus* plants. Parental genotypes were also grown with C_2 plants in a single replicate in the field. Different cultural practices as well as irrigation and application of pesticides were done as and when necessary for the normal growth and development of the plants. Data were taken with respect to plant height, length and width of petal, number of primary branches per plant, pollen fertility (%), siliqua setting (%), number of siliquae per plant, number of viable and sterile seeds per siliqua, siliqua length, beak length, 1000-seed weight, seed yield per plant, days to flowering, and days to maturity. Data were taken from 40 randomly selected C_2 plants and 10 randomly selected plants of each parent. Measurements of mean, range, and coefficients of variation (CV%) of each character were calculated following the formula suggested by Burton [12].

3. Results and Discussion

3.1. Crossing and Study of F_1 Hybrid. Siliqua and seed setting was fairly good in cross between Binasarisha-6 and Alboglabra-1. Out of 106 crosses of Binasarisha-6 as female parent, 67 siliquae were developed with hybrid seeds and gave 63% success. On the other hand, the reciprocal crosses that is, Alboglabra-1 as female parent, were not successful. These results showed an agreement with Malek et al. [13], Choudhary et al. [14], and Sharma et al. [15] who reported the similar performances between crosses and reciprocals in the interspecific crosses within *Brassica* species. Somatic chromosome number in the hybrids ($2n$ = AC) was 19, which showed the equal number of chromosome of amphihaploid between the species, *B. rapa* (n = 10, A) and *B. oleracea* (n = 9, C). The hybrids exhibited vigorous growth with numerous primary as well as secondary branches. Akbar [16] also observed hybrid vigour in the interspecific hybrids of cross between *B. campestris* and *B. oleracea*. Intermediate morphology of F_1 in *Brassica* similar to the present study was also reported earlier by Choudhary et al. [17]. Hybrids flowered abundantly having shriveled, pointed tip and pale colour anthers with reduced filaments. Batra et al. [18] also reported similar morphology of anthers in interspecific hybrids within the genus *Brassica*. Hybrids produced 99-100% sterile pollens. Song et al. [19] also observed high pollen sterility in the F_1 hybrids obtained from all possible combinations of interspecific crosses of the diploid species within the U-triangle. According to Stebbins [20] high pollen sterility as observed in the hybrids of the present study might be due to meiotic irregularities and segregational anomalies as both genomes (A and C) had a single set of chromosome.

3.2. Treatment of Dihaploid Hybrids with Colchicine. It was observed that colchicine produced a drastic effect on growing leaf axils. In general, growth and development was strongly inhibited. The treated auxiliary shoots showed very slow growth and development. After three to four

FIGURE 1: (a) Leaves of Binasarisha-6, F$_1$, synthetic *B. napus*, and Alboglabra-1. (b) Racemes of Binasarisha-6, F$_1$ hybrid, synthetic *B. napus*, and Alboglabra-1. (c) Flowers of Binasarisha-6, F$_1$ hybrid, synthetic *B. napus*, and Alboglabra-1.

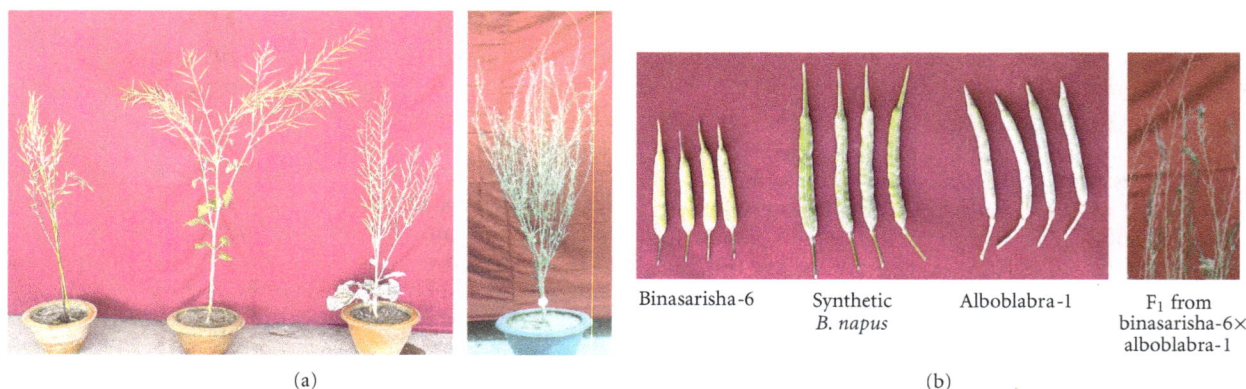

FIGURE 2: (a) Plants of Binasarisha-6, synthetic *B. napus*, Alboglabra-1, and F$_1$, and (b) siliquae of Binasarisha-6, synthetic *B. napus*, and Alboglabra-1, and rachis without siliqua in F$_1$.

weeks of treatment, though growth and development was started, but even then their growth was very slow. The new shoots emerged from the colchicines-treated leaf axils displayed thick and deep green leaves indicating the first symptom of induction of chromosome doubling. The highest chromosome diploidization (84%) was achieved with 0.15% colchicine treatment followed by 0.2% (72%) and 0.1% colchicine gave 60% success. Inhibited growth and development in colchicine treated tissues in *Brassica* hybrids was also reported by Aslam et al. [21]. The results indicated that chromosome diploidization rate differed with the concentration of colchicine, which showed close agreement with the results of Aslam et al. [21]. It has also been reported that success in chromosome doubling differs with method of application, different conditions, duration of treatment, and

at different stages of development [21–23]. Chromosome-doubled shoots produced fertile pollens and seeds in the siliquae.

3.3. C$_2$ B. napus and Its Parents. Colour, shape, and dentition of leaves in C$_2$ plants were intermediate between the parents. Leaf size of C$_2$ plants was larger than that of both parents and F$_1$ (Figure 1(a)). Size of flower buds and flowers of C$_2$ plants was also larger than that of both parents and F$_1$ (Figures 1(b) and 1(c)). The flowers of hybrids and C$_2$ plants had white petals resembling the Alboglabra-1 (Figure 1(c)), which indicates dominance of white petal colour over yellow. Finally, C$_2$ plants showed increased vegetative growth over the parents (Figure 2(a)) which agreed to the earlier results

TABLE 1: Morphological characters, yield attributes, and seed yield of synthetic *B. napus* and its parental genotypes, Alboglabra-1, and Binasarisha-6.

Characters		Alboglabra-1	Synthetic B. napus	Binasarisha-6
Plant height (cm)	Mean	111	143	106
	Range	98–123	132–160	91–116
	CV(%)	6.3	8.3	6.8
Petal length (cm)	Mean	1.9	2.0	1.1
	Range	1.8–2.0	1.7–2.1	1.0–1.2
	CV(%)	2.2	3.0	2.0
Petal width (cm)	Mean	1.17	1.11	0.42
	Range	1.13–1.24	0.95–1.21	0.38–0.47
	CV(%)	3.2	4.6	3.3
Primary branches per plant (no.)	Mean	3.21	4.2	6.8
	Range	2.0–4.0	3–6	5–8
	CV(%)	10.0	13.8	11.3
Pollen fertility (%)	Mean	90	87	91
	Range	87–93	74–94	89–94
	CV(%)	3.0	5.2	3.5
Siliqua setting (%)	Mean	95	93	95
	Range	93–98	71–97	93–99
	CV(%)	2.8	6.4	2.7
Siliqua length (cm)	Mean	5.9	7.7	4.1
	Range	5.3–6.4	6.9–8.1	3.6–4.7
	CV(%)	4.8	10.2	6.0
Beak length (cm)	Mean	0.75	3.02	1.53
	Range	0.69–0.83	2.80–3.23	1.29–1.69
	CV(%)	9.0	12.0	9.9
Siliquae per plant (no.)	Mean	102	77	98
	Range	78–117	61–101	84–110
	CV(%)	9.1	22.4	9.7
Sterile seeds per siliqua (no.)	Mean	—	2.03	—
	Range	—	0.0–3.1	—
	CV(%)	—	11	—
Seeds per siliqua (no.)	Mean	15.6	22.5	22.1
	Range	13–17	17–25	19–25
	CV(%)	7.7	10.4	8.7
1000-seed wt. (g)	Mean	3.2	3.8	3.1
	Range	3.0–3.3	3.7–4.0	3.0–3.3
	CV(%)	3.0	3.1	2.8
Seed yield per plant (g)	Mean	4.9	6.6	6.4
	Range	3.6–6.0	4.3–7.8	5.0–7.4
	CV(%)	11.0	15.8	9.7
Days to 50% flowering	Mean	45	33	35
Days to maturity	Mean	118	95	92

reported by Choudhary et al. [14], Vyas et al. [24], and Chrungu et al. [25].

Data on morphological characters, yield attributes, and seed yield per plant in C_2 plants along with parental genotypes are presented in Table 1. Results revealed that C_2 plants exhibited wider ranges with higher coefficients of variation (CV%) for all the characters studied over the parents. C_2 plants produced taller plants over the parents. Petal length and width were higher than those of the parents and the hybrid. Meng et al. [26] observed taller plant with larger flowers in synthetic *Brassica* hexaploids over their parental genotypes. The vigorous observation of the C_2

plants might be due to larger genome size in polyploids over their parental genotypes.

Average pollen fertility in the C_2 plants was slightly lower than that of both parents, but most of the C_2 plants (73%) had comparatively higher pollen fertility. The fertility of pollens was also reflected in siliqua setting, that is, those C_2 plants that had higher percentages of pollen fertility had also higher percentages of siliqua setting. Some C_2 plants (27%) had lower pollen fertility which might be due to development of aneuploid seeds from C_1. Number of siliquae per plant in C_2 plants usually counted as the most important seed yield component was found to be lower in number than the parents. Number of seeds per siliqua, another important component of yield, was lower than Binasarisha-6 but higher than Alboglabra-1. Mean weight of 1000-seed in C_2 plants was higher than that in the parents. Higher 1000-seed weight was observed in the C_2 plants which might be due to lower number of siliquae per plant as these two component characters are compensating to each other. Finally, C_2 plants produced higher mean seed yield per plant than both parents. Siliqua length as well as beak length in C_2 plants was longer than that of both parents. Although flowering time (50% flowering) in C_2 plants was earlier than both parents, however the days to maturity were little higher over the early maturing *B. rapa* parent. Wider variation for most of the characters in C_2 plants might be due to presence of some aneuploids along with euploids. Richharia [27] and Howard [28] reported lower seed setting in artificially developed *Raphanobrassica*. Tokumasu [29] observed wide variations in F_3 *Raphanobrassica* from a single plant progeny for per cent pollen fertility, per cent siliqua setting, and number of seeds per siliqua. Sarla and Raut [6] observed a wide range of variations for morphological as well as yield contributing characters among 40 C_2 *B. carinata* plants obtained from a single C_1 plant and reported that those wide variations were due to presence of aneuploids along with the euploids in C_2. Formation of univalents or multivalents in C_2 plants may have contributed to unequal segregation at anaphase-I of meiosis and consequently leaded to a decrease in pollen fertility [30, 31]. Aneuploid formation in the synthetic *B. napus* might be occurred due to affinity of allosyndetic pairing between A and C genomes as reported by Inomata [32], Ahmad et al. [33], and Tian et al. [34] resulting in multivalent association at diakinesis and metaphase-I of meiosis [4, 6, 25, 35].

The results of this study clearly showed that it needs further screening and evaluation for the synthesized *B. napus* in the subsequent generations through selection of desirable genotypes having increased pollen fertility as well as high fruit and seed setting resulting in higher seed yield and desired yield contributing characters. However, more research works are needed to stabilize the synthesized *B. napus*.

References

[1] S. Abel, C. Möllers, and H. C. Becker, "Development of synthetic *Brassica napus* lines for the analysis of "fixed heterosis" in allopolyploid plants," *Euphytica*, vol. 146, no. 1-2, pp. 157–163, 2005.

[2] U. Nagaharu, "Genome analysis in *Brassica* with special reference to the experimental formation of *B. napus* and peculiar mode of fertilization," *Japanese Journal of Botany*, vol. 7, pp. 389–452, 1935.

[3] F. Cheung, M. Trick, N. Drou et al., "Comparative analysis between homoeologous genome segments of *Brassica napus* and its progenitor species reveals extensive sequence-level divergence," *Plant Cell*, vol. 21, no. 7, pp. 1912–1928, 2009.

[4] S. Prakash and K. Hinata, "Taxonomy, cytogenetics and origin of crop Brassicas: a review," *Opera Botanica*, vol. 55, pp. 1–57, 1980.

[5] H. C. Becker, G. M. Engqvist, and B. Karlsson, "Comparison of rapeseed cultivars and resynthesized lines based on allozyme and RFLP markers," *Theoretical and Applied Genetics*, vol. 91, no. 1, pp. 62–67, 1995.

[6] N. Sarla and R. N. Raut, "Synthesis of *Brassica carinata* from *Brassica nigra* × *Brassica oleracea* hybrids obtained by ovary culture," *Theoretical and Applied Genetics*, vol. 76, no. 6, pp. 846–849, 1988.

[7] M. E. Schranz and T. C. Osborn, "Novel flowering time variation in the resynthesised polyploid *Brassica napus*," *Journal of Heredity*, vol. 91, no. 3, pp. 242–246, 2000.

[8] W. Albertin, T. Balliau, P. Brabant et al., "Numerous and rapid nonstochastic modifications of gene products in newly synthesized *Brassica napus* allotetraploids," *Genetics*, vol. 173, no. 2, pp. 1101–1113, 2006.

[9] R. T. Gaeta, J. C. Pires, F. Iniguez-Luy, E. Leon, and T. C. Osborn, "Genomic changes in resynthesized *Brassica napus* and their effect on gene expression and phenotype," *Plant Cell*, vol. 19, no. 11, pp. 3403–3417, 2007.

[10] M. Sheidai, M. Arman, A. M. Saeed, and B. Zehzad, "Notes on cytology and seed protein characteristics of Aegilops species in Iran," *The Nucleus*, vol. 43, pp. 118–128, 2000.

[11] A. Gland, "Doubling chromosomes in interspecific hybrids by colchicine treatment," *EUCARPIA Cruciferae NL*, vol. 6, pp. 20–22, 1981.

[12] G. W. Burton, "Quantitative inheritance in grass estimated residual effect was 0.117 indicating that 90% of pea," in *Proceedings of the 6th Grassland Congress*, vol. 1, pp. 277–283, 1952.

[13] M. A. Malek, L. Rahman, M. L. Das, and L. Hassan, "Development of interspecific hybrids between *Brassica carinata* and *B. rapa* (*B. campestris*)," *Bangladesh Journal of Agricultural Sciences*, vol. 33, no. 1, pp. 21–25, 2006.

[14] B. R. Choudhary, P. Joshi, and S. Ramarao, "Interspecific hybridization between *Brassica carinata* and *Brassica rapa*," *Plant Breeding*, vol. 119, no. 5, pp. 417–420, 2000.

[15] S. K. Sharma, S. S. Gosal, and J. L. Minocha, "Effect of growth-regulators on siliqua and seed setting in interspecific crosses of *Brassica* species," *Indian Journal of Agricultural Science*, vol. 67, no. 4, pp. 166–167, 1997.

[16] M. A. Akbar, "Resynthesis of *Brassica napus* aiming for improved earliness and carried out by different approaches," *Hereditas*, vol. 111, no. 3, pp. 239–246, 1989.

[17] B. R. Choudhary, P. Joshi, and S. Rama Rao, "Cytogenetics of *Brassica juncea* × *Brassica rapa* hybrids and patterns of variation in the hybrid derivatives," *Plant Breeding*, vol. 121, no. 4, pp. 292–296, 2002.

[18] V. Batra, S. Prakash, and K. R. Shivanna, "Intergeneric hybridization between *Diplotaxis siifolia*, a wild species and crop brassicas," *Theoretical and Applied Genetics*, vol. 80, no. 4, pp. 537–541, 1990.

[19] K. Song, K. Tang, and T. C. Osborn, "Development of synthetic *Brassica* amphidiploids by reciprocal hybridization and comparison to natural amphidiploids," *Theoretical and Applied Genetics*, vol. 86, no. 7, pp. 811–821, 1993.

[20] G. L. Stebbins, "Chromosomal variation and evolution," *Science*, vol. 152, no. 3728, pp. 1463–1469, 1966.

[21] F. N. Aslam, M. V. Macdonald, P. Loudon, and D. S. Ingram, "Rapid-cycling *Brassica* Species: inbreeding and selection of *B. campestris* for anther culture ability," *Annals of Botany*, vol. 65, no. 5, pp. 557–566, 1990.

[22] L. Currah and D. J. Ockendon, "Chromosome doubling of mature haploid Brussels sprout plants by colchicine treatment," *Euphytica*, vol. 36, no. 1, pp. 167–173, 1987.

[23] S. W. Shi, J. S. Wu, Y. M. Zhou, and H. L. Liu, "Diploidization techniques for microspore-derived haploid plants of rapeseed (*Brassica napus* L.)," *Chinese Journal of Oil Crop Sciences*, vol. 24, no. 1, pp. 1–5, 2002.

[24] P. Vyas, S. Prakash, and K. R. Shivanna, "Production of wide hybrids and backcross progenies between *Diplotaxis erucoides* and crop brassicas," *Theoretical and Applied Genetics*, vol. 90, no. 3-4, pp. 549–553, 1995.

[25] B. Chrungu, N. Verma, A. Mohanty, A. Pradhan, and K. R. Shivanna, "Production and characterization of interspecific hybrids between *Brassica maurorum* and crop brassicas," *Theoretical and Applied Genetics*, vol. 98, no. 3-4, pp. 608–613, 1999.

[26] J. Meng, S. Shi, L. Gan, Z. Li, and X. Qu, "The production of yellow-seeded *Brassica napus* (AACC) through crossing interspecific hybrids of *B. campestris* (AA) and *B. carinata* (BBCC) with *B. napus*," *Euphytica*, vol. 103, no. 3, pp. 329–333, 1998.

[27] R. H. Richharia, "Cytological investigation of *Raphanus sativus*, *Brassica oleracea*, and their F1 and F2 hybrids," *Journal of Genetics*, vol. 34, no. 1, pp. 19–44, 1937.

[28] H. W. Howard, "The fertility of amphidiploids from the cross *Raphanus sativus* × *Brassica oleracea*," *Journal of Genetics*, vol. 36, no. 2, pp. 239–273, 1938.

[29] S. Tokumasu, "The increase of seed fertility of *Brassicoraphanus* through cytological irregularity," *Euphytica*, vol. 25, no. 1, pp. 463–470, 1976.

[30] N. Tel-Zur, S. Abbo, and Y. Mizrahi, "Cytogenetics of semifertile triploid and aneuploid intergeneric vine cacti hybrids," *Journal of Heredity*, vol. 96, no. 2, pp. 124–131, 2005.

[31] W. Qian, X. Chen, D. Fu, J. Zou, and J. Meng, "Intersubgenomic heterosis in seed yield potential observed in a new type of *Brassica napus* introgressed with partial *Brassica rapa* genome," *Theoretical and Applied Genetics*, vol. 110, no. 7, pp. 1187–1194, 2005.

[32] N. Inomata, "Hybrid progenies of the cross, *Brassica campestris* × *B. oleracea*. 1. Cytogenetical studies on F1 hybrids," *Japanese Journal of Genetics*, vol. 55, pp. 189–202, 1980.

[33] H. Ahmad, S. Hasnain, and A. Khan, "Evolution of genomes and genome relationship among the rapeseed and mustard," *Biotechnology*, vol. 1, no. 2, pp. 78–87, 2002.

[34] E. Tian, Y. Jiang, L. Chen, J. Zou, F. Liu, and J. Meng, "Synthesis of a *Brassica* trigenomic allohexaploid (*B. carinata* × *B. rapa*) de novo and its stability in subsequent generations," *Theoretical and Applied Genetics*, vol. 121, no. 8, pp. 1431–1440, 2010.

[35] B. R. Choudhary, P. Joshi, and K. Singh, "Synthesis, morphology and cytogenetics of *Raphanofortii* (TTRR, $2n = 38$): a new amphidiploid of hybrid *Brassica tournefortii* (TT, $2n = 20$)×*Raphanus caudatus* (RR, $2n = 18$)," *Theoretical and Applied Genetics*, vol. 101, no. 5-6, pp. 990–999, 2000.

12

Effects of 1-Methylcyclopropene and Modified Atmosphere Packaging on the Antioxidant Capacity in Pepper "Kulai" during Low-Temperature Storage

Chung Keat Tan,[1] Zainon Mohd Ali,[1] Ismanizan Ismail,[1,2] and Zamri Zainal[1,2]

[1] School of Bioscience and Biotechnology, National University of Malaysia, Selangor, 43600 Bangi, Malaysia
[2] Institute of System Biology (INBIOSIS), National University of Malaysia, Selangor, 43600 Bangi, Malaysia

Correspondence should be addressed to Zamri Zainal, zz@ukm.my

Academic Editors: J. E. Barboza-Corona, A. Mentese, and D. M. Prazeres

The objective of the present study was to simultaneously evaluate the effect of a postharvest treatment on the pepper's antioxidant content and its ability to retain its economical value during the postharvest period. The fruits were pretreated by modified atmosphere packaging (MAP) with or without treatment with 1-methylcyclopropene (1-MCP) before cold storage at 10°C. Changes in the levels of non-enzymatic antioxidants, including the total phenolic, ascorbic acid levels and the total glutathione level, as well as enzymatic antioxidants, including ascorbate peroxidase (APX), glutathione reductase (GR), and catalase (CAT), were determined. Both treatments successfully extended the shelf life of the fruit for up to 25 days, and a high level of antioxidant capacity was maintained throughout the storage period. However, 1-MCP treatment maintained the high antioxidant capacity for a longer period of time. The 1-MCP-treated peppers maintained high levels of phenolic content, a high reduced glutathione (GSH)/oxidised glutathione (GSSG) ratio, decreased levels of ascorbic acid and CAT activity, and increased levels of APX and GR compared with the peppers that were not treated with 1-MCP. The overall results suggested that a combination of 1-MCP and MAP was the most effective treatment for extending shelf life while retaining the nutritional benefits.

1. Introduction

Peppers (*Capsicum annuum* "Kulai") have been consumed for more than three centuries. In fact, most peppers have been used extensively as spices or condiments in Asian foods. The high market demand for peppers is due not only to their natural colours and spices but also to the nutritional benefits of the fruit, particularly because this fruit serves as an excellent source of dietary antioxidants [1]. The total antioxidant activities in peppers, which consist of enzymatic and non-enzymatic antioxidants, are in the highest range among the most commonly consumed vegetables [2]. The intake of foods with high levels of antioxidant constituents promotes health. Dietary antioxidants help protect against the harmful effects of reactive oxygen species (ROS) in our bodies and thus prevent human diseases, including cardiovascular disease and cancer, when adequate amounts are consumed daily

[3]. However, the pepper is a perishable fruit with a short shelf life and, therefore, is highly susceptible to a rapid decrease in quality. Thus, peppers require a modified postharvest environment to reduce the nutritional loss.

Various antioxidants are present in peppers that act as suppressors against oxidative stress. Phenolic compounds, the main contributors to the antioxidant capacity of plants, have been studied often. Phenolic compounds are important for inhibiting lipid autoxidation by acting as radical scavengers [4], which, in turn, protect against the propagation of an oxidative chain. Peppers serve as excellent sources of vitamin C, with one pepper able to satisfy the daily requirement for this vitamin [5]. The high vitamin C content is important for chelating heavy metal ions, scavenging reactive radicals, and suppressing peroxidation, and these actions prevent the effects of degenerative diseases [6, 7]. Glutathione exists in a reduced form (GSH) and an oxidised form

Effects of 1-Methylcyclopropene and Modified Atmosphere Packaging on the Antioxidant Capacity in Pepper
"Kulai" during Low-Temperature Storage

87

(GSSG). In both animals and plants, GSH is the predominant form that acts as an antioxidant and reducing buffer [8]. The enzymatic antioxidant system, which includes ascorbate peroxidase (APX, EC 1.11.1.11), catalase (CAT, EC 1.11.1.6), and glutathione reductase (GR, EC 1.6.4.2), is also synergistically involved with non-enzymatic antioxidants to reduce oxidative stress in plants.

The pepper "Kulai" is mainly harvested and consumed at the mature green stage or at the fully ripened/red stage. However, the market value of peppers is restricted by the postharvest environment due to the short shelf life of peppers [9]. Following the rise in demand from consumers for products of excellent quality and high nutritional value, optimisation of the postharvest conditions to retain the fruit quality is now a focus of current studies. Low-temperature treatment has often been used to prevent physical changes, and the suitable range for pepper storage is between 7°C and 13°C to avoid a chilling injury [10, 11]. Modified atmosphere packaging (MAP) with an optimal gas concentration was able to improve the quality retention of peppers with a low-temperature treatment and to control any postharvest disease [12]. An ethylene receptor blocker, 1-methylcyclopropene (1-MCP), has recently been utilised to extend the shelf life of some climacteric fruits, and a persistent effect was demonstrated [13]. The beneficial effects of 1-MCP include lower lipoxygenase activities and electrolyte leakage, prolonged cold storage life, delayed skin colour changes, and suppression of certain ethylene-induced postharvest physiological disorders [14–16].

Formal postharvest research that is available regarding the effects of a combination treatment (1-MCP and MAP) on the antioxidant capacity of the fruit is still limited, especially in peppers, because 1-MCP is still a new technology. The objective of the present study was to simultaneously evaluate the effect of a postharvest treatment on both the pepper's antioxidant content and the pepper's ability to retain its economical value during the postharvest period. It is anticipated that the results obtained from the current study will help clarify the possible interactions between and roles of the different treatments in regulating antioxidant activity. The findings should be important for developing an effective postharvest treatment to retain the nutritional benefits of the fruit.

2. Materials and Methods

2.1. Materials. The pepper "Kulai" (*Capsicum annuum* cv. Kulai) was harvested at the Bukit Lanchong farm at the mature green and fully ripened/red stages. Fruits that were free from diseases were selected and washed with distilled water. All of the fruits were then air-dried and packed into 13 identical polyethylene bags (thickness: 50 μm) such that each bag contained 100 uniform peppers. Three bags left unsealed served as controls, five sealed bags flushed with CO_2/N_2 mixture (25% CO_2 + 75% N_2) served as the MAP-treated group, and the remaining bags were treated with a combination of MAP (25% CO_2 + 75% N_2) and 90 ppb of gaseous 1-MCP (12 hr, 30°C) from SmartFresh (AgroFresh Inc., Spring House, PA) for 12 h [17]. Following the treatment, both the

MAP-treated and MAP+1-MCP-treated groups were stored at 10°C for up to 25 days. The control group without MAP treatment (unsealed) or 1-MCP treatment was stored at room temperature and given the name UCRT (unsealed control at room temperature). After 21 days of storage, 2 bags from each MAP-treated and MAP+1-MCP-treated group were placed at room temperature for 5 days to simulate commercial shelf handling.

Fifty uniform peppers were sampled at specific intervals during the storage period (as indicated in the figures). The fruits were cut into small pieces, the seeds were discarded, and the fruit pieces were immediately frozen in liquid nitrogen. The tissues were stored at $-80°C$ until further analysis. There were six replicates per time interval with 5 g of peppers per replicate. All of the enzymatic and non-enzymatic assays were performed based on these replicates.

Unless otherwise stated, all solvents, salts, and acids were purchased from Sigma Chemical Co. (St. Louis, USA). All of the reagents were of HPLC grade and of the highest purity available. All aqueous solutions were prepared with distilled water.

2.2. Total Phenolic Content. The total phenolic content in the pepper fruits was measured using the method adapted from Singleton and Rossi [18]. Five grams of fresh weight of the pepper tissues was pulverised using a mortar and pestle with 20 mL of cold 100% methanol, and the samples were then centrifuged at 3,000 rpm and 4°C for 15 min. The extract was appropriately diluted and then oxidised with 100 μL of freshly diluted 50% Folin-Ciocalteu reagent. After 3 min, this reaction was neutralised by adding 2 mL of 2% (w/v) sodium carbonate. After incubating for 30 min at room temperature, the absorbance of the resulting blue-coloured solution was determined at 750 nm using a UV-visible spectrophotometer (UV-160A, Shimadzu, Kyoto, Japan). A standard curve was prepared using the same procedure with gallic acid (10, 20, 50, and 100 mg/L). The total phenolic content of the pepper samples was expressed in milligrams of gallic acid equivalent (GAE) per gram of fresh weight (FW).

2.3. Ascorbic Acid Content. Five grams of the sample was extracted using 25 mL of HPLC-grade water. The homogenate was centrifuged at 13,000 rpm for 15 min. The supernatant was then filtered using a syringe filter (nylon membrane, 0.02 μm). Quantification was achieved using an external standard method adapted from Lim et al. [19] with slight modification. HPLC analysis of ascorbic acid was performed using HPLC system equipped with a diode array detector (Prominence-20A, Shimadzu, Kyoto, Japan). The samples (20 μL) were separated at 40°C on a Waters Symmetry C18 column (3.9 × 150 mm id; 5 μm particle size) (Milford, MA, USA) using a mobile phase of 5% acetic acid at a flow rate of 1 mL/min. The amount of ascorbic acid was calculated from the absorbance at 254 nm using ascorbic acid (20, 40, 60, 80, and 100 mg/L) as a standard. The results are expressed as milligram of ascorbic acid per gram of FW.

2.4. Total Glutathione Content. Five grams of the sample was homogenised in a cold mortar using 15 mL of 5% 5-sulphosalicylic acid. The homogenate was then centrifuged at 9,000 rpm for 15 min. The assay to determine the total glutathione content was based on the method of Anderson [20] with slight modification. The reaction mixture was composed of 700 μL of 0.2 mM NADPH, 100 μL of 6 mM DTNB, and 180 μL of 0.143 M potassium phosphate buffer (pH 7.5). The mixture was incubated in a water bath at 30°C for 5 min before the addition of 20 μL of the supernatant and 1.5 units of glutathione reductase. The change in absorbance at 412 nm during 1 min was monitored using a UV-visible spectrophotometer (UV-160A, Shimadzu, Kyoto, Japan). A standard curve was prepared using the same procedure with GSH equivalents (50, 100, 150, and 200 μM). The total glutathione content of the pepper samples was expressed in μmol per gram of FW.

For the GSSG determination, the supernatant was first diluted 10 times with 0.5 M potassium phosphate buffer (pH 6.5). Then, 20 μL of 2-vinylpyridine was added to 1 mL of the mixture, which was followed by vigorous mixing. After one hour incubation at room temperature, 20 μL of the mixture was removed and used for the glutathione assay described above. A standard curve was plotted using GSSG (25, 50, 75, and 100 μM), and the results were expressed in μmol per gram of FW.

2.5. Enzyme Assays. The APX and CAT assays both used the same extraction procedure. For 5 g of sample tissue, 10 mL of phosphate buffer (0.1 M, pH 7.0), which consisted of 1.0 mM EDTA and 1% polyvinyl polypyrrolidone (PVP), was used as the extraction buffer. The method used for the analysis of the APX activity was adapted from Nakano and Asada [21]. The reaction mixture contained 1.91 mL of phosphate buffer (50 mM, pH 7.0), 0.05 mL of ascorbate (0.5 mM), 0.01 mL of hydrogen peroxide (H_2O_2) (0.1 mM), and 30 μL of enzyme extract. The specific activity of APX was determined by monitoring the decline in the absorbance at 290 nm using a UV-visible spectrophotometer (UV-160A, Shimadzu, Kyoto, Japan) and was expressed as units per gram of fresh weight. One unit of APX was defined as the amount of enzyme that oxidised 1 μmol of ascorbate per min at room temperature.

The CAT activity was estimated according to the method of Beers and Sizer [22]. The reaction mixture consisted of 0.1 mL of enzyme extract, 50 mM phosphate buffer (pH 7.0), and 15 mM H_2O_2. The depletion of H_2O_2 was determined by measuring the change in the absorbance at 240 nm using a UV-visible spectrophotometer (UV-160A, Shimadzu, Kyoto, Japan). One unit of CAT was defined as the amount of enzyme needed to reduce 1 μmol of H_2O_2 in 1 min. The specific activity was expressed as units per gram of FW.

GR was extracted from 5 g samples using 20 mL of cold potassium phosphate extraction buffer, which contained 1 mM EDTA, 0.5% Triton X-100, 0.1 mM 2-mercaptoethanol, and 2% PVP. The assay was adapted from the method of Bergmeyer [23] with a slight modification. The reaction mixture consisted of 0.2 M potassium phosphate buffer (pH 7.5), 2 mM NADPH, 20 mM GSSG, and 80 μL

of enzyme extract. The change in the absorbance was monitored at 340 nm using a UV-visible spectrophotometer (UV-160A, Shimadzu, Kyoto, Japan). One unit of GR was defined as the amount of enzyme needed to oxidise 1 μmol of NADPH in 1 min. The specific activity was expressed as units per gram of FW.

2.6. Statistical Analysis. All measurements were performed on 6 experimental replicates, and the results were reported as the means ± the standard errors. A statistical analysis was performed using the Statistical Analysis System program version 6.12 (SAS Institute Inc., Cary, NC, USA). The data were analysed using analysis of variance (one-way ANOVA). The sources of variation were the types of treatments and the storage duration. The means were compared using the least significant differences (LSD) test at a significance level of 0.05.

3. Results and Discussion

3.1. Total Phenolic Content. Phenolic compounds comprise the largest category of phytochemicals in plants, which include flavonoids, phenolic acids, and phenols. These compounds are excellent antioxidants due to their structure, which allows them to easily donate a hydrogen atom to free radicals, and they are the primary molecules responsible for the antioxidant capacity of fruit [24]. Humans cannot produce phenolic compounds; therefore, the main source of these compounds is the consumption of vegetables and fruits [25]. Peppers are an excellent source of phenolic compounds because they have the highest antioxidant capacity among all of the commonly consumed vegetables [26].

In the current study, mature green and red peppers showed a gradual increase in the level of total phenolic compounds during low-temperature storage (Figure 1). The phenolic content of the MAP- and MAP+1-MCP-treated peppers was lower than that of the UCRT group at an early stage of storage but higher than that of the UCRT group after 3 weeks of storage. The phenolic content in both types of treated groups was approximately 0.15 mg/g FW higher than in the UCRT group at the end of the storage period. Generally, treatment with MAP or MAP+1-MCP resulted in a delayed accumulation of phenolic compounds during the low-temperature storage. These results indicated that treatment with MAP or MAP+1-MCP successfully increased the tolerance of the fruit to low temperatures and consequently led to a lower activity of chilling-induced phenylalanine ammonia lyase (PAL), the main enzyme responsible for the biosynthesis of phenolic compounds. This conclusion was supported by the finding of Lafuente et al. [27], who stated that PAL is a cold-responsive enzyme; appropriate conditioning that induced cold tolerance would lead to a suppressive effect on PAL activity. According to Faragher and Chalmers [28], the biosynthesis of phenolic compounds is also closely correlated with ethylene production, although the details of this correlation remain unknown. Therefore, the inhibitory effect of 1-MCP treatment on ethylene action might be the reason for the lower level of phenolic compounds in the

Effects of 1-Methylcyclopropene and Modified Atmosphere Packaging on the Antioxidant Capacity in Pepper "Kulai" during Low-Temperature Storage

89

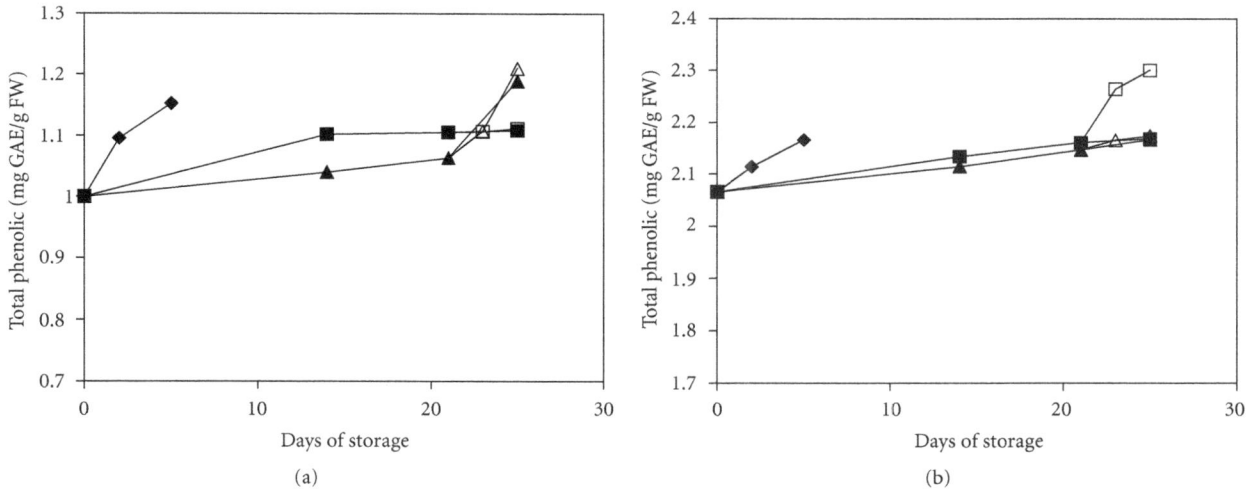

FIGURE 1: Changes in the total phenolic content of mature green peppers (a) and fully ripe/red peppers (b) for the UCRT (◆), MAP (■), and MAP+1-MCP (▲) treatments during storage at 10°C; MAP (□) and MAP+1-MCP (△) values for the fruits transferred to room temperature at day 21. The values are the means of six replicate samples, and their S.E.s are indicated.

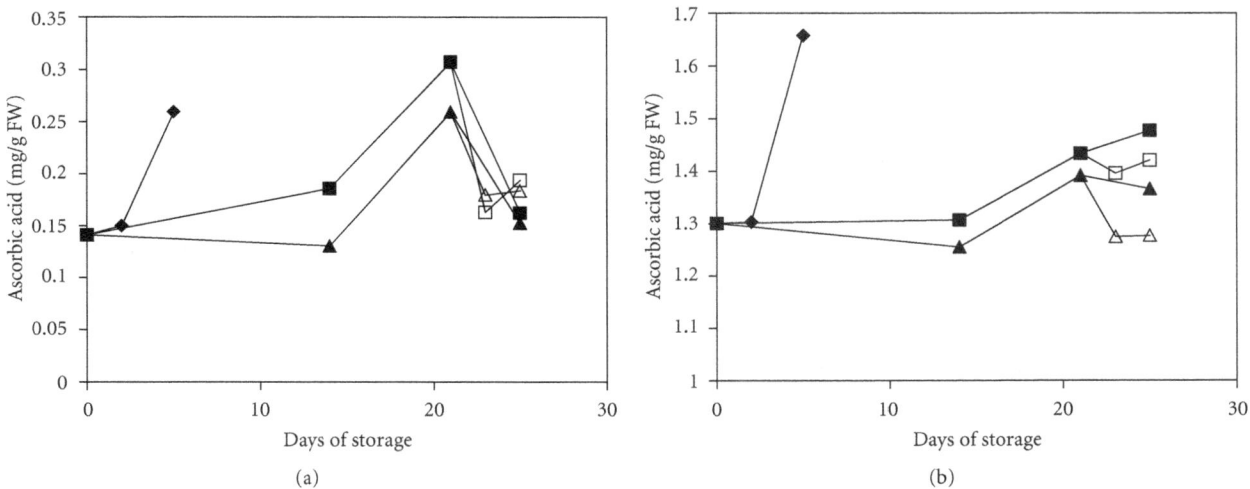

FIGURE 2: Changes in the ascorbic acid content of mature green peppers (a) and fully ripe/red peppers (b) for the UCRT (◆), MAP (■), and MAP+1-MCP (▲) treatments during storage at 10°C; MAP (□) and MAP+1-MCP (△) values for the fruits transferred to room temperature at day 21. The values are the means of six replicate samples, and their S.E.s are shown.

MAP+1-MCP-treated fruit compared with the MAP-treated fruit during storage. However, the differences were not significant ($P < 0.05$).

3.2. Ascorbic Acid Content. Ascorbic acid is a water-soluble antioxidant that neutralises superoxides, hydroxyl radicals, and H_2O_2 [29]. It is a functionally, nutritionally, and biologically active compound in the fruit of a pepper plant [30]. Large variations in ascorbic acid levels have been observed due to differences in the cultivars, harvest stages, postharvest handling, agroclimatic conditions, and analytical methods used [31]. In fact, the stage of harvest, storage conditions and duration of storage are the major factors that determine the ascorbic acid concentration in harvested peppers.

The ascorbic acid content was generally higher after the harvest. However, the accumulation of ascorbic acid was delayed in the MAP- and MAP+1-MCP-treated fruits that were stored at low temperature compared with the UCRT group. The elevation of the ascorbic acid content in green peppers did not persist, and the ascorbic acid content decreased rapidly after 21 days (Figure 2(a)). Conversely, the ascorbic acid content in red peppers was consistently low but exhibited a constant elevation throughout the storage period (Figure 2(b)). Currently, little information is available regarding changes in ascorbic acid concentrations during storage and the mechanisms controlling ascorbic acid production. However, light intensity and temperature are known to be the most important factors in determining the final

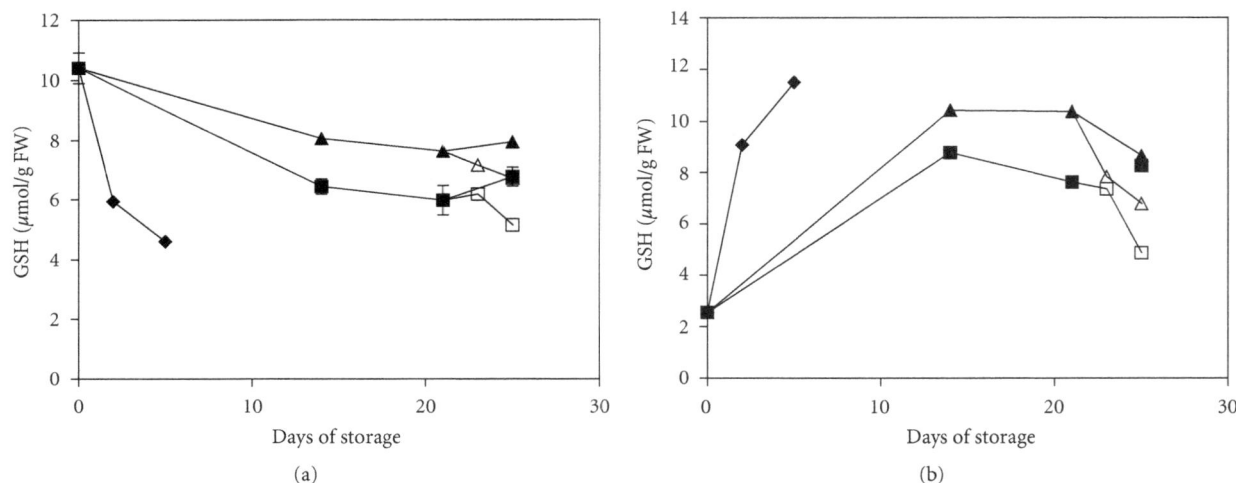

FIGURE 3: Changes in the GSH content of mature green peppers (a) and fully ripe/red peppers (b) for the UCRT (◆), MAP (■), and MAP+1-MCP (▲) treatments during storage at 10°C; MAP (□) and MAP+1-MCP (△) values for the fruits transferred to room temperature at day 21. The values are the means of six replicate samples, and their S.E.s are shown.

ascorbic acid content [32]. The accumulation of ascorbic acid during the first 3 weeks of storage suggested that high ascorbic acid levels might be a self-protective response against chilling stress, and the delay in the accumulation rate of ascorbic acid proved that the development of chilling stress was successfully suppressed as a result of the treatment. Separately, the storage duration might be the cause of the rapid decrease in ascorbic acid content at the later stages of storage [33]. Nevertheless, Win et al. [34] reported that a high concentration of 1-MCP might have a suppressive effect on the ascorbic acid content in fruit, which suggested a possible reason for the significant difference ($P < 0.05$) observed between the MAP-treated and the MAP+1-MCP-treated green peppers. Furthermore, the difference between the treatment groups might also be due to the additional protection effect of the 1-MCP treatment against chilling stress. Although the fruits were protected from the effect of oxidative stress, the nutritional values were also depleted at the same time due high consuming rate of ascorbic acid in scavenging activities.

3.3. Total Glutathione Content.

Reduced glutathione, GSH, is a tripeptide molecule that exists interchangeably with oxidised glutathione, GSSG. GSH plays a key role in many biological mechanisms, including amino acid transport, the biosynthesis of DNA and proteins, and, most importantly, the protection of cells from oxidation. Glutathione is directly involved in the APX-GR system to remove reactive oxygen species (ROS) and maintains a highly reduced intracellular environment [35].

Our results showed that the GSH content was generally decreased in green peppers but increased in red peppers regardless of the storage temperature or treatment (Figure 3). Clearly, treatment with 1-MCP results in higher levels of GSH in both pepper types compared with the MAP treatment alone, and the difference between these two treated

groups was significant ($P < 0.05$) (Table 1). In plants, GSH is the predominant form of glutathione and contains up to 90% of the total glutathione [36]. High ratios of GSH to GSSG are particularly important for a strong defensive mechanism against oxidative stress and to minimise the harm caused by H_2O_2 at a cellular level. Conversely, GSSG production was decreased during storage in general (Figure 4). According to an ANOVA analysis, the GSSG content in the MAP+1-MCP-treated group was significantly lower ($P < 0.05$) than that in the MAP-treated group for green peppers. The GSSG content in green pepper was reduced by half by the end of the storage period. However, the difference between the two treated groups in red peppers was not significant. Additionally, the results indicated that the GSH/GSSG ratio in the MAP-treated fruit decreased by 5- to 10-fold; however, in the MAP+1-MCP-treated fruit, the ratio increased by approximately 9- to 18-fold (Table 2). This result implied that the redox status shifted significantly towards a reduced state as a result of the treatment with 1-MCP, which then suppressed oxidative stress and its effect on the cells. A high ratio of GSH/GSSG proved to be important in raising the resistance of fruits against chilling injury [37, 38]. Apart from that, consumption of fruits with high GSH content, such as the MAP+1-MCP-treated pepper, had been proved to be important in enhancing the immune function in human [39].

3.4. Enzymatic Antioxidants.

The enzymatic antioxidant system, which includes ascorbate peroxidase (APX, EC 1.11.1.11), glutathione reductase (GR, EC 1.6.4.2), and catalase (CAT, EC 1.11.1.6), plays a key role in regulating the defensive response against oxidative stress. The activities of these enzymes in cells were mainly influenced by metabolite specificities, inherent characteristics of the cells, and, most importantly, the environmental factors to which the cells were exposed, such as the level of ROS or the presence of chemicals. Furthermore, these enzymes also exhibit

Effects of 1-Methylcyclopropene and Modified Atmosphere Packaging on the Antioxidant Capacity in Pepper "Kulai" during Low-Temperature Storage

91

TABLE 1: Effect of MAP and MAP+1-MCP treatment on antioxidant constituents in mature green peppers and fully ripe/red peppers during low-temperature storage.

Green peppers	Treatments			
	MAP+RT*	MAP	MAP+1-MCP+RT	MAP+1-MCP
(1) Content				
Total phenolic	1.09 ± 0.09a	1.08 ± 0.08a	1.08 ± 0.08a	1.07 ± 0.08a
Ascorbic acid	0.19 ± 0.01a	0.20 ± 0.01a	0.16 ± 0.01b	0.16 ± 0.01b
GSH	6.83 ± 0.32c	7.39 ± 0.54bc	7.99 ± 0.25ab	8.51 ± 0.36a
GSSG	0.95 ± 0.08a	1.07 ± 0.09a	0.95 ± 0.08a	1.04 ± 0.09a
(2) Activity				
APX	2.83 ± 0.12a	2.84 ± 0.11a	3.21 ± 0.15a	3.01 ± 0.16a
GR	167.05 ± 12.02b	152.19 ± 13.04b	174.19 ± 12.34a	177.72 ± 15.23a
CAT	42.44 ± 2.63ab	45.34 ± 3.56a	36.32 ± 3.28ab	32.83 ± 2.98b
Red peppers	Treatments			
	MAP+RT	MAP	MAP+1-MCP+RT	MAP+1-MCP+RT
(1) Content				
Total phenolic	2.19 ± 0.28a	2.13 ± 0.15a	2.13 ± 0.15a	2.12 ± 0.17a
Ascorbic acid	1.37 ± 0.08a	1.38 ± 0.09a	1.29 ± 0.07a	1.33 ± 0.06a
GSH	6.23 ± 0.24a	6.33 ± 0.35a	7.03 ± 0.62a	7.30 ± 0.41a
GSSG	0.72 ± 0.06ab	0.75 ± 0.03a	0.59 ± 0.05b	0.65 ± 0.08ab
(2) Activity				
APX	4.25 ± 0.02a	4.27 ± 0.03a	4.08 ± 0.05a	4.08 ± 0.03a
GR	579.31 ± 26.25b	576.16 ± 31.25b	610.56 ± 32.16a	618.83 ± 38.92a
CAT	41.35 ± 2.61a	37.29 ± 3.12a	35.08 ± 2.81a	36.21 ± 3.15a

Values are means ± SE of 5 measurements. Different letters indicate significant differences (LSD test, $P < 0.05$) for the means of any antioxidant constituent across the rows.

* RT represents the fruits that were transferred to room temperature at day 21.

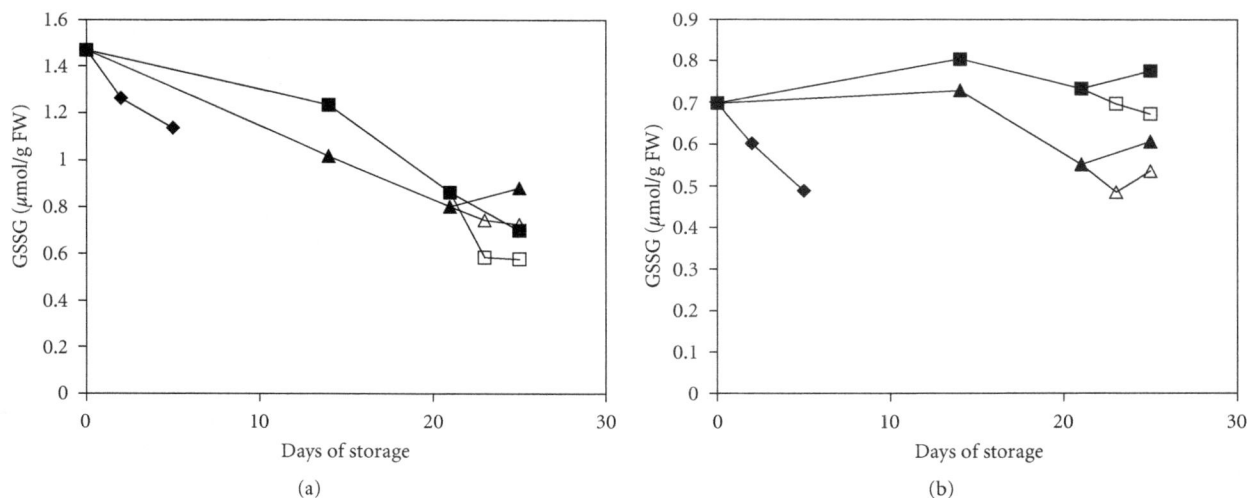

FIGURE 4: Changes in the GSSG of mature green peppers (a) and fully ripe/red peppers (b) for the UCRT (◆), MAP (■), and MAP+1-MCP (▲) treatments during storage at 10°C; MAP (□) and MAP+1-MCP (△) values for the fruits transferred to room temperature at day 21. The values are the means of six replicate samples, and their S.E.s are shown.

a synergistic interaction with non-enzymatic antioxidants to maintain a reduced environment. The APX-GR system is one of the best examples of this type of interaction.

APX and GR form a closely associated system that effectively removes H_2O_2. APX breaks down the H_2O_2 that has escaped from the CAT activity or was generated during respiration [40], and GR is responsible for GSH production and ascorbate regeneration [41]. The APX activity was generally lower in the treated fruit compared with that in the UCRT group fruit, especially during the first 2 weeks of cold storage

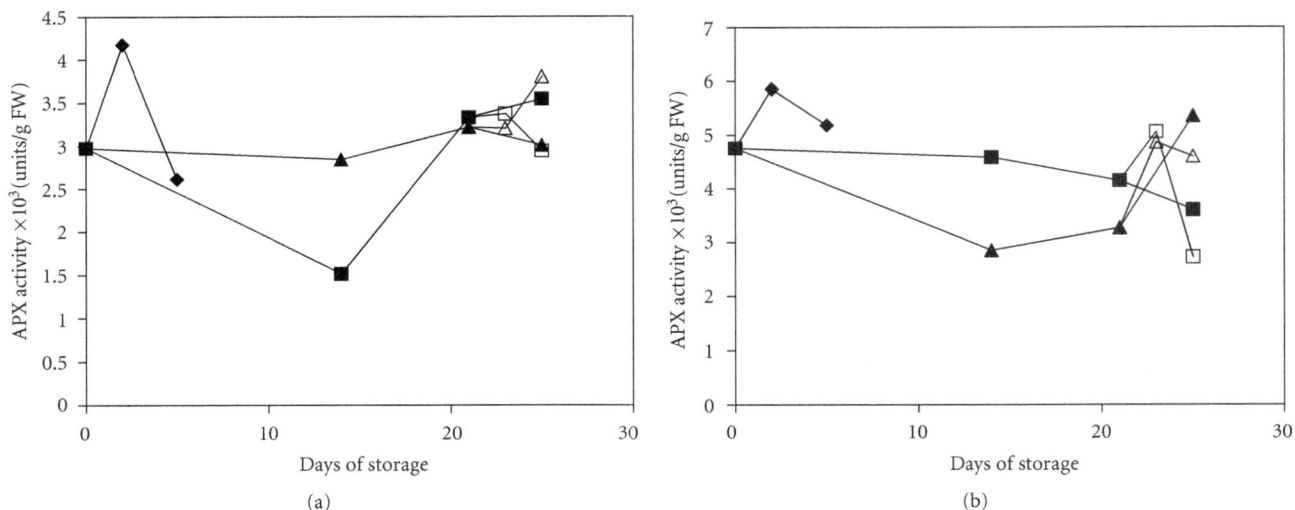

FIGURE 5: Changes in the APX activity of mature green peppers (a) and fully ripe/red peppers (b) for the UCRT (◆), MAP (■), and MAP+1-MCP (▲) treatments during storage at 10°C; MAP (□) and MAP+1-MCP (△) values for the fruits transferred to room temperature at day 21. The values are the means of six replicate samples, and their S.E.s are shown.

TABLE 2: Change in GSH/GSSG ratio in mature green peppers and fully ripe/red peppers stored at low temperature as affected by the MAP and MAP+1-MCP treatments.

Days	Green peppers (GSH)/(GSSG)	Red peppers (GSH)/(GSSG)
UCRT: 0	7.092	3.654
UCRT: 2	4.699	15.107
UCRT: 5	4.05	23.632
MAP: 14	5.214	10.929
MAP: 21	6.917	10.407
MAP: 25	9.683	10.677
MAP: 23 (RT)*	10.583	10.583
MAP: 25 (RT)*	8.922	7.265
MAP/1-MCP: 14	7.912	14.352
MAP/1-MCP: 21	9.512	18.848
MAP/1-MCP: 25	9.003	14.289
MAP/1-MCP: 23 (RT)*	9.644	16.188
MAP/1-MCP: 25 (RT)*	9.278	12.655

*RT represents the fruits that were transferred to room temperature at day 21.

(Figure 5). After two weeks of storage at 10°C, the APX activity in the MAP+1-MCP-treated fruit showed a sustainable increase and eventually reached a higher level than that observed in the UCRT fruit. The resulting APX activity in the MAP+1-MCP-treated fruit was 3.8 kUnits/g FW and 5.8 kUnits/g FW for the mature green peppers and red peppers, respectively, at the end of storage. This finding is in agreement with the study by Singh and Dwivedi [42], which suggested that treatment with 1-MCP provided an additive and sustainable effect to elevate the APX activity. High levels of APX activity might have a suppressive effect on the ascorbic acid levels in the MAP+1-MCP-treated fruit, as was noted earlier in this study. Such observations indicate that the

APX activity in the MAP+1-MCP-treated fruit was actively involved in slowing the development of chilling stress during the later stages of storage. Conversely, the APX activity in the MAP-treated red pepper showed a variable change and decreased after two weeks of storage. This result implied that the effect of the MAP treatment alone was not sufficient to sustain the APX activity until the end of storage.

The GR activity in the treated fruit was likely inconsistent with the change of the GSH content during storage (Figure 6). The GR activity of the MAP- and MAP+1-MCP-treated groups was constantly maintained at a higher level than that of the UCRT group throughout the storage period, especially in the MAP+1-MCP-treated fruit. This higher level was a predictable result because GR is important for sustaining a highly reduced cellular environment by maintaining a high GSH/GSSG ratio [43]. Furthermore, the 1-MCP treatment again showed an additive effect on increasing the GR activity in response to the chilling stress conditions. The difference between the MAP-treated fruit and the MAP+1-MCP-treated fruit was significantly evident ($P < 0.05$) in the red pepper. A similar finding was also observed in cotton plants under stress conditions [44]. Overall, these findings demonstrated that 1-MCP treatment played an important role in regulating the ascorbate-glutathione cycle against oxidative stress.

CAT catalyses the downstream scavenging system by breaking down H_2O_2 [45]. The results obtained in the present study revealed that changes in CAT activity in the treated fruit followed a trend similar to that observed in the UCRT fruit, with both exhibiting an increase in activity at the early stages followed by a decrease (Figure 7). This result is in contrast to the changes in the APX activity, which suggested that the effective removal of harmful substances, such as H_2O_2, occurred via a cooperative mediation between APX and CAT. Such modulation is mainly dependent on the levels of substrate (H_2O_2) and reductant (ascorbate). CAT has a high

Effects of 1-Methylcyclopropene and Modified Atmosphere Packaging on the Antioxidant Capacity in Pepper "Kulai" during Low-Temperature Storage

93

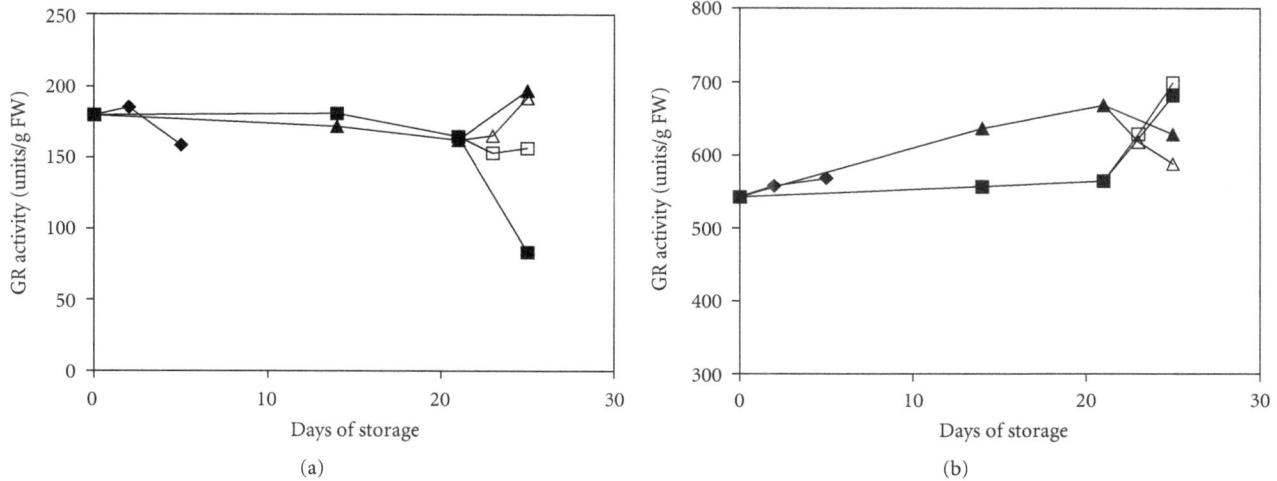

FIGURE 6: Changes in the GR activity of mature green peppers (a) and fully ripe/red peppers (b) for the UCRT (♦), MAP (■), and MAP+1-MCP (▲) treatments during storage at 10°C; MAP (□) and MAP+1-MCP (△) values for the fruits transferred to room temperature at day 21. The values are the means of six replicate samples, and their S.E.s are shown.

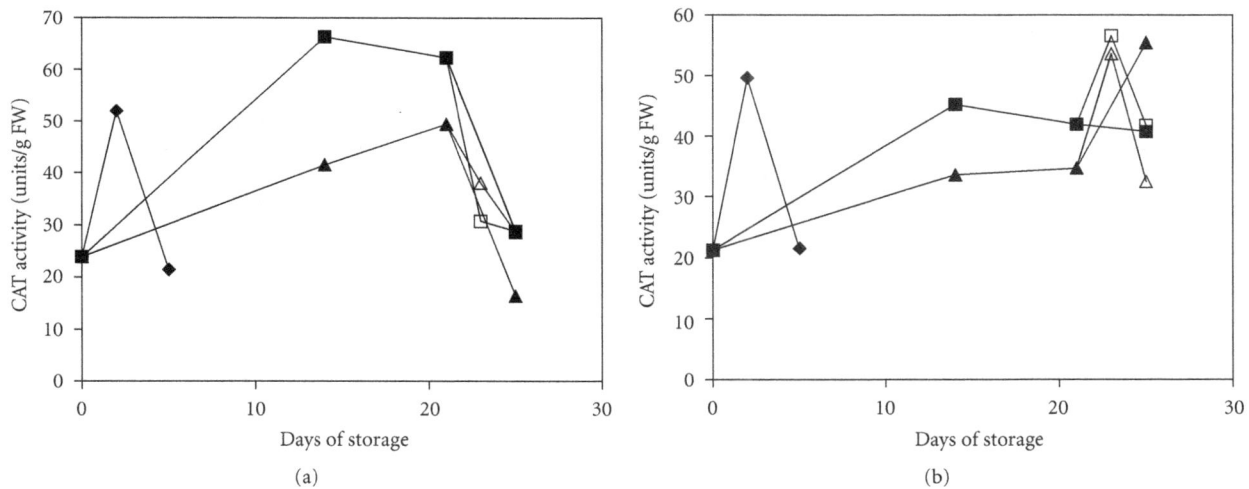

FIGURE 7: Changes in the CAT activity of mature green peppers (a) and fully ripe/red peppers (b) for the UCRT (♦), MAP (■), and MAP+1-MCP (▲) treatments during storage at 10°C; MAP (□) and MAP+1-MCP (△) values for the fruits transferred to room temperature at day 21. The values are the means of six replicate samples, and their S.E.s are shown.

catalytic rate but a low affinity towards the substrate, whereas APX has a much higher affinity but requires a sufficient amount of reductant to be activated [21, 46]. Generally, CAT activity in the MAP-treated fruit was higher compared with the MAP+1-MCP-treated fruit. However, in the MAP+1-MCP-treated red pepper, the CAT activity was consistently increased throughout the storage period, which indicated that the 1-MCP treatment might be potentially effective in retaining the CAT activity. A similar finding was also reported in a study regarding plum fruit [47].

4. Conclusion

The postharvest conditions for commonly consumed fruits and vegetables are a topic of concern, especially for highly perishable fruits such as the pepper "Kulai" (*Capsicum annuum* cv. Kulai). The present study showed that treatment with MAP or MAP+1-MCP can effectively delay the chilling injury development at low temperatures and extend the shelf life of a pepper by up to 25 days while retaining the nutritional quality of the pepper. The accumulation of the total phenolic and ascorbic acid contents were delayed as a result of the treatments given. The ascorbic acid content and the CAT activity were lower in the MAP+1-MCP treated fruit. However, treatment with 1-MCP did show an impressive effect by upregulating the APX-GR system, which can contribute to the reduction in oxidative stress caused by storage at low temperatures. The GR activity was consistent with the GSH content in maintaining a high GSH/GSSG ratio, which is important to enhance human immunity.

More importantly, the effect of 1-MCP was sustained even after the fruits were transferred to storage at room temperature. The commercial utilisation of this technology is likely to have a dramatic impact on the improvement of the storage and handling of horticultural products at the current stage. With slight modification, this postharvest application of 1-MCP can be beneficial for extending the shelf life and market quality of other vegetables and fruits.

Acknowledgments

This work was supported by Research University Grant Scheme (OUP) through UKM-GUP-KPB-08-33-135 and UKM-OUP-KPB-33-169/2011. C. K. Tan is one of the recipients of MyBrain15 Sponsorship Program (MyPhD) provided by the Higher Education Ministry.

References

[1] L. R. Howard, S. T. Talcott, C. H. Brenes, and B. Villalon, "Changes in phytochemical and antioxidant activity of selected pepper cultivars (*Capsicum* species) as influenced by maturity," *Journal of Agricultural and Food Chemistry*, vol. 48, no. 5, pp. 1713–1720, 2000.

[2] B. Ou, D. Huang, M. Hampsch-Woodill, J. A. Flanagan, and E. K. Deemer, "Analysis of antioxidant activities of common vegetables employing oxygen radical absorbance capacity (ORAC) and ferric reducing antioxidant power (FRAP) assays: a comparative study," *Journal of Agricultural and Food Chemistry*, vol. 50, no. 11, pp. 3122–3128, 2002.

[3] P. M. Bramley, "Is lycopene beneficial to human health?" *Phytochemistry*, vol. 54, no. 3, pp. 233–236, 2000.

[4] H. Qian and V. Nihorimbere, "Antioxidant power of phytochemicals from Psidium guajava leaf," *Journal of Zhejiang University*, vol. 5, no. 6, pp. 676–683, 2004.

[5] J. J. Otten, J. P. Hellwig, and L. D. Meyers, *Diatery References Intakes: The Essential Guide to Nutrient Requirments*, The National Academies Press, Washington, DC, USA, 2006.

[6] B. H. J. Bielski, H. W. Richter, and P. C. Chan, "Some properties of the ascorbate free radical," *Annals of the New York Academy of Sciences*, vol. 258, pp. 231–237, 1975.

[7] J. R. Harris, *Subcellular Biochemistry, Ascorbic Acid: Biochemistry and Biomedical Cell Biology*, Plenum, New York, NY, USA, 1996.

[8] A. Pompella, A. Visvikis, A. Paolicchi, V. De Tata, and A. F. Casini, "The changing faces of glutathione, a cellular protagonist," *Biochemical Pharmacology*, vol. 66, no. 8, pp. 1499–1503, 2003.

[9] J. J. Polderdijk, H. A. M. Boerrigter, E. C. Wilkinson, J. G. Meijer, and M. F. M. Janssens, "The effects of controlled atmosphere storage at varying levels of relative humidity on weight loss, softening and decay of red bell peppers," *Scientia Horticulturae*, vol. 55, no. 3-4, pp. 315–321, 1993.

[10] R. E. Hardenburg, A. E. Watada, and C. Y. Wong, "The commercial storage of fruits, vegetables, and florist and nursery stocks," *USDA Agriculture Handbooks*, vol. 66, pp. 130–142, 1986.

[11] R. E. Paull, "Chilling injury of crops of tropical and subtropical origin," in *Chilling Injury of Horticultural Crops*, C. Y. Wang, Ed., CRC Press, Boca Raton, Fla, USA, 1990.

[12] K. S. Lee, K. L. Woo, and D. S. Lee, "Modified atmosphere packaging for green chili peppers," *Packaging Technology and Science*, vol. 7, no. 1, pp. 51–58, 1994.

[13] E. C. Sisler, T. Alwan, R. Goren, M. Serek, and A. Apelbaum, "1-Substituted cyclopropenes: effective blocking agents for ethylene action in plants," *Plant Growth Regulation*, vol. 40, no. 3, pp. 223–228, 2003.

[14] E. Bassetto, A. P. Jacomino, A. L. Pinheiro, and R. A. Kluge, "Delay of ripening of "Pedro Sato" guava with 1-methylcyclopropene," *Postharvest Biology and Technology*, vol. 35, no. 3, pp. 303–308, 2005.

[15] S. L. Chae, M. K. Seong, L. C. Jeoung, K. C. Gross, and A. B. Woolf, "Bell pepper (*Capsicum annuum* L.) fruits are susceptible to chilling injury at the breaker stage of ripeness," *HortScience*, vol. 42, no. 7, pp. 1659–1664, 2007.

[16] A. Manenoi, E. R. V. Bayogan, S. Thumdee, and R. E. Paull, "Utility of 1-methylcyclopropene as a papaya postharvest treatment," *Postharvest Biology and Technology*, vol. 44, no. 1, pp. 55–62, 2007.

[17] Z. S. Ilić, R. Trajković, Y. Perzelan, S. Alkalai-Tuvia, and E. Fallik, "Influence of 1-methylcyclopropene (1-MCP) on postharvest storage quality in green bell pepper fruit," *Food and Bioprocess Technology*, 2011.

[18] V. L. Singleton and J. A. Rossi, "Colorimetry of total phenolics with phosphomolybdic phosphotungstic acid reagents," *American Journal of Enology and Viticulture*, vol. 16, pp. 144–158, 1965.

[19] Y. Y. Lim, T. T. Lim, and J. J. Tee, "Antioxidant properties of several tropical fruits: a comparative study," *Food Chemistry*, vol. 103, no. 3, pp. 1003–1008, 2007.

[20] M. E. Anderson, "Determination of glutathione and glutathione disulfide in biological samples," *Methods in Enzymology*, vol. 113, pp. 548–555, 1985.

[21] Y. Nakano and K. Asada, "Hydrogen peroxide is scavenged by ascorbate-specific peroxidase in spinach chloroplasts," *Plant and Cell Physiology*, vol. 22, no. 5, pp. 867–880, 1981.

[22] R. F. Beers Jr. and I. W. Sizer, "A spectrophotometric method for measuring the breakdown of hydrogen peroxide by catalase," *The Journal of Biological Chemistry*, vol. 195, no. 1, pp. 133–140, 1952.

[23] H. U. Bergmeyer, *Methods of Enzymatic Analysis*, VCH, Weinheim, Germany, 3rd edition, 1987.

[24] K. W. Lee, Y. J. Kim, D. O. Kim, H. J. Lee, and C. Y. Lee, "Major phenolics in apple and their contribution to the total antioxidant capacity," *Journal of Agricultural and Food Chemistry*, vol. 51, no. 22, pp. 6516–6520, 2003.

[25] M. Materska and I. Perucka, "Antioxidant activity of the main phenolic compounds isolated from hot pepper fruit (*Capsicum annuum* L.)," *Journal of Agricultural and Food Chemistry*, vol. 53, no. 5, pp. 1750–1756, 2005.

[26] Y. F. Chu, J. Sun, X. Wu, and R. H. Liu, "Antioxidant and antiproliferative activities of common vegetables," *Journal of Agricultural and Food Chemistry*, vol. 50, no. 23, pp. 6910–6916, 2002.

[27] M. T. Lafuente, M. A. Martinez-Téllez, G. González-Aguilar et al., "Physiological and biochemical responses associated with chilling sensitivity of "Fortune" mandarins and temperature conditioning," *CIHEM-Options Mediterraneennes*, vol. 1, pp. 125–134, 1995.

[28] J. D. Faragher and D. J. Chalmers, "Regulation of anthocyanin synthesis in apple skin. III. Involvement of phenylalanine ammonia-lyase," *Australian Journal of Plant Physiology*, vol. 4, pp. 133–141, 1977.

Effects of 1-Methylcyclopropene and Modified Atmosphere Packaging on the Antioxidant Capacity in Pepper "Kulai" during Low-Temperature Storage

95

[29] A. Podsedek, "Natural antioxidants and antioxidant capacity of Brassica vegetables: a review," *LWT - Food Science and Technology*, vol. 40, no. 1, pp. 1–11, 2007.

[30] I. M. C. M. Rietjens, M. G. Boersma, L. D. Haan et al., "The pro-oxidant chemistry of the natural antioxidants vitamin C, vitamin E, carotenoids and flavonoids," *Environmental Toxicology and Pharmacology*, vol. 11, no. 3-4, pp. 321–333, 2002.

[31] A. Mozafar, *Plant Vitamins: Agronomic, Physiological and Nutritional Aspects*, CRC Press, Boca Raton, Fla, USA, 1994.

[32] S. K. Lee and A. A. Kader, "Preharvest and postharvest factors influencing vitamin C content of horticultural crops," *Postharvest Biology and Technology*, vol. 20, no. 3, pp. 207–220, 2000.

[33] W. Kalt, "Effects of production and processing factors on major fruit and vegetable antioxidants," *Journal of Food Science*, vol. 70, no. 1, pp. R11–R19, 2005.

[34] T. O. Win, V. Srilaong, J. Heyes, K. L. Kyu, and S. Kanlayanarat, "Effects of different concentrations of 1-MCP on the yellowing of West Indian lime (*Citrus aurantifolia*, Swingle) fruit," *Postharvest Biology and Technology*, vol. 42, no. 1, pp. 23–30, 2006.

[35] K. Tanaka, T. Sano, K. Ishizuka, K. Kitta, and Y. Kawamura, "Comparison of properties of leaf and root glutathione reductases from spinach," *Physiologia Plantarum*, vol. 91, no. 3, pp. 353–358, 1994.

[36] H. Rennenberg, "Glutathione metabolism and possible biological roles in higher plants," *Phytochemistry*, vol. 21, no. 12, pp. 2771–2781, 1980.

[37] G. Kocsy, G. Szalai, A. Vágújfalvi, L. Stéhli, G. Orosz, and G. Galiba, "Genetic study of glutathione accumulation during cold hardening in wheat," *Planta*, vol. 210, no. 2, pp. 295–301, 2000.

[38] M. A. Walker and B. D. McKersie, "Role of the ascorbateglutathione antioxidant system in chilling resistance of tomato," *Journal of Plant Physiology*, vol. 141, pp. 234–239, 1993.

[39] W. Dröge and R. Breitkreutz, "Glutathione and immune function," *Proceedings of the Nutrition Society*, vol. 59, no. 4, pp. 595–600, 2000.

[40] C. H. Foyer, P. Descourvie'res, and K. J. Kunert, "Protection against oxygen radicals: an important defence mechanism studied in transgenic plants," *Plant, Cell and Environment*, vol. 17, pp. 507–523, 1994.

[41] H. Saruyama and M. Tanida, "Effect of chilling on activated oxygen-scavenging enzymes in low temperature-sensitive and -tolerant cultivars of rice (*Oryza sativa* L.)," *Plant Science*, vol. 109, no. 2, pp. 105–113, 1995.

[42] R. Singh and U. N. Dwivedi, "Effect of ethrel and 1-methylcyclopropene (1-MCP) on antioxidants in mango (*Mangifera indica* var. Dashehari) during fruit ripening," *Food Chemistry*, vol. 111, no. 4, pp. 951–956, 2008.

[43] P. E. Gamble and J. J. Burke, "Effect of water stress on the chloroplast antioxidant system. Alterations in glutathione reductase activity," *Plant Physiology*, vol. 76, pp. 615–621, 1984.

[44] M. K. Eduardo, M. O. Derrick, and L. S. John, "Effect of 1-MCP on antioxidants, enzymes, membrane leakage, and protein content of drought-stressed cotton plants," *Summaries of Arkansas Cotton Research*, pp. 102–107, 2007.

[45] C. Bowler, M. Van Montagu, and D. Inzé, "Superoxide dismutase and stress tolerance," *Annual Review of Plant Physiology and Plant Molecular Biology*, vol. 43, no. 1, pp. 83–116, 1992.

[46] H. Willekens, D. Inze, M. Van Montagu, and W. Van Camp, "Catalases in plants," *Molecular Breeding*, vol. 1, no. 3, pp. 207–228, 1995.

[47] J. Kan, J. Che, H. Y. Xie, and C. H. Jin, "Effect of 1-methylcyclopropene on postharvest physiological changes of "Zaohong" plum," *Acta Physiologiae Plantarum*, vol. 12, pp. 56–67, 2010.

13

Effect of Irrigation to Winter Wheat on the Radiation Use Efficiency and Yield of Summer Maize in a Double Cropping System

Li Quanqi,[1] Chen Yuhai,[2] Zhou Xunbo,[2] Yu Songlie,[2] and Guo Changcheng[3]

[1] *College of Water Conservancy and Civil Engineering, Shandong Agricultural University, Tai'an 271018, China*
[2] *State Key Laboratory of Crop Biology, Shandong Key Laboratory of Crop Biology, Shandong Agricultural University, Tai'an 271018, China*
[3] *Tianjin Key Laboratory of Water Resources and Environment, Tianjin Normal University, Tianjin 300074, China*

Correspondence should be addressed to Chen Yuhai, yhchen@sdau.edu.cn

Academic Editors: U. Feller and G. Humphreys

In north China, double cropping of winter wheat and summer maize is a widely adopted agricultural practice, and irrigation is required to obtain a high yield from winter wheat, which results in rapid aquifer depletion. In this experiment conducted in 2001-2002, 2002-2003, and 2004-2005, we studied the effects of irrigation regimes during specific winter wheat growing stage with winter wheat and summer maize double cropping systems; we measured soil moisture before sowing (SMBS), the photosynthetic active radiation (PAR) capture ratio, grain yield, and the radiation use efficiency (RUE) of summer maize. During the winter wheat growing season, irrigation was applied at the jointing, heading, or milking stage, respectively. The results showed that increased amounts of irrigation and irrigation later in the winter wheat growing season improved SMBS for summer maize. The PAR capture ratio significantly (LSD, $P < 0.05$) increased with increased SMBS, primarily in the 3 spikes leaves. With improved SMBS, both the grain yield and RUE increased in all the treatments. These results indicate that winter wheat should be irrigated in later stages to achieve reasonable grain yield for both crops.

1. Introduction

In north China, the most important crops are winter wheat and summer maize; thus, a double cropping practice has been widely adopted [1, 2]. Evapotranspiration during the winter wheat growing season is approximately 400–500 mm, but annual precipitation typically does not exceeded 200 mm [3, 4]. Therefore, additional irrigation is required to achieve a satisfactory winter wheat grain yield. The summer maize growing season (June to September) is during the rainy season in north China. The average annual precipitation during the summer maize growing season is approximately 325 mm, which meets the crop's water consumption requirements. However, due to the influence of seasonal winds, drought often occurs during the summer maize seedling stage; therefore, soil moisture before sowing (SMBS) has become an important parameter to obtain a stable yield.

Previously, Li et al. [5] studied the winter wheat and summer maize double cropping system of winter wheat and summer maize in north China, including the influence of irrigation during the winter wheat growing season on water use and physiological characteristics of summer maize. The results showed that irrigation during the winter wheat growing season could increase soil moisture accumulation before summer maize sowing. With increased SMBS, the water use efficiency (WUE) of summer maize could increase in dry and moderate years, but not in wet years. Irrigation during the winter wheat growing season also affected the photosynthesis rate, transpiration, stomatal conductance, and leaf temperature of summer maize [6]. Thus, irrigation during the winter wheat growing season not only influenced the water consumption characteristics of winter wheat, but also the following summer maize crop. Given the excessive exploitation of groundwater resources for irrigation in north China

[7–9], in a double cropping system, the combined grain yield and WUE for both winter wheat and summer maize should be considered.

Biomass production occurs when leaves intercept incoming photosynthetic active radiation (PAR) and the plant transforms the intercepted radiation into energy [10, 11]. Throughout the world, many researchers have studied the relationship between crop grain yield and RUE of crops [12–14]. To support agricultural water conservation efforts in north China, many researchers have studied the relationship between deficit irrigation and RUE. Li et al. [4] indicated that irrigating at the jointing and heading or jointing, heading, and milking stages could help increase the PAR capture ratio later in the winter wheat growing season. Han et al. [15] reported that varietal and deficit irrigation effects on the RUE and grain yield of winter wheat resulted from PAR modifications in the winter wheat canopies. Li et al. [16] showed that in north China, a furrow planting pattern should be used in combination with deficit irrigation to increase both the RUE and grain yield of winter wheat. However, all of these studies focused only on one crop, and did not examine deficit irrigation, RUE, and the planting system together.

This study aimed to determine whether irrigation regimes during specific winter wheat growing stages in a double cropping system affect summer maize leaf area characteristics responsible for intercepting incoming radiation and RUE.

2. Materials and Methods

2.1. Experimental Site. The experiments were conducted using 12 irrigation plots at Yucheng Comprehensive Experimental Station (36°57′N, 116°38′E), Chinese Academy of Science, during the years 2001-2002 and 2002-2003, and at Tai'an Experimental Station (36°10′N, 117°09′E), Agronomy College, Shandong Agricultural University, in 2004-2005. The plot areas at Yucheng and Tai'an were 6.7 and 9.0 m², respectively, at a depth of 1.5 m; the plots were enclosed with a concrete wall, and the bottom surfaces of the plots were not sealed. The plot surfaces were 15 cm above ground level on all sides to prevent runoff, run-on, and subsurface water movement between the plots. The 2 experimental stations are located in Shandong province, north China. The mean annual precipitation at the Yucheng Comprehensive Experimental Station is 590 mm, of which approximately 62% falls between June and September—the summer maize growing season. The soil at the experimental site is classified as sandy loam with an organic matter content of approximately 0.5–0.6%, pH of approximately 8.5, and field capacity and wilting point of 25.1% and 8.0% by volume, respectively. The mean annual precipitation at the Tai'an Experimental Station is 700 mm, of which approximately 65% falls between June and September. The soil at this experimental site is classified as loam with an organic matter content of approximately 1.4%, pH of approximately 6.9, and field capacity and wilting point of 25.8% and 7.7% by volume, respectively. The precipitation under natural conditions for the summer maize growing seasons in 2002, 2003, and 2005 were 133.6, 308.4, and 576.4 mm, respectively. Base on annual mean precipitation during the summer maize growing season in the studied regions, the year 2002 was a very dry year, 2003 was a moderate year, and 2005 was a wet year.

2.2. Experimental Design. The experiments were conducted in triplicate using a randomized block design during 2001-2002, 2002-2003, and 2004-2005; the following 4 irrigation amounts were applied throughout the entire growth cycle of winter wheat: no supplemental irrigation (T0); irrigated only at the jointing stage (T1); irrigated at the jointing and heading stages (T2); irrigated at the jointing, heading, and milking stages (T3). In 2002, irrigation was applied on March 23, April 12, and May 9; in 2003, on April 6, April 30, and May 16; in 2005, on April 7, April 27, and May 14. All irrigation applications consisted of 60 mm of water supplied from a pump outlet to the plots via plastic pipes; a flow meter was used to measure the amount of water applied.

2.3. Cultural Procedures and Measurements. Winter wheat was planted on October 4 in 2001 and 2002, and on October 6 in 2004. Before sowing, 300 kg·hm⁻² of triple superphosphate, 300 kg·hm⁻² of urea, and 75 kg·hm⁻² of potassium chloride were applied. The wheat plants were harvested on June 6 in 2002 and 2003, and June 7 in 2005. The maize cultivar "Nongda108" was manually planted immediately after harvesting in all 3 years. At the beginning of July, urea was applied at a rate of 140 kg·hm⁻² depending on the rainfall. When the maize plants were at the 5-leaf stage, their density was fixed at 6.6×10^4 plants·hm⁻². The maize plants were harvested on September 24, September 26, and September 28 in 2002, 2003, and 2005, respectively. After air-drying, the dry weight of the grain was measured.

Beginning at 28 days after sowing, the leaf area index (LAI) was measured every 7 days using a LI-3000 Portable Area Meter (Li-Cor Co.Ltd, Lincoln, Nebraska, USA) at Yucheng Comprehensive Experimental Station in 2002 and 2003. At the Agronomy Station of Shandong Agricultural University, the LAI was measured using a 1.5 m long linear sensor (SunScan) every 10 days from emergence to maturity.

To measure transmitted radiation, the linear sensor (SunScan) was placed parallel to the row direction of each plot in the middle of each summer maize row. The average of these measurements was considered to be the radiation transmitted by the canopy. Moreover, the transmitted radiation at the 3 spike leaves (including the leaves on, above, and below the spike), above the 3 spike leaves, and below the 3 spike leaves were measured, respectively. The incoming solar radiation above the crop canopy was also monitored. We determined the amount of solar radiation intercepted by the canopy by calculating the difference between the above-canopy and soil surface solar radiation as measured by the SunScan [16].

RUE during the summer maize growing season was calculated using the approach proposed by Quanqi et al. [17]:

$$E\% = \frac{\Delta W \cdot H}{\Sigma S} \times 100\%, \tag{1}$$

TABLE 1: Soil moisture accumulation in 0–1.2 m soil profiles before summer maize sowing (mm).

Treatments	2002	2003	2005	Mean
T0	221.4^c	194.0^d	247.2^d	220.9^d
T1	231.9^c	217.3^c	266.7^c	238.6^c
T2	263.7^b	225.2^b	315.4^b	268.1^b
T3	296.8^a	255.4^a	349.3^a	300.5^a

Means in each column with in each year followed by the same letter are not significantly different at $P < 0.05$ based on LSD test.

TABLE 2: Effect of irrigation during the winter wheat growing season on the PAR reflection ratio, PAR penetration ratio, and PAR capture ratio in summer maize canopy (%).

Treatments	PAR reflection ratio	PAR penetration ratio	PAR capture ratio
T0	3.8^a	20.4^a	75.8^d
T1	3.7^a	16.1^b	80.2^c
T2	3.3^{ab}	14.5^c	82.2^b
T3	2.9^c	11.7^d	85.4^a

The data was the average values on August 20, Aug 27, and Septemper 5, 2002; and Aug 24, and Sep 9, 2003; and Aug 19, Aug 20, and Aug 22, 2005. Values followed by a different letter are significantly different at 5% probability level.

where ΔW is dry matter weight at each growing stage after drying at 80°C for constant weight; $H = 17.782 \, \text{KJ} \cdot \text{g}^{-1}$ is the energy conversion coefficient; and ΣS is global incoming radiation for each summer maize growing season, which were obtained from weather stations located near the experiment area at Yucheng and within 0.5 km of the experimental site at Tai'an.

2.4. Statistical Analysis. The treatments were run as an analysis of variance (ANOVA). For ANOVA, $\alpha = 0.05$ was set as the level of significance to determine whether differences existed among treatments means. The multiple comparisons were done for significant effects with the least significant difference (LSD) test at $\alpha = 0.05$.

3. Results

3.1. Soil Moisture Accumulation before Summer Maize Sowing. Irrigation during the winter wheat growing season has the effect of increasing the soil moisture accumulation before summer maize sowing (Table 1). With more irrigation, soil moisture in the 1.2 m soil profiles increased significantly (LSD, $P < 0.05$), by 10.5, 42.3, and 75.4 mm at T1, T2, and T3, respectively, compared to T0 in 2002: by 23.3, 31.2, and 61.4 mm, respectively, in 2003; by 19.5, 68.2, and 102.1 mm, respectively, in 2005. The average soil moisture after the first, second, and third irrigations improved by 17.7, 47.2, and 79.6 mm, respectively. These results indicated that irrigation later in the winter wheat growing season and increased irrigation amounts improved SMBS for summer maize.

However, the effect of irrigation at each soil layer was different; soil moisture was increased significantly (LSD,

TABLE 3: PAR capture ratio in the canopy of summer maize (%).

Treatment	Above the 3 spike leaves	The 3 spike leaves	Below the 3 spike leaves
T0	12.8^d	29.4^d	33.6^a
T1	15.1^c	43.7^c	21.4^b
T2	20.6^b	50.2^b	11.4^c
T3	22.0^a	54.3^a	9.1^d

The data was the average values on Aug 20, Aug 27, and Sep 5, 2002; and Aug 24, and Sep 9, 2003, and Aug 19, Aug 20, and Aug 22, 2005. Values followed by a different letter are significantly different at 5% probability level.

$P < 0.05$) for the soil layer below 20 cm, but not above it (Figure 1).

3.2. Leaf Area Index. Figure 2 shows dynamic LAI variations during the summer maize growing season in 2002. The values corresponding to 2003 and 2005 are not shown because they are very similar to the values shown for 2002. As shown in Figure 2, LAI increased with growth stages. At approximately 56 days after sowing, LAI for each treatment group reached the maximum value. After that point, LAI decreased as the growing season continued. During the summer maize growing season, LAI was highest at T3, with a maximum value of 7.9, which was higher than those at T2, T1, and T0 by 3.2, 1.7, and 0.5, respectively. For the summer maize growing season overall, LAI at T3 was higher than at other treatments, which could have significantly affected the complete capture and utilization of PAR.

3.3. PAR Capture Ratio. Table 2 shows the PAR reflection ratio, PAR penetration ratio, and PAR capture ratio later in the summer maize growing season. The PAR reflection ratio was not significantly (LSD, $P < 0.05$) different at T0, T1, and T2, all of which were significantly (LSD, $P < 0.05$) higher than at T3. The PAR penetration ratio and PAR capture ratio were significantly (LSD, $P < 0.05$) different between treatments. With increased summer maize SMBS, the PAR penetration ratio significantly (LSD, $P < 0.05$) decreased for all treatment groups; however, the PAR capture ratio significantly (LSD, $P < 0.05$) increased, primarily due to improved LAIs associated with the increased SMBS (Figure 2).

The PAR capture ratios at the 3 spike leaves were contrary to those obtained below the 3 spike leaves (Table 3). However, the PAR capture ratios below the 3 spike leaves were consistent with the PAR penetration ratios, that is, more PAR captured below the 3 spike leaves resulted in more PAR penetration. The PAR capture ratio at T3 was higher than at T2, T1, and T0 by 3.2%, 5.2%, and 9.6%, respectively. However, at the 3 spike leaves, the PAR capture ratio at T3 was higher than at T2, T1, and T0 by 4.1%, 10.6%, and 24.9%, respectively. Thus, although irrigation during the winter wheat growing season clearly altered the LAI, it had little effect on the PAR interception amount; however, it had significantly (LSD, $P < 0.05$) affected the vertical distribution of PAR in the canopy. With increased SMBS, the PAR capture ratio in the 3 spike leaves improved.

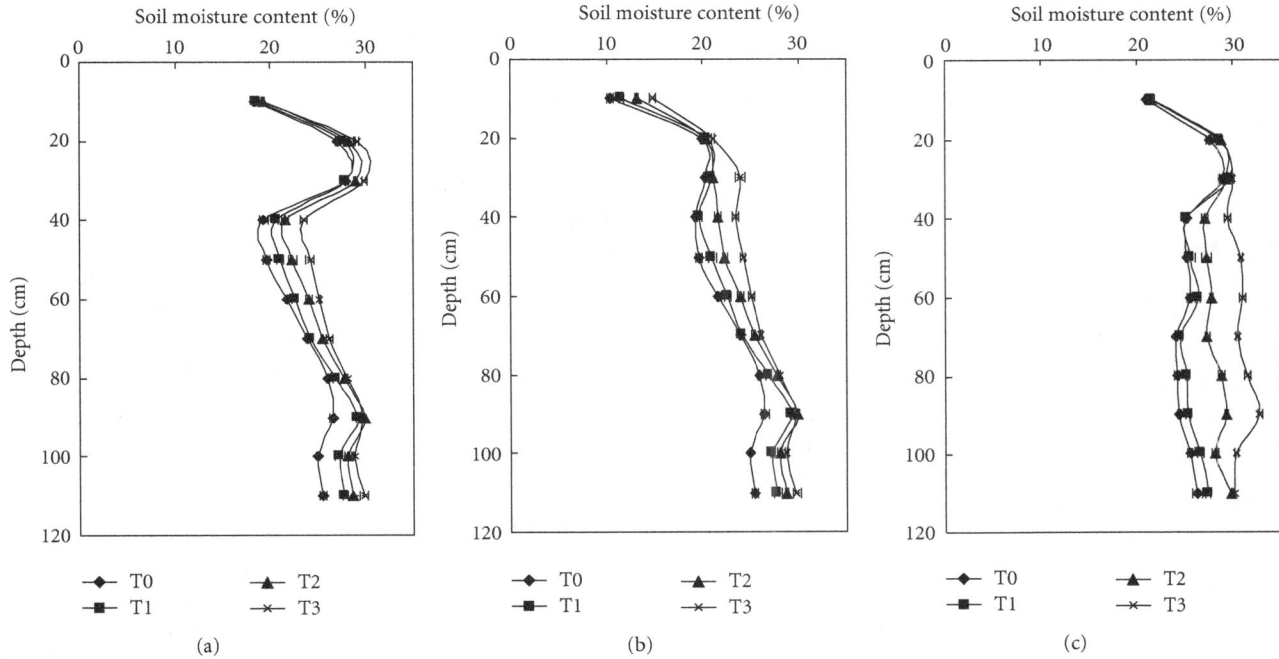

FIGURE 1: Soil moisture status before summer maize sowing in 2002 (a), 2003 (b), and 2005 (c). Horizontal bars are standard errors.

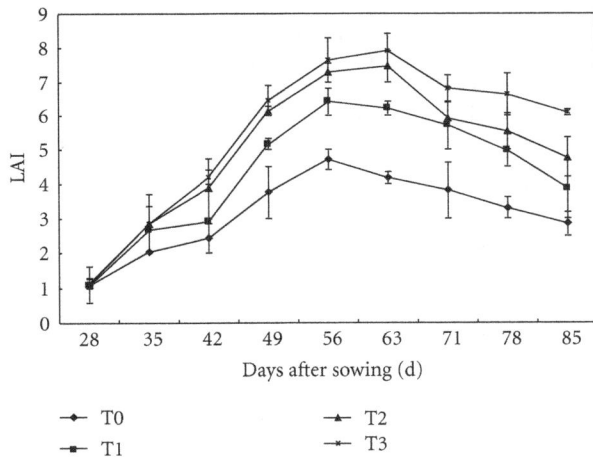

FIGURE 2: The dynamic variation of LAI in 2002 at Yucheng. Vertical bars are standard errors.

3.4. Dry Matter Accumulation at Maturity and Grain Yield.

Table 4 shows the dry matter accumulation at maturity and grain yield for summer maize. The average dry matter accumulation at maturity at T0 was 14587.2 kg·hm^{-2}, which was significantly (LSD, $P < 0.05$) lower than at T2 and T3 by 303.2 and 316.3 kg·hm^{-2}, respectively. The average grain yield at T0 was only 6710.1 kg·hm^{-2}, which was significantly (LSD, $P < 0.05$) lower than at T2 and T3 by 586.2 and 890.7 kg·hm^{-2}, respectively. Thus, it is apparent that with increased SMBS, the dry matter accumulation and grain yield also increase consistently.

TABLE 4: Effect of irrigation during the winter wheat growing season on the dry matter accumulation at maturity and grain yield of summer maize (kg · hm^{-2}).

Treatment	Dry matter accumulation	Grain yield
T0	14587.2[b]	6710.1[c]
T1	14616.7[ab]	7016.0[bc]
T2	14890.4[a]	7296.3[ab]
T3	14903.5[a]	7600.8[a]

The data was the average values in 2002, 2003, and 2005. Values followed by a different letter are significantly different at 5% probability level.

TABLE 5: Effect of irrigation during the winter wheat growing season on the RUE of summer maize at different growth stages (%).

Treatment	Jointing	Large bell-mouth	Tasseling	Milking	Maturity
T0	0.6[d]	1.1[b]	3.1[c]	3.7[b]	2.2[b]
T1	0.7[c]	1.2[b]	3.2[bc]	3.9[ab]	2.3[b]
T2	0.8[b]	1.3[ab]	3.4[ab]	4.0[ab]	2.4[ab]
T3	0.9[a]	1.5[a]	3.6[a]	4.1[a]	2.6[a]

The data was the average values in 2002, 2003, and 2005. Values followed by a different letter are significantly different at 5% probability level.

3.5. Radiation Use Efficiency.

Table 5 shows the RUE of summer maize at different growth stages. During the summer maize growing season, the highest RUE was observed at the milking stage. At jointing, large bell-mouth, tasseling, milking, and maturity stages, the RUE at T3 was significantly (LSD, $P < 0.05$) higher than at T0. Except at the jointing stage, the RUE at T3 was not significantly (LSD, $P < 0.05$)

different than at T2. Furthermore, the RUE at T0 was lower than at T1, T2, and T3 by 4.5%, 9.1%, and 18.2%, respectively. Hence, in a winter wheat and summer maize double cropping system, RUE was enhanced during the summer maize growing season by increased irrigation amounts and irrigation later in the winter wheat growing season.

4. Discussion

In north China, approximately 70% of the total cultivated land is planted with winter wheat and summer maize in a double cropping system [18]. The experiment showed that the timing and amount of irrigation during the winter wheat growing season could affect the SMBS for summer maize in a double cropping system. Although the irrigation water applied during the winter wheat growing season was not absorbed completely, some of it was utilized by the summer maize crop. The results indicated that irrigation later in the winter wheat growing season and increased irrigation amounts improved SMBS for summer maize. With increased SMBS, the grain yield and RUE for summer maize improved consistently. Former studies by the authors [3] showed that during the winter wheat growing season in north China, irrigation applied at the jointing and heading stages or jointing and milking stages had no significantly (LSD, $P < 0.05$) differences between the grain yield and WUE for winter wheat. Therefore, by adopting effective measures, such as irrigation at the jointing and milking stages during the winter wheat growing season, satisfactory grain yields for winter wheat and summer maize can be obtained in a double cropping system. The experiment also showed that in winter wheat and summer maize double cropping systems, research studies that focused only on winter wheat water consumption was not sufficient. Researchers should consider the effect of winter wheat irrigation regimes on the grain yield, RUE, WUE, and economic benefits of the entire planting system. Similarly, in order to conserve agricultural water in north China, irrigation regimes should be customized based on specific planting systems.

Most of the green organ photosynthetic matter was produced by the 3 spike leaves [19], hence, an increase in the green organ's PAR capture ratio would aid in the accumulation and transportation of photosynthetic products later in the summer maize growing season. Therefore, the improved PAR capture ratio and transformational ability in these plant parts were very important for increasing the quantity of photosynthetic matter. The results presented in this experiment show that the PAR capture ratio associated with the high yielding treatment (T3) at the 3 spike leaves was 54.3%. This result is consistent with Fang's study of winter wheat [20]. It is apparent that increasing SMBS could optimize PAR distributions in the summer maize canopy, which will result in a highly efficient photosynthetic colony.

The amount of incoming PAR that is absorbed by the canopy primarily depends on LAI and crop geometry [21]. The differences in the PAR capture ratios for the summer maize canopies were not only due to dynamic LAI variations, but also due to alterations in vertical distributions. Although vertical LAI distributions were not included in this paper, the topic will be addressed in a future study.

Acknowledgments

This work was supported in part by the National Natural Science Foundation of China (31101114), by the Public Welfare Industry (Agricultural) Scientific Funds for Scientific Research Project (200903003), by the State Key Program of Basic Research of China (2009CB118602), by the Development of Science and Technology Plan Projects in Shandong Province of China (2011GNC21101), by the Natural Science Foundation of Shandong Province of China (ZR2010CQ009), and by the Opening Fund of Tianjin Key Laboratory of Water Resources and Environment (52XS1014).

References

[1] Q. Fang, Q. Yu, E. Wang et al., "Soil nitrate accumulation, leaching and crop nitrogen use as influenced by fertilization and irrigation in an intensive wheat-maize double cropping system in the North China Plain," *Plant and Soil*, vol. 284, no. 1-2, pp. 335–350, 2006.

[2] Z. X. Zhu, B. A. Stewart, and X. J. Fu, "Double cropping wheat and corn in a sub-humid region of China," *Field Crops Research*, vol. 36, no. 3, pp. 175–183, 1994.

[3] Q. Li, Y. Chen, M. Liu, X. Zhou, S. Yu, and B. Dong, "Effects of irrigation and straw mulching on microclimate characteristics and water use efficiency of winter wheat in North China," *Plant Production Science*, vol. 11, no. 2, pp. 161–170, 2008.

[4] Q. Li, M. Liu, J. Zhang, B. Dong, and Q. Bai, "Biomass accumulation and radiation use efficiency of winter wheat under deficit irrigation regimes," *Plant, Soil and Environment*, vol. 55, no. 2, pp. 85–91, 2009.

[5] Q. Li, Y. Chen, M. Liu, X. Zhou, B. Dong, and S. Yu, "Effect of irrigation to winter wheat on the soil moisture, evapotranspiration, and water use efficiency of summer maize in north China," *Transactions of the ASABE*, vol. 50, no. 6, pp. 2073–2080, 2007.

[6] Q. Q. Li, Y. H. Chen, H. F. Han et al., "Effects of different soil moisture before sowing on physiologic character and yield of summer maize," *Chinese Agricultural Science Bulletin*, vol. 20, no. 6, pp. 116–119, 2004.

[7] C. Liu, J. Yu, and E. Kendy, "Groundwater exploitation and its impact on the environment in the North China Plain," *Water International*, vol. 26, no. 2, pp. 265–272, 2001.

[8] H. Yang and A. Zehnder, "China's regional water scarcity and implications for grain supply and trade," *Environment and Planning A*, vol. 33, no. 1, pp. 79–95, 2001.

[9] J. Xia, M. Y. Liu, and S. F. Jia, "Water security problem in North China: research and perspective," *Pedosphere*, vol. 15, no. 5, pp. 563–575, 2005.

[10] R. M. Gifford, J. H. Thorne, W. D. Hitz, and R. T. Giaquinta, "Crop productivity and photoassimilate partitioning," *Science*, vol. 225, no. 4664, pp. 801–808, 1984.

[11] D. M. Whitfield and C. J. Smith, "Effects of irrigation and nitrogen on growth, light interception and efficiency of light conversion in wheat," *Field Crops Research*, vol. 20, no. 4, pp. 279–295, 1989.

[12] M. F. Dreccer, A. H. C. M. Schapendonk, G. A. Slafer, and R. Rabbinge, "Comparative response of wheat and oilseed

rape to nitrogen supply: absorption and utilisation efficiency of radiation and nitrogen during the reproductive stages determining yield," *Plant and Soil*, vol. 220, no. 1-2, pp. 189–205, 2000.

[13] J. R. Kiniry, C. A. Jones, J. C. O'toole, R. Blanchet, M. Cabelguenne, and D. A. Spanel, "Radiation-use efficiency in biomass accumulation prior to grain-filling for five grain-crop species," *Field Crops Research*, vol. 20, no. 1, pp. 51–64, 1989.

[14] K. Opoku-Ameyaw and P. M. Harris, "Intercropping potatoes in early spring in a temperate climate. 2. Radiation utilization," *Potato Research*, vol. 44, no. 1, pp. 63–74, 2001.

[15] H. Han, Z. Li, T. Ning, X. Zhang, Y. Shan, and M. Bai, "Radiation use efficiency and yield of winter wheat under deficit irrigation in North China," *Plant, Soil and Environment*, vol. 54, no. 7, pp. 313–319, 2008.

[16] Q. Li, Y. Chen, M. Liu, X. Zhou, S. Yu, and B. Dong, "Effects of irrigation and planting patterns on radiation use efficiency and yield of winter wheat in North China," *Agricultural Water Management*, vol. 95, no. 4, pp. 469–476, 2008.

[17] L. Quanqi, C. Yuhai, L. Mengyu, Y. Songlie, Z. Xunbo, and D. Baodi, "Effects of planting patterns on biomass accumulation and yield of summer maize," in *Ecosystems and Sustainable Development*, E. Tiezzi, J. C. Marques, C. A. Brebbia, and S. E. Jorgensen, Eds., pp. 437–445, WIT Transactions on Ecology and the Environment, 2007.

[18] X. Zhang, D. Pei, S. Chen, H. Sun, and Y. Yang, "Performance of double-cropped winter wheat-summer maize under minimum irrigation in the North China Plain," *Agronomy Journal*, vol. 98, no. 6, pp. 1620–1626, 2006.

[19] T. R. Sinclair and T. Horie, "Leaf nitrogen, photosynthesis and crop radiation-use efficiency: a review," *Crop Science*, vol. 29, pp. 90–98, 1989.

[20] Q. X. Fang, Y. H. Chen, Q. Q. Li et al., "Effects of soil moisture on radiation utilization during late growth stages and water use efficiency of winter wheat," *Acta Agronomica Sinica*, vol. 32, no. 6, pp. 861–866, 2006.

[21] D. Plénet, A. Mollier, and S. Pellerin, "Growth analysis of maize field crops under phosphorus deficiency. II. Radiation-use efficiency, biomass accumulation and yield components," *Plant and Soil*, vol. 224, no. 2, pp. 259–272, 2000.

14

Crop Row Detection in Maize Fields Inspired on the Human Visual Perception

J. Romeo,[1] G. Pajares,[1] M. Montalvo,[2] J. M. Guerrero,[1] M. Guijarro,[1] and A. Ribeiro[3]

[1] *Department of Software Engineering and Artificial Intelligence, Faculty of Informatics, University Complutense, 28040 Madrid, Spain*
[2] *Department of Computer Architecture and Automatic, Faculty of Informatics, University Complutense, 28040 Madrid, Spain*
[3] *Artificial Perception Group, Center for Automation and Robotics (CAR), CSIC-UPM, 28500, Arganda del Rey, Madrid, Spain*

Correspondence should be addressed to G. Pajares, pajares@fdi.ucm.es

Academic Editors: C. Dell and A. Garcia y Garcia

This paper proposes a new method, oriented to image real-time processing, for identifying crop rows in maize fields in the images. The vision system is designed to be installed onboard a mobile agricultural vehicle, that is, submitted to gyros, vibrations, and undesired movements. The images are captured under image perspective, being affected by the above undesired effects. The image processing consists of two main processes: image segmentation and crop row detection. The first one applies a threshold to separate green plants or pixels (crops and weeds) from the rest (soil, stones, and others). It is based on a fuzzy clustering process, which allows obtaining the threshold to be applied during the normal operation process. The crop row detection applies a method based on image perspective projection that searches for maximum accumulation of segmented green pixels along straight alignments. They determine the expected crop lines in the images. The method is robust enough to work under the above-mentioned undesired effects. It is favorably compared against the well-tested Hough transformation for line detection.

1. Introduction

1.1. Problem Statement. The increasing development of robotics equipped with machine vision sensors applied to Precision Agriculture is demanding solutions for several problems. The robot navigates and acts over a site-specific area of a larger farm [1], where one important part of the information is supplied by the vision system.

An important issue related with the application of machine vision methods is that concerning the crop row and weed detection, which has attracted numerous studies in this area [2–6]. This will allow site-specific treatments trying to eliminate weeds and to favor the growth of crops.

The robot navigates on a real terrain presenting irregularities and roughness. This produces vibrations and also swinging in the pitch, yaw, and roll angles. Moreover, the spacing of crop rows is also known. Because of the above, the crop rows are not projected on the expected locations in the image. On the other hand the discrimination of crops and weeds in the image is a very difficult task because their red, green, and blue spectral components display similar values. This means that no distinction is possible between crops and weeds based on the spectral components. Thus the problem is to locate the crop rows in the image. To achieve this goal, in this paper we propose a new strategy that exploits the specific arrangement of crops (maize) in the field and also applies the knowledge of perspective projection based on the camera intrinsic and extrinsic parameters. This method is inspired by the human visual perception and like humans applies a similar reasoning for locating crop rows in the images, although it exploits the camera system geometry, because it is available. As we will see in the next section, the crop row location is not new and has been considered already in the literature; the method proposed in this paper gains advantage over existing approaches because it has been designed to achieve high effectiveness in real-time applications. This makes the main contribution of this paper. The method does not include a segmentation step, which is found in most other methods for plant detection. This tries to avoid time consumption as compared to other strategies.

The segmentation step has been replaced by a simple thresholding method, where the threshold is previously established by applying a learning-based fuzzy clustering strategy. Cluster centers for green textures are obtained during an off-line learning phase, and then this knowledge is exploited during the online decision phase. Crop row detection is an easy step where simple straight lines are traced, based on perspective projection, looking for specific pixels alignments defining crop rows.

Moreover it applies downsampling for reducing image sizes. All above steps are oriented to gain time reduction during the computational process. The proposed approach is favorably compared against some existing approaches in both effectiveness and time reducing.

1.2. Revision of Methods. Several strategies have been proposed for crop row detection. Fontaine and Crowe [7] tested the abilities of fourth-line detection algorithms to determine the position and the angle of the camera with respect to a set of artificial rows with and without simulated weeds. These were stripe analysis, blob analysis, linear regression, and Hough transform.

(1) Methods Based on the Exploration of Horizontal Strips. Søgaard and Olsen [8] apply RGB color image transformation to grayscale. This is done by first dividing the color image into its red, green, and blue spectral channels and then by applying the well-tested methods to extract living plant tissue described in [9]. After this, the greyscale image is divided into horizontal strips where maximum grey values indicate the presence of a candidate row, each maximum determines a row segment, and the center of gravity of the segment is marked at this strip position. Crop rows are identified by joining marked points through a similar method to the one utilized in the Hough transform or by applying linear regression. Sainz-Costa et al. [6] have developed a strategy based on analysis of video sequences for identifying crop rows. Crop rows persist along the directions defined by the perspective projection with respect the 3D scene in the field. Exploiting this fact, they apply greyscale transformation, and then the image is binarized by thresholding. Each image is divided into four horizontal strips. Rectangular patches are drawn over the binary image to identify patches of crops and rows. The gravity centers of these patches are used as the points defining the crop rows, and a line is adjusted considering these points. The first frame in the sequence is used as a lookup table that guides the full process for determining positions where the next patches in subsequent frames are to be identified. Hague et al. [10] transform the original RGB image to gray scale. The transformed image is then divided into eight horizontal bands. The intensity of the pixels across these bands exhibits a periodic variation, due to the parallel crop rows. Since the camera characteristics, pose and the crop row spacing are known a priori, the row spacing in image pixels can be calculated for each of the horizontal bands using a pinhole model of the camera optics. A bandpass filter can then be constructed which will enhance this pattern and has a given frequency domain response. Sometimes horizontal patterns are difficult to extract because crops and weeds form a unique patch.

(2) Methods Based on the Hough Transformation. According to Slaughter et al. [11], one of the most commonly used machine vision methods for identifying crop rows is based upon the Hough [12] transform. It was intended to deal with discontinues lines, where the crop stand is incomplete with gaps in crop rows due to poor germination or other factors that result in missing crop plants in the row. It has been intended for real-time automatic guidance of agricultural vehicles [13–16]. It is applied to binary images, which are obtained by applying similar technique to the ones explained above, that is, RGB image transformation to grayscale and binzarization [3, 4, 17]. Gée et al. [18] apply a double Hough transform under the assumption that crop rows are the only lines of the image converging to the vanishing point, the remainder lines are rejected, additional constraints such as interrow spacing and perspective geometry concepts help to identify the crop rows. It is required to determine the threshold required by the Hough transform to determine maximum peaks values [19, 20] or predominant peaks [21]. Depending on the crop densities, several lines could be feasible, and a posterior merging process is applied to lines with similar parameters [3, 4, 17]. Ji and Qi [22] report that Hough transform is slow due to the huge computation; they propose a randomized Hough transform to reduce computational time. Some modifications have been proposed to improve the Hough transformation such as the one proposed in Asif et al. [23], which apply the Hough only to those points which are edge points along the crops. But this requires the application of techniques for edge extraction. Also the randomized Hough transformation has been proposed with this goal [22]. It is intended to avoid redundant computations in the Hough transform. It operates iteratively by randomly sampling a set of points to compute a single localization in the Hough space. Since two pixels are trivially collinear, the parameters of the line on which they lie can be estimated. These parameters are used to increment the accumulator cell in the Hough space. In summary, the Hough transform is computationally expensive and the randomized Hough transform requires selecting pairs of points to be considered as a unique line, that is, pairs of points belonging to a crop row. If we apply this technique in images where edge points have been extracted, the selection of those pairs becomes more complex. Furthermore, the computational cost of Hough-based algorithms is very sensitive to the image resolution after down-sampling, but also when weeds are present and irregularly distributed, this our case, this is could cause the failure detection. Moreover, as the weed density increases the crop row detection becomes more and more difficult.

(3) Vanishing Point Based. Pla et al. [24] propose an approach which identifies regions (crops/weeds and soil) by applying color image segmentation. They use the skeleton of each defined region as a feature to work out the lines which

define the crop. The resulting skeletons of each region can be used as curves which define the underlying structure of the crop and to extract the straight lines where the plants, and soil rows lie. Segments in the skeletons are defined as chains of connected contour points and they must be of a defined length. This allows selecting candidate lines for crop row detection, among all candidates the ones that meet the vanishing point. The vanishing point is detected using previous information about the vanishing point found in the previous images, performing a sort tracking on the vanishing point. This process is highly dependent on the skeletons, which are not always easy to extract and isolate, particularly considering that crops and weeds patches appear overlapped among them.

(4) Methods Based on Linear Regression. Some of the techniques above apply this approach. Billingsley and Schoenfisch [25] reported a crop detection system that is relatively insensitive to additional visual "noise" from weeds. They used linear regression in each of three crop row segments and a cost function analogous to the moment of the best-fit line to detect lines fitted to outliers (i.e., noise and weeds) as a means of identifying row guidance information. As mentioned above, Søgaard and Olsen [8] also apply linear regression, which is a feasible approach if weed density is low and pixels belonging to crop rows are well separated. Otherwise it is highly affected by pixels belonging to weeds because of their strong contribution to line estimation.

(5) Stereo-Based Approach. Kise et al. [26] and Kise and Zhang [27] developed a stereovision-based agricultural machinery crop row tracking navigation system. Stereoimage processing is used to determine 3D locations of the scene points of the objects of interest from the obtained stereoimage. Those 3D positions, determined by means of stereoimage disparity computation, provide the base information to create an elevation map which uses a 2D array with varying intensity to indicate the height of the crop. This approach requires crops with significant heights with respect the ground. Because in maize fields, during the treatment stage, the heights are not relevant, it becomes ineffective in our application. Rovira-Más et al. [28] have applied and extended stereovision techniques to other areas inside Precision Agriculture. Stereo-based methods are only feasible if crops or weeds in the 3D scene display a relevant height and the heights differ in both kind of plants.

(6) Methods Based on Blob Analysis. This method finds and characterizes regions of contiguous pixels of the same value in a binarized image [7]. The algorithm searches for white blobs (interrow spaces) of more than 200 pixels, as smaller blobs could represent noise in the crop rows. Once the blobs are identified, the algorithm determines the angle of their principal axes and the location of their center of gravity. For a perfectly straight white stripe, the center of gravity of the blob was over the centerline of the white stripe, and the angle was representative of the angle of the interrow spaces. The algorithm returned the angle and center of gravity of the blob

closest to the centre of the image. Identification of blobs in images infested with weeds in maize fields becomes a very difficult task, because weeds and crops under overlapping in localized areas produce wide blobs.

(7) Methods Based on the Accumulation of Green Plants. Olsen [29] proposed a method based on the consideration that along the crop row an important accumulation of green parts in the image appears. The image is transformed to gray scale, where green parts appear clearer that the rest. A sum curve of gray levels is obtained for a given rectangular region exploring all columns in the rectangle. It is assumed that vertical lines follow this direction in the image. Images are free of perspective projection because they are acquired with the camera in orthogonal position. A sinusoidal curve is fitted by means of least squares to the sum curve previously obtained. Local maxima of the sinusoid provide row centers locations. This is a simple and suitable method, which can be still simplified but it is not of our interest because of the fact that the images we work with are taken from the tractor under perspective projection but not orthogonal. In this paper we exploit the idea of green plant accumulation under a simpler strategy.

(8) Methods Based on Frequency Analysis. Because crop rows are vertical in the 3D scene, they are mapped under perspective projection onto the image displaying some behavior in frequency domain. Vioix et al. [30] exploit this feature and apply a bidimensional Gabor filter, defined as a modulation of a Gaussian function by a cosine signal. The frequency parameter required by the Gabor filter is empirically deduced from the 2D-Fast Fourier Transform [31]. Bossu et al. [32] apply wavelets to discriminate crop rows based on the frequency analysis. They exploit the fact that crop rows are well localized in the frequency domain; thus selecting a mother wavelet function with this frequency the crop rows can be extracted. In maize fields where the experiments are carried out, crops do not display a clear frequency content in the Fourier space, therefore the application of filters based on the frequency becomes a difficult task.

1.3. Motivational Research and Design of the Proposed Strategy. Our work is focused on crop row detection in maize for specific treatments requiring discrimination among crops and weeds. This means that crop rows must be identified and located in the image conveniently. Some of the requirements proposed by Astrand [33] and reported in [11] for guidance systems can be considered for crop row detection; the problem is essentially similar. Therefore, our system is designed to be

   (i) able to locate crop rows with the maximum accuracy as possible,

   (ii) able to work on real-time,

   (iii) able to work on sown crops, not manually planted, which means that weeds and crops grow simultaneously displaying, at the early growth stage of the treatment, similar heights and also similar spectral signatures. This means that discrimination between crops and weed cannot be made by height or spectral signatures only,

(iv) able to work when plants are missing in the row,

(v) able to work when there is high weed pressure,

(vi) able to work under different weather (luminance) conditions,

(vii) able to locate crop rows with the least assumptions and constraints.

The aim of this study is to present a general method for identifying crop rows in maize fields from the images. We exploit the advantages of the existing methods introduced above, extracting the main ideas, and design a new strategy for crop row detection inspired on the human visual perception abilities which is able to work in real time.

This method is also dedicated to be applied in maize with crop row spacing and also to deal with and without seedling spacing. It is summarized in the two main steps as follows.

(a) Segmentation of green plants (crop and weeds).

(b) Crop row identification.

2. Materials and Methods

2.1. Images. The images used for this study correspond to maize crops. They were captured with a Canon EOS 400D camera during April/May 2011 in a 1.7 ha experimental field of maize in La Poveda Research Station, Arganda del Rey, Madrid. All acquisitions were spaced by five/six days, that is, they were obtained under different conditions of illumination and different growth stages. The digital images were captured under perspective projection and stored as 24-bit color images with resolutions of 5 MP saved in RGB (red, green, and blue) color space in the JPG format. The images were processed with MATLAB R2009a [34] under Windows 7 and Intel Core 2 Duo CPU, 2.4 GHz, 2.87 GB RAM. A set of 350 images were available for processing.

With the aim of testing the robustness and performance of the proposed approach, we have worked with images captured under different conditions, including different number of crop rows; these conditions have been identified in the real fields as possible and also those that could cause problems during the detection process in normal operation. The following is a list of representative images from the set of available images, illustrating some of such conditions:

(a) different brightness due to different weather conditions, Figures 1(a) and 1(b);

(b) different growth stages, Figures 2(a) and 2(b);

(c) different camera orientations, that is, different yaw, pitch and roll angles, and heights from the ground, Figures 3(a) and 3(b);

(d) different weed densities, Figures 4(a) and 4(b).

2.2. Image Segmentation: Green Plants Identification. For real-time applications is of great relevance to simplify this process as much as possible. Instead of using vegetation indices [9, 35], which require an image transformation from RGB color space to gray scale, we used a learning-based approach with the goal of obtaining the percentage of the green spectral component with respect to the remainder, which allowed us to consider a pixel belonging to a green plant. This relative percentage is intended to deal with illumination variability so that it determines relative values among the three spectral RGB components that identify green plants. This is carried out by applying a *fuzzy clustering* approach. Under this approach there is a learning phase which is applied to during offline activity for computing the relative percentage or threshold and a decision phase where the threshold is applied without additional computation.

The learning phase was designed as follows. From the set of available images we randomly extracted n training samples, stored in X, that is, $X = \{\mathbf{x}_1, \mathbf{x}_2, \dots, \mathbf{x}_n\} \in \mathfrak{R}^d$, where d is the data dimensionality. Each sample vector $\mathbf{x}_i$ represents an image pixel, where its components are the three RGB spectral components of that pixel at the original image location (x, y). This means that in our experiments the data dimensionality is $d = 3$. Each sample is to be assigned to a given cluster w_j, where the number of possible clusters is c, that is, $j = 1, 2, \dots, c$. In the proposed approach c is set to 2 because we were only interested in two types of textures, that is, green plants (crop/weeds) and the remainder (soil, debris, stones).

The samples in X are to be classified based on the well-known fuzzy clustering approach that receives the input training samples $\mathbf{x}_i$ and establishes a partition, assuming the number of clusters c is known. The process computes for each $\mathbf{x}_i$ at the iteration t, its degree of membership in the cluster $w_j(\mu_i^j)$ and updates the cluster centers $\mathbf{v}_j$ as follows [36]:

$$\mu_i^j(t+1) = \frac{1}{\sum_{r=1}^{c} \left(d_{ij}(t)/d_{ir}(t) \right)^{2/(b-1)}};$$

$$\mathbf{v}_j(t+1) = \frac{\sum_{i=1}^{n} \left[\mu_i^j(t) \right]^b \mathbf{x}_i}{\sum_{i=1}^{n} \left[\mu_i^j(t) \right]^b}. \tag{1}$$

$d_{ij}^2 \equiv d^2(\mathbf{x}_i, \mathbf{v}_j)$ is the squared Euclidean distance. The number b is called the exponential weight [37, 38], $b > 1$. The stopping criterion of the iteration process is achieved when $\|\mu_i^j(t+1) - \mu_i^j(t)\| < \varepsilon$ for all ij or a number $t_{\max}$ of iterations is reached.

The method requires the initialization of the cluster centers, so that (1) can be applied at the iteration $t = 1$. For this purpose, we applied the pseudorandom procedure described in Balasko et al. [39].

(1) Perform a linear transform $Y = f(X)$ of the training sample values so that they range in the interval $[0, 1]$.

(2) Initialize $\mathbf{v} = 2D\overline{\mathbf{M}} \circ \mathbf{R} + D\overline{\mathbf{m}}$, where $\overline{\mathbf{m}}$ is the mean vector for the transformed training samples values in

Figure 1: Different brightness due to different weather conditions: (a) darker; (b) clearer.

Figure 2: Different crop growth stages: (a) low; (b) high.

Figure 3: Different yaw, pitch and roll angles, and heights from the ground.

FIGURE 4: Different weed densities: (a) low; (b) high.

Y and $\overline{\mathbf{M}} = \max(\mathrm{abs}(Y - \overline{\mathbf{m}}))$, both of size $1 \times d$; $D = [1 \ \dots \ 1]^T$ with size $c \times 1$; $\mathbf{R}$ is a $c \times d$ matrix of random numbers in $[0, 1]$; the operation $\circ$ denotes the element by element multiplication.

Once the learning process is finished we obtain two cluster centers $\mathbf{v}_1$ and $\mathbf{v}_2$ associated to clusters w_1 and w_2. Without loss of generality, let $\mathbf{v}_1 \equiv \{v_{1R}, v_{1G}, v_{1B}\}$ the one associated to the green plants. It is a 3-dimensional vector where its components v_{1R}, v_{1G}, and v_{1B} represent the averaged values for the corresponding RGB spectral components; thus the threshold value for discriminating among green plants and the remainder ones is finally set to $T_G = v_{1G}/(v_{1R} + v_{1G} + v_{1B})$.

Once T_G is available, the green parts on the images are identified assuming the corresponding RGB pixels contain the G spectral value greater than T_G. Therefore, during the online identification process only is required the logical comparison.

2.3. Crop Row Identification.

Once green parts were extracted in the image, next step was crop row identification. For such purpose we make use of the following constraints, based on the image perspective projection and the general knowledge about the maize field.

(i) The number of crop lines (L) to be detected is known and also the approximate x position or image column at the bottom of the image where every crop line starts. This assumption is based on the system geometry and the image perspective projection.

(ii) We are going to detect crop lines that start from the bottom of the image and end at the top of the image. Lines starting from both left and right sides of the image and vanishing at the top are rejected. This is because image geometry allows to consider this situation. An extension of this algorithm could be done to detect those lines with its corresponding computing time cost.

(iii) All the images have been acquired with a camera onboard a tractor and pointing in the same direction as the crop lines, therefore images are mapped under in perspective projection and the crop lines converge in the well-known vanishing point. This constraint is inspired by methods based on the vanishing point, as described in the introduction. Though crop lines are parallel, distance between crop lines seems to be greater at the bottom of the image than at the top, due to perspective. This algorithm works considering that crop lines are going to have that appearance in the image with a range of tolerance that can be adjusted depending on the stability of the tractor and the evenness of the ground. We assume that crop lines starting from the left bottom of the image have a clockwise slope and lines starting from the right bottom of the image have an anticlockwise slope. The bigger the range of tolerance the higher the computing time. In this paper we have used a 15% of tolerance which means that crop lines may vary from one image to the next one a 15% of the width of the image.

The algorithm works as follows.

(1) From every pixel in the bottom row we trace all the possible lines starting on that pixel and ending on every pixel of the top line, that is, if the image has N-columns, we will trace N^2 lines from every pixel of the bottom row, which means that we finally trace N^2 lines. This number of calculated lines is the highest number of lines in case we make no constraints. Nevertheless, as we will see in step 4 and 5 important constrains can be applied to reduce this number. Figure 5 shows the bean of lines traced for two pixels placed at the bottom row of the image. For illustrating purposes we have traced broad beams of lines, but the number of lines to cope with all possible situations, but this number could be considerably reduced by applying previous knowledge, like the slope. This is applied in this work as described below, reducing the computational cost.

(2) For every traced line starting on a pixel of the bottom row, we count the number of "green" pixels that belong to that line. This is possible because the image

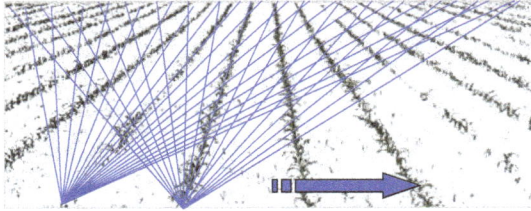

FIGURE 5: Lines traced from every pixel of the bottom row in the image.

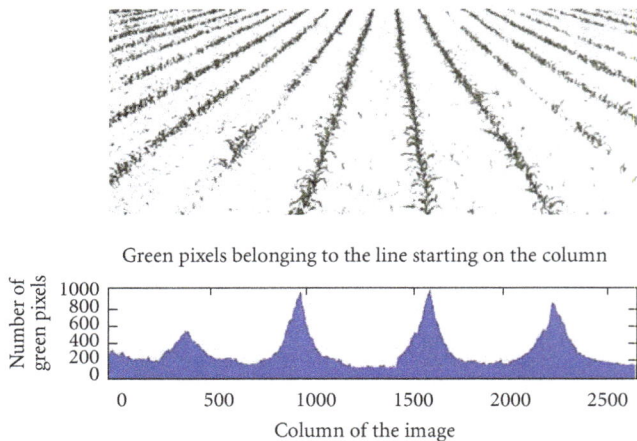

FIGURE 6: Number of green pixels found for the best line of every pixel of the bottom line.

has been already segmented and pixels belonging to green parts have been identified. Now, the line with the highest number of green pixels is the candidate line to represent the crop row for that pixel of the bottom row.

(3) We repeat the same procedure for all the pixels of the bottom row, and finally we obtain c-candidates lines, that is, one for every pixel of the bottom row. As we can see in Figure 6, every pixel location in the bottom row of the image has a value for its best line. Those values become higher as the represented line approaches the real crop line. They are the peaks in the lower part in Figure 6.

(4) Since we know the number of crop lines to be detected and also where they roughly start at the bottom row, we can choose the closer and highest values, which are identified by peaks in the accumulator. With such a figure, assume the number of crop rows to be detected is four; so we look for four peaks that are conveniently spaced because of the crop rows arrangement in maize fields and also based on camera system geometry. This idea is inspired in methods based on the accumulation of green plants, described in the introduction.

The following are three considerations that can be applied to speed up the computational process from the point of view of a real-time application.

(1) For each selected line we store the start and end points, obtaining the corresponding equation for the straight line.

(2) Because of the perspective of the image it is not necessary to trace all the lines to the top row (as mentioned in step 1) but only those whose slope is according to what we expect. That is, if we are dealing with left pixels of the image we would trace only lines with a slope clockwise and without reaching the end of the right side of the image. For right pixels we would search for anticlockwise slopes starting from the right side and without reaching the left side of the image. This idea is based on the vanishing point concept, applied in some approaches as described in the introduction.

(3) In addition to it, it is not necessary to trace lines pixel by pixel. Depending on the image resolution a "pixel step" can be used without affecting final result and reducing considerably the computational cost.

(4) Notice that there are some values to be adjusted before the algorithm runs. These values depend on the images we are dealing with and on the stability of the camera. The higher the image resolution the higher the "pixel step" for lines calculation. Furthermore, the higher the stability of the tractor the thinner the range of pixels of the top row for tracing lines.

3. Results

Our proposed crop row detection (CRD) method consists on a first stage or learning phase where the threshold T_G is obtained for segmenting green plants. With such purpose we have processed 200 images, selected from the set of 350 images available, from which we have randomly extracted 40.000 training samples. The selected images cover the broad range of situations, that is, different number of rows, weather conditions, weed concentrations and growth states according to Figures 1 to 4.

With these training samples, we apply the fuzzy clustering procedure described in Section 2.2, from which we compute two cluster centers identifying both green plants (v_1) and soil or other components (v_2). Our interest is only focused on segmenting green plants, therefore, from v_1 we compute the threshold T_G defined in Section 2.2 as the percentage of the green component in v_1, that is, $T_G = 0.37$. This is the threshold finally used for image segmentation.

Table 1 displays both cluster centers v_1 and v_2 and the percentage of the highest value in the spectral components associated to each cluster center. As we can see green and soil pixels can be identified by the corresponding percentages, each one applied over the green and red spectral components. This was the general behavior observed for the set of images analyzed.

As mentioned during the introduction, the Hough transform has been applied in several methods for crop row detection, hence we compare the performance of our CRD approach against the Hough (HOU) transform. We have

TABLE 1: Percentage of the green spectral component for green plants and for other components (soil, debris, stones).

	Spectral component values	Percentage of the highest spectral component
$\mathbf{v}_1$ (green plants)	{137.80 140.68 106.07}	0.37 (green)
$\mathbf{v}_2$ (soil and other components)	{188.49 177.71 153.53}	0.36 (red)

TABLE 2: Performances of HOU and CRD approaches measured in terms of percentage of effectiveness and processing times.

Image resolution (pixels)	Percentage of effectiveness		Processing time (seconds)	
	HOU	CRD	HOU	CRD
162×216	86.3	97.1	1,088	0,580
194×259	89.4	97.3	1,305	0,737
243×324	89.1	97.3	2,120	0,928
324×432	90.9	97.4	4,752	1,667
486×648	91.1	97.5	8,153	3,216

applied identical conditions to the Hough transform than the ones applied in our CRD approach, so it works in terms of comparability; they are synthesized as follows.

(a) Search for lines arising from the bottom of the image and ending at the top, that is, suspicious useless lines are not explored.

(b) Only are allowed lines with slopes close to the ones expected at each side of the image. Horizontal lines and many others that do not meet the above are rejected.

(c) The Hough transform is implemented to work under the normal representation, polar coordinates [40], with unit increments in the parameter representing the angle.

The comparison is established in terms of effectiveness and processing times. The effectiveness is measured based on the expert human criterion, where a line, which has been detected, is considered as correct if it overlaps with the real crop row alignment. Over the set of 350 images analyzed, we compute the average percentage of coincidences for both CRD and HOU. Also, because the main goal of the proposed approach is its profit for real time applications, we measure computational times. Also, with the goal of real-time, we have tested these performances for different image resolutions. As we can see image resolutions differ from the ones in the original images, these resolutions have been obtained by applying a down-sampling process to the original image.

This is intended under the idea that it is possible to reduce the image dimension retaining the main information without affecting the effectiveness and reducing the processing time. Table 2 displays the results. The first column contain different image resolutions, which are obtained by selecting large regions of interest in each image, with horizontal and vertical sizes of 1940×2590 pixels, and these regions contains different number of crop rows under different configurations provided by the images displayed

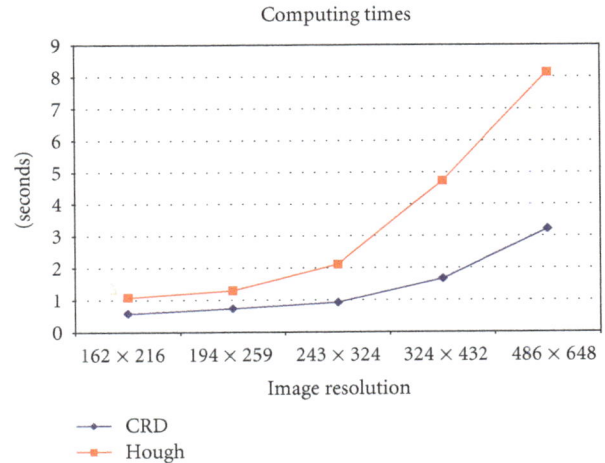

FIGURE 7: Times in seconds against the different image resolutions.

in Figures 1 to 4. These large regions are split by 10, 8, 6, and 4, which are, respectively the ones represented in Table 2. We have chosen this set of values because with them we obtain similar performances in terms of effectiveness with acceptable processing times. The effectiveness for higher resolutions is similar, but the processing times increase considerably. Below the lower resolution, the effectiveness decreases considerably.

The second and third columns contain the percentage of effectiveness and columns fourth and fifth the processing times measured in seconds. All these measurements represent average values over the set of 350 images processed. For clarity, Figure 7 represents the processing times in Table 2, for the four resolutions studied.

From results in Table 2 we can infer that CRD outperforms HOU in terms of effectiveness, with a near constant value regardless the image resolution. With lower resolutions, that is, with image divisions above 12, this percentage decreases drastically, achieving values below 85%. This is because for low resolutions some important information in the images is lost. Thus, from values in Table 2 and because the processing time is lower with small image resolutions, from a real-time point of view, a suitable resolution with acceptable performance is the lowest, that is, the one for 162×216. From Figure 7, one can see that the increasing of time is not linear. For resolutions above 243×324 time differences are more pronounced.

The worst performance obtained for HOU can be explained because crops and weeds concentration produces a high density of values, representing peaks, in the cell accumulator. These values do not display a high clear value, theoretically representing a unique crop line. Moreover, the absolute maximum value around the expected crop line most times does not represent the correct line. Thus, it is necessary to define a patch selecting different high peak values for each expected line, which are averaged, to obtain the final value. Because this patch has not clear limits, its selection becomes a difficult task and errors in the selection produce errors in the crop row localization, which explain the worst performance of HOU against CRD.

4. Conclusions

We developed a new method for crop row detection that improves Hough-based methods in terms of effectiveness and computing time. The goal is its application to real-time implementations.

Furthermore, our approach has been proved to be robust enough to different images typologies.

The proposed method is robust enough to work in images under perspective projection. It can detect any number of crop lines with any slope converging in a vanishing point. It works with either high or low image resolutions.

Future works must be oriented toward weeds detection by establishing cells around crop lines and calculating the percentage of greenness of every cell. This should be intended for posterior actuations to kill weeds.

Acknowledgments

The research leading to these results has been funded by the European Union's Seventh Framework Programme [FP7/2007–2013] under Grant Agreement no. 245986 in the Theme NMP-2009-3.4-1 (Automation and robotics for sustainable crop and forestry management). The authors wish also to acknowledge to the Project AGL-2008-04670-C03-02, supported by the Ministerio de Ciencia e Innovación of Spain within the Plan Nacional de I+D+i.

References

[1] G. Davies, W. Casady, and R. Massey, "Precision agriculture: an introduction. Water Quality Focus Guide. WQ450," http://extension.missouri.edu/explorepdf/envqual/wq0450.pdf.

[2] C. M. Onyango and J. A. Marchant, "Segmentation of row crop plants from weeds using colour and morphology," *Computers and Electronics in Agriculture*, vol. 39, no. 3, pp. 141–155, 2003.

[3] A. Tellaeche, X. BurgosArtizzu, G. Pajares, A. Ribeiro, and C. Fernández-Quintanilla, "A new vision-based approach to differential spraying in precision agriculture," *Computers and Electronics in Agriculture*, vol. 60, no. 2, pp. 144–155, 2008.

[4] A. Tellaeche, X. P. Burgos-Artizzu, G. Pajares, and A. Ribeiro, "A vision-based method for weeds identification through the Bayesian decision theory," *Pattern Recognition*, vol. 41, no. 2, pp. 521–530, 2008.

[5] X. P. Burgos-Artizzu, A. Ribeiro, A. Tellaeche, G. Pajares, and C. Fernández-Quintanilla, "Improving weed pressure assessment using digital images from an experience-based reasoning approach," *Computers and Electronics in Agriculture*, vol. 65, no. 2, pp. 176–185, 2009.

[6] N. Sainz-Costa, A. Ribeiro, X. Burgos-Artizzu, M. Guijarro, and G. Pajares, "Mapping wide row crops with video sequences acquired from a tractor moving at treatment speed," *Sensors*, vol. 11, no. 7, pp. 7095–7109, 2011.

[7] V. Fontaine and T. G. Crowe, "Development of line-detection algorithms for local positioning in densely seeded crops," *Canadian Biosystems Engineering*, vol. 48, pp. 7.19–7.29, 2006.

[8] H. T. Søgaard and H. J. Olsen, "Determination of crop rows by image analysis without segmentation," *Computers and Electronics in Agriculture*, vol. 38, no. 2, pp. 141–158, 2003.

[9] D. M. Woebbecke, G. E. Meyer, K. von Bargen, and D. A. Mortensen, "Shape features for identifying young weeds using image analysis," *Transactions of the American Society of Agricultural Engineers*, vol. 38, no. 1, pp. 271–281, 1995.

[10] T. Hague, N. Tillett, and H. Wheeler, "Automated crop and weed monitoring in widely spaced cereals," *Precision Agriculture*, vol. 1, no. 1, pp. 95–113, 2006.

[11] D. C. Slaughter, D. K. Giles, and D. Downey, "Autonomous robotic weed control systems: a review," *Computers and Electronics in Agriculture*, vol. 61, no. 1, pp. 63–78, 2008.

[12] P. V. C. Hough, "A method and means for recognizing complex patterns," U.S. Patent Office No. 3069654, 1962.

[13] J. Marchant, "Tracking of row structure in three crops using image analysis," *Computers and Electronics in Agriculture*, vol. 15, no. 2, pp. 161–179, 1996.

[14] T. Hague, J. A. Marchant, and D. Tillett, "A system for plant scale husbandry," *Precision Agriculture*, pp. 635–642, 1997.

[15] B. Åstrand and A. J. Baerveldt, "A vision based row-following system for agricultural field machinery," *Mechatronics*, vol. 15, no. 2, pp. 251–269, 2005.

[16] V. Leemans and M. F. Destain, "Application of the Hough transform for seed row localisation using machine vision," *Biosystems Engineering*, vol. 94, no. 3, pp. 325–336, 2006.

[17] A. Tellaeche, G. Pajares, X. P. Burgos-Artizzu, and A. Ribeiro, "A computer vision approach for weeds identification through Support Vector Machines," *Applied Soft Computing Journal*, vol. 11, no. 1, pp. 908–915, 2011.

[18] Ch. Gée, J. Bossu, G. Jones, and F. Truchetet, "Crop/weed discrimination in perspective agronomic images," *Computers and Electronics in Agriculture*, vol. 60, no. 1, pp. 49–59, 2008.

[19] G. Jones, Ch. Gée, and F. Truchetet, "Modelling agronomic images for weed detection and comparison of crop/weed discrimination algorithm performance," *Precision Agriculture*, vol. 10, no. 1, pp. 1–15, 2009.

[20] G. Jones, Ch. Gée, and F. Truchetet, "Assessment of an inter-row weed infestation rate on simulated agronomic images," *Computers and Electronics in Agriculture*, vol. 67, no. 1-2, pp. 43–50, 2009.

[21] F. Rovira-Más, Q. Zhang, J. F. Reid, and J. D. Will, "Hough-transform-based vision algorithm for crop row detection

of an automated agricultural vehicle," *Journal Automobile Engineering, Part D*, vol. 219, no. 8, pp. 999–1010, 2005.

[22] R. Ji and L. Qi, "Crop-row detection algorithm based on Random Hough Transformation," *Mathematical and Computer Modelling*, vol. 54, no. 3-4, pp. 1016–1020, 2011.

[23] M. Asif, S. Amir, A. Israr, and M. Faraz, "A vision system for autonomous weed detection robot," *International Journal of Computer and Electrical Engineering*, vol. 2, no. 3, pp. 486–491, 2010.

[24] F. Pla, J. M. Sanchiz, J. A. Marchant, and R. Brivot, "Building perspective models to guide a row crop navigation vehicle," *Image and Vision Computing*, vol. 15, no. 6, pp. 465–473, 1997.

[25] J. Billingsley and M. Schoenfisch, "The successful development of a vision guidance system for agriculture," *Computers and Electronics in Agriculture*, vol. 16, no. 2, pp. 147–163, 1997.

[26] M. Kise, Q. Zhang, and F. Rovira Más, "A stereovision-based crop row detection method for tractor-automated guidance," *Biosystems Engineering*, vol. 90, no. 4, pp. 357–367, 2005.

[27] M. Kise and Q. Zhang, "Development of a stereovision sensing system for 3D crop row structure mapping and tractor guidance," *Biosystems Engineering*, vol. 101, no. 2, pp. 191–198, 2008.

[28] F. Rovira-Más, Q. Zhang, and J. F. Reid, "Stereo vision three-dimensional terrain maps for precision agriculture," *Computers and Electronics in Agriculture*, vol. 60, no. 2, pp. 133–143, 2008.

[29] H. J. Olsen, "Determination of row position in small-grain crops by analysis of video images," *Computers and Electronics in Agriculture*, vol. 12, no. 2, pp. 147–162, 1995.

[30] J.-B. Vioix, J.-P. Douzals, F. Truchetet, L. Assémat, and J.-P. Guillemin, "Spatial and spectral method for weeds detection and localization," *Eurasip Journal on Advances in Signal Processing*, vol. 7, pp. 679–685, 2004.

[31] J. Bossu, Ch. Gée, J. P. Guillemin, and F. Truchetet, "Development of methods based on double Hough transform and Gabor filtering to discriminate crop and weeds in agronomic images," in *Proceedings of the SPIE 18th Annual Symposium Electronic Imaging Science and Technology*, vol. 6070, no. 23, San Jose, Calif, USA, 2006.

[32] J. Bossu, Ch. Gée, G. Jones, and F. Truchetet, "Wavelet transform to discriminate between crop and weed in perspective agronomic images," *Computers and Electronics in Agriculture*, vol. 65, no. 1, pp. 133–143, 2009.

[33] B. Astrand, *Vision based perception or mechatronic weed control*, Ph.D. thesis, Chalmers and Halmstad Universities, Sweden, Stockholm, 2005.

[34] TheMatworks, http://www.mathworks.com/, 2011.

[35] A. Ribeiro, C. Fernández-Quintanilla, J. Barroso, and M. C. García-Alegre, "Development of an image analysis system for estimation of weed," in *Proceedings of the 5th European Conference on Precision Agriculture (ECPA '05)*, vol. 1, no. 1, pp. 169–174, 2005.

[36] H. J. Zimmermann, *Fuzzy Set Theory and its Applications*, Kluwer Academic, Norwell, Mass, USA, 1991.

[37] J. C. Bezdek, *Pattern Recognition with Fuzzy Objective Function Algorithms*, Plenum Press, New York, NY, USA, 1981.

[38] R. O. Duda, P. E. Hart, and D. S. Stork, *Pattern Classification*, John Wiley & Sons, New York, NY, USA, 2000.

[39] B. Balasko, J. Abonyi, and B. Feil, *Fuzzy Clustering and Data Analysis Toolbox for Use with Matlab*, Veszprem University, Hungary, Budapest, 2008.

[40] R. C. Gonzalez, R. E. Woods, and S. L. Eddins, *Digital Image Processing Using MATLAB*, Prentice Hall, Upper Saddle River, NJ, USA, 2009.

15

Storage Insects on Yam Chips and Their Traditional Management in Northern Benin

Y. L. Loko,[1] A. Dansi,[1,2] M. Tamo,[3] A. H. Bokonon-Ganta,[4] P. Assogba,[2] M. Dansi,[1] R. Vodouhè,[5] A. Akoegninou,[6] and A. Sanni[7]

[1] Laboratory of Agricultural Biodiversity and Tropical Plant breeding (LAAPT), Faculty of Sciences and Technology (FAST-Dassa), University of Abomey-Calavi (UAC), P.O. Box 526, Cotonou, Benin
[2] Crop, Aromatic and Medicinal Plant Biodiversity Research and Development Institute (IRDCAM), 071 BP 28 Cotonou, Benin
[3] Institut International d'Agriculture Tropicale (IITA), 08 BP 0932 Cotonou, Benin
[4] Service de la Protection des Végétaux et du Contrôle Phytosanitaire, Direction de l'Agriculture, BP 58 Porto-Novo, Benin
[5] Bioversity International, Office of West and Central Africa, 08 BP 0932 Cotonou, Benin
[6] National Herbarium, Department of Botany and Plant Biology, Faculty of Sciences and Technology (FAST), University of Abomey-Calavi (UAC), P.O. Box 526, Cotonou, Benin
[7] Laboratory of Biochemistry and Molecular Biology, Faculty of Sciences and Technology (FAST), University of Abomey-Calavi (UAC), P.O. Box 526, Cotonou, Benin

Correspondence should be addressed to A. Dansi; adansi2001@gmail.com

Academic Editors: G. E. Brust, A. Ferrante, and J. R. Qasem

Twenty-five villages of Northern Benin were surveyed to identify the constraints of yam chips production, assess the diversity of storage insects on yam chips, and document farmers' perception of their impacts on the stocks and their traditional management practices. Damages due to storage insects (63.9% of responses) and insufficiency of insect-resistant varieties (16.7% of responses) were the major constraints of yam chips production. Twelve insect pest species were identified among which *Dinoderus porcellus* Lesne (Coleoptera, Bostrichidae) was by far the most important and the most distributed (97.44% of the samples). Three predators (*Teretrius nigrescens* Lewis, *Xylocoris flavipes* Reuter, and *Alloeocranum biannulipes* Montrouzier & Signoret) and one parasitoid (*Dinarmus basalis* Rondani) all Coleoptera, Bostrichidae were also identified. The most important traditional practices used to control or prevent insect attack in yam chips were documented and the producers' preference criteria for yam cultivars used to produce chips were identified and prioritized. To further promote the production of yam chips, diversification of insect-resistant yam varieties, conception, and use of health-protective natural insecticides and popularization of modern storage structures were proposed.

1. Introduction

In economic terms, yams (*Dioscorea* spp.) are the world's fourth most important tuber crop after potatoes, cassava, and sweet potatoes [1]. They are cultivated in most tropical countries but especially in West Africa, where over 95% of the world's output is produced [1, 2]. They are the main source of carbohydrate for millions of people. In West Africa, many yam species are cultivated but the African domesticates known as the Guinea yams (*D. cayenensis* Lam.-*D. rotundata* Poir. complex) are, however, the most important, most preferred, and widely planted [3].

In Benin, the fourth yam-producing country behind Nigeria, Ivory Coast, and Ghana, yam is among the most important food crops [1] and has economic and sociocultural importance [4]. Yam is seasonal and the fresh tubers are highly perishable. Postharvest losses are very high, ranging from 30% to 85% of the total production [4]. In order to overcome this high perishability of the tubers and the irregularity of its availability throughout the year, yams are traditionally processed into dried chips or cossettes [5–7], hence reinforcing food security [7]. Unfortunately, yam chips are often severely attacked by insects, which sometimes reduce whole yam stocks into powder in very few months [8, 9]. Very little

TABLE 1: Administrative localisation of the ethnic areas and sites surveyed.

N°	Ethnic areas	Districts	Number of sites	Selected villages
1	Yom	Djougou	6	Déwa, Alfapara, Pélébina, Mone, Gangamou, Dangoussar
2	Lokpa	Djougou	4	Ouarlgou, Yarakéou, Pohomto, Niagba-kabia
3	Ani	Bassila	2	Penessoulou, Saramanga
4	Nago	Bassila	7	Modogui, Ouanou, Papané, Agramarou, Koko, Agbassa, Wari-Maro
5	Taneka	Copargo	3	Kataban, Setrah, Foungou
6	Bariba	Tchaourou	2	Woria, Badékparou
7	Peulh	Tchaourou	1	Gakpenou

research attention has been given to storage insects attack on yam chips and traditional management practices in Benin. Gnonlonfin et al. [10] reported the existence of many species of insects but focused their study mainly on their population's dynamics in stored yams chips. Consequently, the diversity of the insect species in the yam chips producing zone is still unknown, and farmers' perception of the importance of insect damages in the stocks has never been assessed. Traditional management practices (including the storage structures) used to prevent or control insect infestations have also not been documented. Yam chips are produced from tubers of single-harvest varieties, locally known as "Kokoro" characterized by their numerous small-sized tubers. Within Kokoro yams, many varieties of different agronomic and technological characteristics exist [11] but the yam chips producers' variety preference or selection criteria have never been studied. Knowledge of farmers' selection criteria will be useful in designing concrete breeding programmes that could facilitate the adoption of improved varieties [12].

We report in this paper a survey conducted in the most important yam chips producing zone of Benin in order to

(i) identify and prioritise the constraints related to the production of yam chips in Benin and farmers' propositions for overcoming these constraints,

(ii) document farmers' perceptions about insect pests on stored yam chips and traditional management practices,

(iii) assess the diversity and the importance of the storage insects species in the most important yam chips production zone of Benin,

(iv) Identify and prioritize the producers' variety preference or selection criteria across study zones and ethnic groups for popularization and breeding purposes.

2. Material and Methods

2.1. Study Area. The study was conducted in five districts (Djougou, Copargo, Ouaké, Bassila, and Tchaourou) of the Departments of Donga and Borgou in northern Benin. These districts are known to be the major yam chips production zones of Benin [11, 13]. The inhabitants are members of seven ethnic groups (Figure 1; Table 1) (Ani, Bariba, Lokpa, Nago, Peulh, Taneka, and Yom) and have a very long tradition in processing Kokoro yam tubers into chips or cossettes (Figure 2). The departments of Donga and Borgou

are located in a semiarid agroecological zone characterized by unpredictable and irregular rainfall (800–1300 mm/year) with only one rainy season (May to October) and a dry season lasting for more than 5 months sometimes [14]. Mean annual temperatures range from 26°C to 28°C and may exceptionally reach 35°C [15, 16]. Yam production in this area is intensive and essentially based on Kokoro yams [13], which have very variable yields from one season to another due to climatic hazards [13].

2.2. Site Selection and Survey. Twenty-five villages (Table 1; Figure 1) were randomly selected throughout the study area and its ethnic zones for the survey. Data were collected from the different sites during expeditions through the application of Participatory Research Appraisal tools and techniques such as granary visits, direct observation, focus group discussions, and individual interviews using a questionnaire and the help of translators from each area following Dansi et al. [12]. In each site, local farmers' associations were involved in the study to facilitate the identification of the households to survey and the data collection. Within villages, 10 to 12 households were randomly selected for individual interviews using the transect method described by Dansi et al. [17]. In each household, the interviewee (head of household or his wife or one of his wives in case of polygamy) was selected by mutual agreement with the hosting couple according to Christinck et al. [18]. Apart from the socioeconomic data such as age, gender, and educational level of the interviewees, data collected included the farmers' perceptions of the constraints of yam chips production, the cossette storage structures and practices, the importance of damages caused by insects, the time of the infestation, the farmers' knowledge of the insect species, the traditional management practices on the infested stocks, and the farmers' preference criteria of kokoro varieties used in the production of the chips. Preference criteria of kokoro varieties were identified and prioritized using the matrix scoring technique described by Defoer et al. [19], Adoukonou-Sagbadja et al. [20], and Dansi et al. [12]. In each village, samples (500 g) of infested yam chips were collected directly from two to three randomly selected yam chips storage structures following Mendesil et al. [21] and Koradaa et al. [22]. Initial weights of the samples to be collected were taken using the numerical balance (model SF-400). Infested samples collected were preserved in plastic containers with perforated lids to allow for ventilation. With the aid of a plant taxonomist at the national herbarium of the University of Abomey-Calavi, insecticides and/or insect repulsive plants

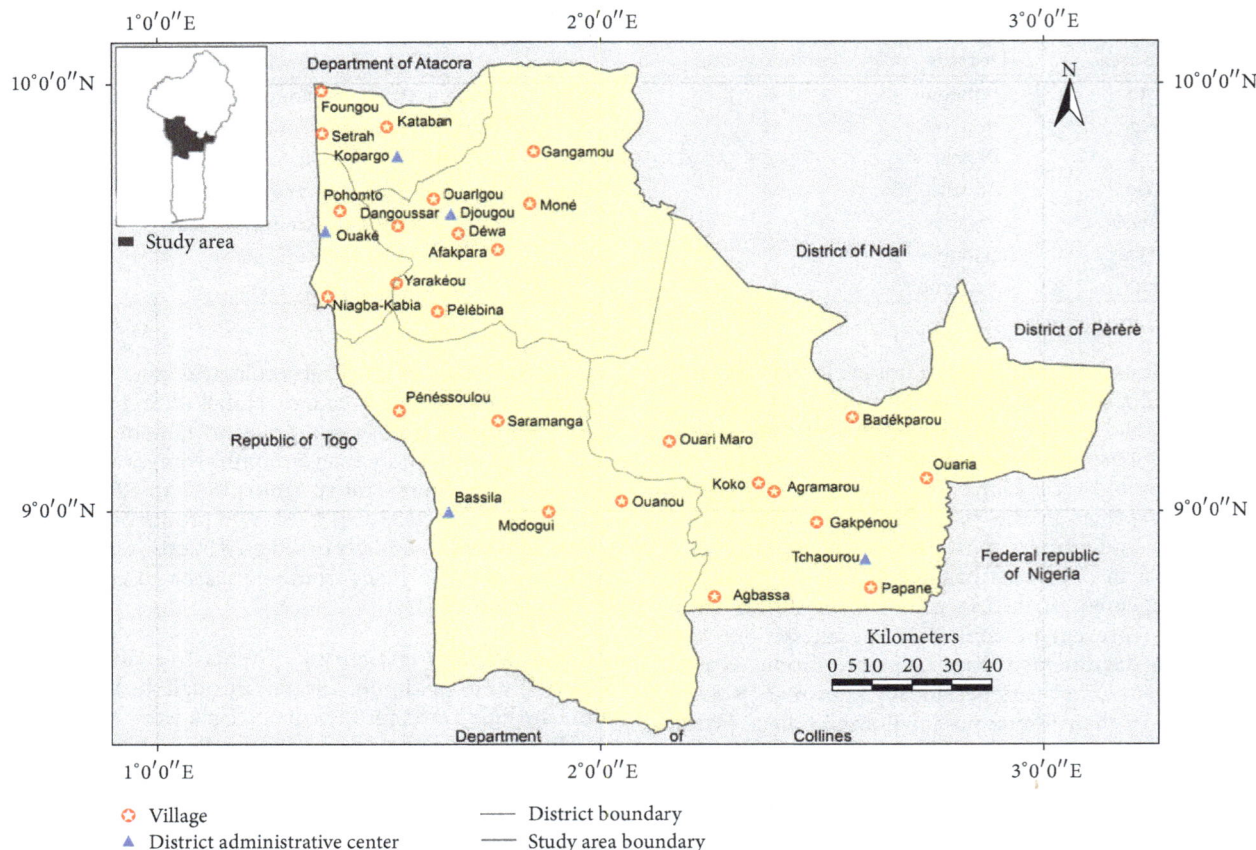

FIGURE 1: Map of Benin showing the geographical position of the surveyed villages.

used in processing reported by interviewees we sampled and their scientific names were determined using the analytical flora of Benin [16].

2.3. Incubation of the Samples and Isolation and Identification of the Insects. The labeled plastic containers containing the samples of the infested chips were incubated for three months under laboratory conditions at temperature of 25–27°C and 70%–80% relative humidity, following the method described by Eze et al. [23]. After the incubation period, the samples were broken into particles of less than 0.5 cm using a hand mortar and the insects were recovered through a 0.25 mm nylon net sieve [24]. Recovered insects were counted and conserved in a flask containing 70% alcohol for safeguarding and identification. Species' identification was done at the Biodiversity Center of the International Institute of Tropical Agriculture (IITA-Benin).

2.4. Statistical Analysis. Data were analyzed through descriptive statistics (frequencies, percentages, means, etc.) to generate summaries and tables at different (villages, individuals) levels using SAS software [25].

3. Results

3.1. Characteristics of the Respondents. The respondents were in majority (98%) women. Sixty-three (63%) are illiterate and

47% attended primary school only. Their ages varied from 17 to 60 years with an average of 37 years. In all the households surveyed, yam chips were produced for either home consumption and for the market (95.2% of the respondents) or for home consumption only (4.8%).

3.2. Constraints of Yam Chips Production. Six constraints (Table 2) related to yam chips production in Benin were recorded. They were all directly or indirectly linked to the storage of the chips. Among them, damages due to storage insects were the most important (63.9% of responses), followed by insufficiency of insect-resistant varieties (16.7% of responses) and the lack of natural human health preserving insecticides (10.2% of responses). The other three constraints (lack of organised markets, low availability of fresh kokoro yam tubers, and the lack of appropriate and specific storage structures) were of very low importance (only 1.1% to 4.5% of responses). The majority of the yam chips producers (72.12%) estimated at 40%–60% the importance of the damage caused by the storage insects on the yam ships (Figure 2(a)). This however depends on the variety used, the conservation structure and the drying level of the chips. For the great majority (92.94%) of the respondents, the infestation of the cossettes in stock occurred during the first two months (Figure 2(b)). In order to minimize these constraints, yam chips producers proposed six key solutions

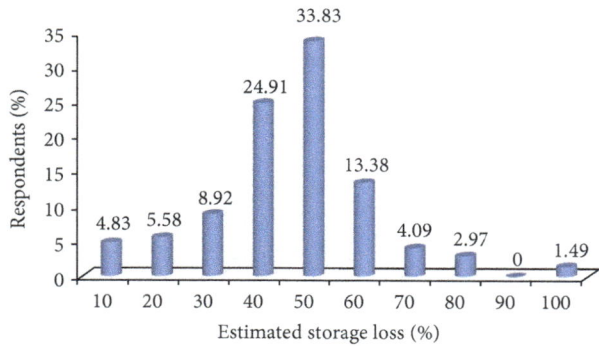

(a)

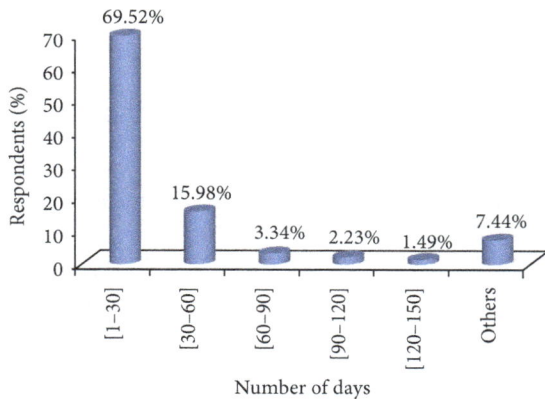

(b)

FIGURE 2: Farmers' perception of (a) storage loss due to stored yam chips insect pests, (b) the period of infestation of the yam chips.

FIGURE 3: Yam chips with insect infestation in Benin.

TABLE 2: Yam chips production constraints in Benin.

Constraints	Percentage of responses
Damages caused by storage insects	63.9
Insufficiency of insect-resistant varieties	16.7
Lack of natural human health preserving insecticides	10.2
Lack of appropriate and specific storage structures	4.5
Insufficient availability of fresh kokoro yam tubers	3.6
Lack of organised markets	1.1

TABLE 3: Solutions for the constraints and their importance as proposed by the interviewees in the study area.

Solutions	Importance (% of responses)
Diversification of good storage insect-resistant kokoro yam	30.2
Development of a natural human health preserving insecticides	24.2
Enhancement of the production of kokoro yam	21.2
Development of fast drying areas for the yam chips	19.5
Development of efficient and specific yam chips storage structures	3.3
Establishment of a well-organised yam chips good market	1.6

(Table 3) including diversification of good storage insect-resistant kokoro yam (30.2% of responses), development of a natural human health preserving insecticides (24.2% of responses), enhancement of the production of kokoro yam (21.2% of responses), and development of fast drying areas for the yam chips (19.5% of responses).

3.3. Farmers' Knowledge of the Insect Pests Damaging Yam Chips in Stocks and Diversity Assessment.

In the study area, all the storage insects were traditionally classified in a single group named *Benonkpé* in Ani, *Kokolibo* in Nago, *Doridji* in Peulh, *Gbénénoukokonou* in Bariba, *Dresse* in Yom, and *Poucasse* in Lokpa and Taneka. All these six vernacular names literally mean beetles. Farmers reported that these beetles act by penetrating the chips and drastically reducing their internal parts into powdery waste (Figure 3). Although interviewees recognized storage insects as major constraint in yam chips production, only 47% of them were able to differentiate some species. The few respondents who attempted to identify yam chips insect pests based their identification mainly on the colour (45.12% of responses) and the relative size (32.18% of responses) of the insects and on the symptoms of the damage they caused (22.7% of responses).

The diversity analysis conducted on the total of 78 samples collected and incubated revealed 12 species of insects belonging to four orders (Table 3) which are Coleoptera (eight species), Hemiptera (two species), Hymenoptera (one species), and Psocoptera (undetermined species). Species of the order of Coleoptera were the most numerous and the most represented in the samples. On average 223 insects of the order Coleoptera were counted by 500 g of yam chips against 11.4 for all the other orders put together (Table 4). Among the species identified *Dinoderus porcellus* Lesne was the most represented. It was found in 97.44% of the collected samples and was also the most abundant in all the samples in which it was found (Table 4). This was followed by the species *Tribolium castaneum* (Herbst), detected in 52.56% of the samples (Table 4). The other species were found in only 2 to 10 samples out of the 78 samples collected and in

TABLE 4: Results of the samples incubated at the laboratory showing the species of insects used and their relative abundance.

Types of insects	Infested samples (%)	Average count for 500 g	Percentage of abundance (500 g)	Rank
Dinoderus porcellus	76 (97.44)	208.72	89.03	1
Psocoptera spp.	19 (24.36)	8.83	3.77	2
Tribolium castaneum	41 (52.56)	6.12	2.61	3
Lasioderma serricorne	4 (5.13)	3.3	1.41	4
Sitophilus zeamais	10 (12.82)	2.57	1.09	5
Xylocoris flavipes	10 (12.82)	2.46	1.05	6
Cryptolestes pusillus	10 (12.82)	1.26	0.54	7
Carpophilus dimidiatus	7 (8.97)	0.85	0.36	8
Teretrius nigrescens	5 (6.41)	0.13	0.05	9
Carpophilus binotatus	3 (3.85)	0.09	0.04	10
Alloeocranum biannulipes	5 (6.41)	0.09	0.04	11
Dinarmus basalis	2 (2.56)	0.03	0.01	12

very few numbers. Among the 12 species of insects identified, three (*Teretrius nigrescens* Lewis, *Xylocoris flavipes* Reuter, and *Alloeocranum biannulipes* Montrouzier & Signoret) were predators and one (*Dinarmus basalis* Rondani) was a parasitoid. *Xylocoris flavipes* was found in 12.82% of the samples and appeared to be the most abundant predator.

3.4. Traditional Yam Chips Storage Systems and Duration of the Conservation. In all the households surveyed, yam chips were stored inside houses and rooms in various containers. The great majority (97.77%) of producers used maize bags (made with synthetic materials) of various sizes as containers, depending on the quantity of chips to be conserved. Only few producers (2.23%) preferred to store in large-sized and hermetically closed plastic buckets, jars, or barrels to prevent insect infestations. No specific structure was dedicated to storage of yam chips. Interviewees reported that storage period varied from 1 to 13 months with an average of 8 months. For 49.82%, 31.97%, and 18.21% of the respondents, storage duration of yam chips varied between 1 and 5 months, 5 and 10 months, and 10 and 15 months, respectively.

3.5. Traditional Control Systems of Yam Chips Storage Insects Pests. Under traditional storage conditions, interviewees used seven strategies to reduce losses due to insects (Table 5). Among those, the most important were regular inspection and exposure of chips to sunlight to repel insects (35.93% of responses), use of insect-resistant varieties (26.80% of responses), and use of insecticide and/or insect's repulsive plants during preparation (26.45% of responses). The other strategies such as shaking of the yam chips to remove insects along powdery waste, use of insecticides, treatment with pepper powder, and minimising frequent opening of storage structures to avoid entrance of the insects were poorly used (Table 5). According to interviewees, severely infested stocks of yam chips were sold (66.17% of responses), used for home consumption only (24.54% of responses), or simply thrown away (9.29% of responses).

TABLE 5: Farmers' management practices for the control of yam chips insect pests.

Management practices	Importance (% of responses)
Exposure of the infested yam chips to sun	35.93
Use of insect-resistant varieties	26.80
Use of insecticide and/or insect's repulsive plants during preparation	26.45
Sifting of yam chips to remove insects along powdery waste	8.59
Use of insecticides	1.12
Treatment with pepper powder	0.74
Minimising frequent opening of storage structures to avoid entrance of the insects	0.37

The study revealed that eight species of plant were used to prevent infestation of the yam chips or to control insect pests (Table 6). Among these species, three (*Bridelia ferruginea*, *Blighia sapida*, and *Khaya senegalensis*) were reported to be insecticide while four (*Piliostigma thonningii*, *Lophira lanceolata*, *Tectona grandis*, and *Sorghum bicolor*) were said to be dyes and insect repulsive (Table 6). Cassava leaves were also used during the parboiling to harden the chips. *Piliostigma thonningii* and *Sorghum bicolor* were known and used across all the ethnic groups while the other species apart from *Tectona grandis* were each used in only one ethnic group. The number and types of species of plants varied from one ethnic group to another. Yom and Peulh used only two species of plant, Lokpa and Ani used four species, and Nago used five. For the different plant species identified, the plant parts (leaves or bark) used, the application or treatment methods (infusion or fumigation), and the frequencies of utilization across ethnic groups are summarized in Table 6.

Throughout the study zone, 37 kokoro yam cultivars used to produce chips were listed as tolerant to storage insects. The number of cultivars reported varies across ethnic areas. Eight cultivars were reported with the Nago, Bariba, and Lokpa,

TABLE 6: List of plants used to protect yam chips against storage insect pests and their utilisation methods.

Species	Part used	Role	Method of application	Percentage of farmers using the plants across ethnic groups						
				Peulh	Nago	Ani	Taneka	Bariba	Lokpa	Yom
Piliostigma thonningii	Leaf/bark	Dye	Infusion	42.86	19.40	15	30.30	16.67	26.47	30.30
Lophira lanceolata	Leaf	Dye	Infusion	—	7.46	—	—	—	—	—
Blighia sapida	Leaf	Insecticide	Infusion	—	5.97	—	—	—	—	—
Bridelia ferruginea	Leaf/bark	Insecticide	Infusion	—	—	5	—	—	—	—
Khaya senegalensis	Bark	Insecticide	Fumigate	—	—	—	—	11.11	—	—
Tectona grandis	Leaf	Dye	Infusion	—	19.40	5	12.12	—	14.71	—
Manihot esculentus	Leaf	Hardening of the yam chips	Infusion	—	—	—	—	—	2.94	—
Sorghum bicolor	Stem/oil cakes	Dye	Infusion	57.14	47.77	75	57.58	72.22	55.88	69.70

seven with the Yom, four with the Taneka, and only two with the Ani (Table 7). In each ethnic area, certain cultivars were more common. With the Nago ethnic group, Oguidigbo, Adakada, Tabané, and Omonya were the most important while with the Bariba ethnic group, the most listed cultivars were Otoukpannan, Tchakatchaka, and Yakanougo (Table 6).

3.6. Farmer's Preferences Criteria for Kokoro Yam Cultivars for Chips Production.

Throughout the study zone, kokoro yam cultivars, used for chips production, were selected among the existing diversity based on eight criteria. Among them, the quality of the paste made with the flour, the storage aptitude of the chips, and the quality of the Wassa-Wassa (local couscous made with yam chips' flour) were the most important and represent altogether 74.08% of the responses (Table 8). The number and importance of the criteria also varied across ethnic groups. With the Nago and the Peulh, storage aptitude was the most important criterion while with the other ethnic groups the quality of the paste came at the first position (Table 8). The quality of Wassa-Wassa, which was the third most important criterion among the Lokpa, Ani and Yom people, was not even mentioned by the Bariba people. Similarly, the fast drying quality of the cultivars, which was important to the Nago, the Taneka, and the Ani, was not listed among the Peulh people and had very low values with the other ethnic groups. While all the eight criteria were listed by the Lokpa ethnic group, all but one was recorded with the Taneka and the Yom and only four were identified with the Peulh (Table 8).

4. Discussion

The respondents were in majority women. This can be explained by the fact that in all ethnic groups of the study area, women were the sole processors of kokoro yam tubers into chips. The few males interviewed responded on behalf of their wives, who gave way to them out of respect. The culture of the people was also evident in the yam production system where tasks had been traditionally divided according to gender. Men were in charge of the most important activities in terms of labour requirements, while foods processing and transformation of the yam tubers into chips, among other activities, were devoted to women [4]. According to Bricas

and Vernier [26], the commercialization of yam chips is by far more economically profitable than the one of fresh tubers. This could justify the importance that women in the study area gave to commercialization as a means of substantially improving their household income.

Among the constraints related to yam chips production in the study area, damages caused by storage insects stood out as the most important. Similar results were reported by Osuji [27] and Adedire and Gbaye [28] in Nigeria. The importance of the damages raised by the respondents is key indicator of the urgent necessity to develop control strategies against the storage insects.

Sun drying of infested chips was the major method used by farmers to control these insect pests. This method, which is the oldest known technique of conservation of the agro-alimentary products, also presents several disadvantages. Chalal et al. [29] reported that sun drying directly exposes the products to dust and to ultraviolet rays which can cause the deterioration of food vitamins. Among the solutions proposed by farmers were diversification of good storage insect-resistant kokoro yams and development of natural human health preserving insecticides. These two propositions, which call the attention of plant geneticist and breeders on one hand and industrial chemists on the other, indicate that producers are very concerned about their health. The numerous cases of food poisoning that were associated with the use of cotton insecticides on yam chips recorded these last years in the country and which led to the death of many persons may have contributed to this health consciousness. In Nigeria, Adedoyin et al. [30] and Adeleke [31] also reported poisoning due to the consumption of yam flour (treated with insecticide) in some families in Ilorin and Kano.

Our study revealed that in the different samples of infested yam chips collected and analyzed, *Dinoderus porcellus* was the most represented. This species which is known to be mostly associated with dried yams [32] has already been reported as the most important pest of stored yam chips in Nigeria [27, 28, 32]. *Dinoderus porcellus* particularly infests well-dried chips [27, 28]. Therefore, it is possible that the few samples, in which it was absent, were not well dried or had relatively higher moisture contents. The presence of *Psocoptera* spp., *Carpophilus dimidiatus*, and *Carpophilus Binotatus* in the samples without *Dinoderus*

TABLE 7: Kokoro yam cultivars tolerant to storage insect pests and their importance across ethnic areas.

Ethnic areas	Insect-resistant varieties	Importance (number of farmers)
Ani	Demkpenai	14
	Awanawou	11
Bariba	Otoukpannan	18
	Tchakatchaka	16
	Yakanougo	15
	Omonya	8
	Singor	6
	Ankakorouwoura	1
	Gaboubaba	1
	Kourakourogouroko	1
Lokpa	Azowi	17
	Iootchra	14
	Moghoun	12
	Kounto	10
	Kparokoumè	8
	Soprova	6
	Tougbana	4
	Tédoman	1
Nago	Oguidigbo	54
	Adakada	45
	Tabané	41
	Hounbonon	33
	Kokorogbambe	26
	Kokorolakolako	18
	Kokoroagbalè	13
	Adjawoungbo	5
Taneka	Atawouraï	27
	Souwoukou	19
	Gréé	7
	Djèssoumè	4
Yom	Koutonouman	51
	Biboï	44
	Assinakpeina	39
	Adjôgba	17
	Mouhame	3
	Ayè	2
	Satchila	2

porcellus supports this hypothesis as they are known as insects associated with wet food products [33]. Tribolium castaneum and Psocoptera spp. were also found in not negligible number of the samples. The red flour beetle, Tribolium castaneum, is a common and one of the most important stored product pests associated with a wide range of durable commodities (barley, bran, cacao, ginger, maize, millet, cassava chips, nutmeg, peanut, pepper, rice, sorghum, etc.) and food-processing facilities worldwide [34–36] were also found. Its presence in

the samples examined is not surprising as it has already been reported by Vernier et al. [9], Soldati et al. [37], and Oni and Omoniyi [32]. In some yam chips samples collected outside our study area, Vernier et al. [9, 38] and Gnonlonfin et al. [10] identified five other species which were not found in our studies. These included Dinoderus bifoveolatus (Wollaston), Palorus subdepressus (Wollaston), Rhyzopertha dominica (F.), C. quadricollis (Guérin-Méneville), Gnatocerus maxillosus (F.), and Prostephanus truncatus. In order to have an exhaustive list of all the stored-products insect pests associated with yam chips in Benin and map their geographical distribution, further studies need to be conducted by including the remaining part of the country. A good knowledge of the diversity of the species will be of great utility for the yam breeders who may like to select kokoro cultivars producing tubers that are tolerant to storage insect pests. For example, The Laboratory of Agricultural Biodiversity and Tropical Plant Breeding of the University of Abomey-Calavi (Benin) and the Global Crop Diversity Trust (Rome, Italy) are currently introducing yam chips technology to the arid zone of the department of Atakora (far northwest of Benin), where the environment is quite suitable for fast drying of the chips. To succeed, however, it will be necessary to reckon with kokoro cultivars tolerant to storage insect pests.

To control insect pests and diseases in crops, Integrated Pest Management (IPM) is recommended [39]. IPM promotes biological control based on the use of the natural enemies of pests (predators and parasitoids) and the genetic control through growing of pest and disease tolerant or resistant cultivars [40]. Among the natural enemies encountered in the infested yam chips, Xylocoris flavipes is known as an effective polyphagous predator of eggs, larvae and chrysalis of coleopteran insects [41]. This natural enemy is also a predator of larvae of T. castaneum [42] and is frequently associated with the insects of cereals stocks [43]. According to Helbig [44], X. flavipes has some interesting biological characteristics that make it a potential control agent of storage insect pests. Unfortunately, it was reported that X. flavipes only eliminates populations of small-sized insects, but not larger insects or insects with internal feeding such as D. porcellus [45]. A. biannulipes on the other hand (Montrouzier & Signoret) is known as a predator of the large-sized storage insect pests including L. serricorne (F.) and Tribolium castaneum [46–48]. A biological control program, combining these predators, will be useful in eliminating various types of insects and will help control the insect pests' complex associated with yam chips.

It appeared from our study that yam chips producers also used diverse plants to protect chips against insect attacks. Phytochemical studies conducted by Dumaine et al. [49] revealed that none of the four plant species (L. lanceolata, T. grandis, P. thonningii, and B. ferruginea) used in the study areas has insecticidal or insect repulsing effects. However, Akinpelu and Obuotor [50] found that P. thonningii bark extract has a bactericidal activity which is also important for improving the sanitary conditions of the chips. Among the plants used, Blighia sapida, Bridelia ferruginea, and Khaya senegalensis are even believed, and rightly so, to have insecticide properties by the Nago, the Ani, and the Bariba people (Table 6). In fact, the bark extract of K. senegalensis has been

TABLE 8: Famer's preference criteria of good kokoro yam cultivar for chips production in the study area and across ethnic groups.

Preferences criteria	Study area (% of responses)	Ethnic groups (% of responses)						
		Nago	Peulh	Bariba	Taneka	Lokpa	Yom	Ani
Quality of the paste	35.62	23.03	21.43	41.38	38.15	31.25	35.65	32.14
Storage aptitude of the chips	26.5	31.99	57.14	37.93	24.5	23.96	22.61	28.57
Quality of Wassa-Wassa	11.96	4.11	7.14	0	3.54	22.92	13.04	17.86
Colour of the paste	9.67	12.33	14.29	13.79	2.7	6.25	13.04	3.57
Flour richness of the yam chips	3.05	2.74	—	—	7.9	7.29	0	—
Crushing facility of the chips	1.02	—	—	—	2.6	2.08	0.87	—
Taste of the paste	4.83	—	—	3.45	—	5.21	10.44	3.57
Fast drying	7.35	25.8	—	3.45	20.61	1.04	4.35	14.29

proved to be antifungal [51], antibacterial [52, 53], and insect antifeedant [54]. Mitchell and Ahmad [55] reported that *B. sapida* has acaricide and insecticide properties. Similarly it has been shown that all the fruit components (skin, aril and granulates, oil) of this plant have repulsive properties against stock insects such as *Callosobruchus maculatus*, *Cryptolestes ferrugineus*, *T. castaneum*, and *S. zeamais* [55–58]. Experiments should be conducted to assess the effects of the extract of these three species on insect pests that damage stored yam chips.

Chips producers reported that the importance of the damages is a function of the yam cultivars used and listed 37 kokoro yam landraces producing tubers rarely attacked by the storage insect pests. Due to the existence of numerous synonymies in farmer-named yam cultivars [11], these listed landraces may not all correspond to 37 different genotypes. Therefore, agromorphological characterization coupled with molecular analysis should be carried out to identify duplicates and establish the equivalence between recorded names following Tamiru et al. [59] and Kombo et al. [60]. Moreover, and as recommended by Vernier et al. [38], it will be also important to assess by a well-elaborated trial the effectiveness of the tolerance of the chips derived from the tubers of these varieties to storage insect pests. The use of resistant varieties remains the most economically profitable and the best healthy method of combating chips storage insect pests. Because of this, kokoro yams in the chips production zone should be strengthened with more high yielding cultivars that are suitable for chips and resistant to storage insect pests. According to Dansi et al. [11], such cultivars exist in the traditional agriculture and could be identified through participatory evaluation. Within the existing diversity, cultivars to be used for the chips are selected based on diverse criteria, among which those related to the quality of the foods (Wassa-Wassa; paste) made with the yam chips flour and the technological characteristics of the chips are the most important (66.66% of the responses). This result is expected because in Benin, chips are only made and used for food purposes. In the preference criteria identified, aspects related to conservation come in second position indicating that producers really give particular importance to insect damages. The variation of the preference criteria noted across ethnic groups is frequent and has been already reported in many crops such as cowpea [61], banana and plantain [62], maize [63], telf [64], sorghum

[65], yam [11], and even fonio [12]. The fast drying criteria importantly raised by the Nago, the Taneka, and the Ani ethnic groups should be seriously considered as it influences the hygienic quality of the chips and their market value. Nago, Taneka, and Ani people mostly produced chips for economic purposes through commercialization. One understands therefore how important fast drying could be to them.

5. Conclusion

This study has allowed us to identify several constraints that hamper yam chips production in northern Benin. Attacks by storage insects were the major constraints identified. Yam chips were infested by various insects, of which the most important was *Dinoderus porcellus*. Several plants are traditionally used to fight these insects. Following farmers' requests, efforts should be directed towards diversification of good kokoro cultivars which are tolerant to storage insects. In this framework and to identify such cultivars, we recommend the participatory evaluation of existing kokoro yam, the identification of duplicates, and clarification of synonymies and the assessment of the tolerance of the chips manufactured with tubers produced by the identified varieties.

Acknowledgments

This research was sponsored by the Ministry of Higher Education and Scientific Research of Benin. The authors are grateful to Dr. Goergen Georg, who is in charge of the Insect Museum of the International Institute of Tropical Agriculture (Cotonou, Benin) for his assistance during the identification of the insects and to Dr. H. Yédomonhan (National Herbarium, Department of Botany and Plant biology, University of Abomey-Calavi) for the identification of the plant species. They thank all the chips manufacturers and agricultural technicians they met for fruitful discussions during the survey.

References

[1] FAO, *FAOSTAT Database*, Food and Agriculture Organization, Roma, Italy, 2010, http://www.fao.org/.

[2] J. M. Babajide, O. Q. Bello, and S. O. Babajide, "Quantitative determination of active substances (preservatives) in *Piliostigma thonningii* and *Khaya ivorensis* leaves and subsequent

transfer in dry-yam," *African Journal of Food Science*, vol. 4, no. 6, pp. 382–388, 2010.

[3] H. D. Mignouna and A. Dansi, "Yam (*Dioscorea* ssp.) domestication by the Nago and Fon ethnic groups in Benin," *Genetic Resources and Crop Evolution*, vol. 50, no. 5, pp. 519–528, 2003.

[4] M. N. Baco, S. Tostain, R. L. Mongbo, O. Dainou, and C. Agbangla, "Gestion dynamique de la diversité variétale des ignames cultivées (*Dioscorea cayenensis—D. rotundata*) dans la commune de Sinendé au nord Bénin," *Plant Genetic Resources Newsletter*, vol. 139, pp. 18–24, 2004.

[5] N. Akissoé, D. J. Hounhouigan, N. Bricas, P. Vernier, C. M. Nago, and O. A. Olorunda, "Physical, chemical and sensory evaluation of dried yam (*Dioscorea rotundata*) tubers, flour and amala, a flour-derived product," *Tropical Science*, vol. 41, no. 3, pp. 151–155, 2001.

[6] D. J. Hounhouigan, A. P. Kayode, N. Bricas, and C. M. Nago, "Desirable culinary and sensory characteristics of yams in urban Benin," *Benin Journal of Agricultural Sciences*, vol. 21, no. 12, pp. 2815–2820, 2003.

[7] J. M. Babajide, O. O. Atanda, T. A. Ibrahim, H. O. Majolagbe, and S. A. Akinbayode, "Quantitative effect of 'abafe' (*Piliostigma thionnigii*) and 'agehu' (*Khaya ivorensis*) leaves on the microbial load of dry-yam 'gbodo'," *African Journal of Microbiology Research*, vol. 2, pp. 292–298, 2008.

[8] E. Ategbo, N. Bricas, J. Hounhouigan et al., "Le développement de la filière chips d'igname pour l'approvisionnement des villes au Nigeria, au Bénin et au Togo," in *Actes du Séminaire International Cirad-Inra-Orstom-Coraf*, pp. 339–341, Montpellier, France, Juin 1997.

[9] P. Vernier, G. Goergen, R. Dossou, P. Letourmy, and J. Chaume, "Utilization of biological insecticides for the protection of stored yam chips," *Outlook on Agriculture*, vol. 34, no. 3, pp. 173–179, 2005.

[10] G. J. B. Gnonlonfin, K. Hell, A. B. Siame, and P. Fandohan, "Infestation and population dynamics of insects on stored cassava and yams chips in Benin, West Africa," *Journal of Economic Entomology*, vol. 101, no. 6, pp. 1967–1973, 2008.

[11] A. Dansi, H. D. Mignouna, J. Zoundjihékpon, A. Sangare, R. Asiedu, and F. M. Quin, "Morphological diversity, cultivar groups and possible descent in the cultivated yams (*Dioscorea cayenensis—D. rotundata*) complex in Benin Republic," *Genetic Resources and Crop Evolution*, vol. 46, no. 4, pp. 371–388, 1999.

[12] A. Dansi, H. Adoukonou-Sagbadja, and R. Vodouhè, "Diversity, conservation and related wild species of Fonio millet (*Digitaria* spp.) in the Northwest of Benin," *Genetic Resources and Crop Evolution*, vol. 57, no. 6, pp. 827–839, 2010.

[13] Y. L. Loko, A. Dansi, C. Linsoussi et al., "Current status and spatial analysis of Guinea yam (*Dioscorea cayenensis* Lam. -*D. rotundata* Poir. complex) diversity in Benin," *Genetic Resources and Crop Evolution*. In press.

[14] S. Adam and M. Boko, *Le Bénin*, Les éditions du Flamboyant / EDICEF, 1993.

[15] A. C. Adomou, *Vegetation patterns and environmental gradients in Benin: implications for biogeography and conservation [Ph.D. thesis]*, University of Wageningen, Wageningen, The Netherlands, 2005.

[16] A. Akoegninou, W. J. van der Burg, and L. J. G. van der Maesen, *Flore Analytique du Bénin*, Backhuys Publishers, 2006.

[17] A. Dansi, A. Adjatin, H. Adoukonou-Sagbadja et al., "Traditional leafy vegetables and their use in the Benin Republic," *Genetic Resources and Crop Evolution*, vol. 55, no. 8, pp. 1239–1256, 2008.

[18] K. V. Christinck, B. Kshirsagar, E. Weltzien, and P. J. Bramel-Cox, "Participatory methods for collecting germplasm : experiences with famers of Rajasthan, India," *Plant Ressources Newsletter*, vol. 121, p. 129, 2000.

[19] T. Defoer, A. Kamara, and H. De Groote, "Gender and variety selection: farmers' assessment of local maize varieties in Southern Mali," *African Crop Science Journal*, vol. 5, no. 1, pp. 65–76, 1997.

[20] H. Adoukonou-Sagbadja, A. Dansi, R. Vodouhè, and K. Akpagana, "Indigenous knowledge and traditional conservation of fonio millet (*Digitaria exilis, Digitaria iburua*) in Togo," *Biodiversity & Conservation*, vol. 15, no. 8, pp. 2379–2395, 2006.

[21] E. Mendesil, C. Abdeta, A. Tesfaye, Z. Shumeta, and H. Jifar, "Farmers' perceptions and management practices of insect pests on stored sorghum in Southwestern Ethiopia," *Crop Protection*, vol. 26, no. 12, pp. 1817–1825, 2007.

[22] R. R. Koradaa, S. K. Naskara, and S. Edison, "Insect pests and their management in yam production andstorage: a world review," *International Journal of Pest Management*, vol. 56, no. 4, pp. 337–349, 2010.

[23] S. C. Eze, J. E. Asiegbu, B. N. Mbah, G. C. Orkwor, and R. Asiedu, "Effects of four agrobotanical extracts and three types of bags on the control of insect pests and moulds of stored yam chips," *Journal of Agriculture, Food, Environment and Extension*, vol. 5, no. 1, pp. 8–12, 2006.

[24] A. Delobel, "Les cossettes de manioc, un important réservoir d'insectes des denrées stockées en Afrique centrale," *Revue de Zoologie Africaine*, vol. 106, no. 1, pp. 17–25, 1992.

[25] SAS, "User's guide," SAS Institute, Cary, NC, USA, 1996.

[26] N. Bricas and P. Vernier, Dossier: la transformation de l'igname. Bulletin du réseau TPA N° 18, 5–12, 2000, http://infotpa.gret.org/fileadmin/bulletin/bulletin18.pdf.

[27] F. N. C. Osuji, "Observations on beetles attacking dried yams and yam flour from three Nigerian markets," *Tropical Stored Products Information*, vol. 39, pp. 35–38, 1980.

[28] C. O. Adedire and O. A. Gbaye, "Seasonal prevalence and life history of the yam beetle, *Dinoderus porcellus* (Lesne) (Coleoptera: Bostrichidae)," *Nigerian Journal of Experimental and Applied Biology*, vol. 3, pp. 323–329, 2002.

[29] N. Chalal, A. Bellhamri, and L. Bennamoun, "Étude d'un séchoir solaire fonctionnant en mode direct et indirect," *Revue des Energies Renouvelables*, pp. 117–126, 2008, Séminaire Maghrébin sur les Sciences et les Technologies de Séchage (SMSTS '08).

[30] O. T. Adedoyin, A. Ojuawo, O. O. Adesiyun, F. Mark, and E. A. Anigilaje, "Poisoning due to yam flour consumption in five families in Ilorin, Central Nigeria," *West African Journal of Medicine*, vol. 27, no. 1, pp. 41–43, 2008.

[31] S. I. Adeleke, "Food poisoning due to yam flour consumption in Kano (Northwest) Nigeria," *Online Journal of Health and Allied Sciences*, vol. 8, no. 2, p. 10, 2009.

[32] M. O. Oni and A. O. Omoniyi, "Studies on temperature influence on oviposition and development of immature stages of the yam beetle *Dinoderus porcellus* Lesne. Coleoptera: Bostrichidae on dried yam species," *Journal of Agricultural Science*, vol. 4, no. 2, pp. 213–218, 2012.

[33] R. J. Bartelt and M. S. Hossain, "Chemical ecology of *Carpophilus* sap beetles (Coleoptera: Nitidulidae) and development of an environmentally friendly method of crop protection," *Terrestrial Arthropod Reviews*, vol. 3, no. 1, pp. 29–61, 2010.

[34] J. M. Turaki, B. M. Sastawa, B. G. J. Kabir, and N. E. S. Lale, "Susceptibility of flours derived from various cereal grains to infestation by the rust-red flour beetle (*Tribolium castaneum* Herbst) (Coleoptera: Tenebrionidae) in different seasons," *Journal of Plant Protection Research*, vol. 47, no. 3, pp. 279–288, 2007.

[35] N. E. S. Lale and B. A. Yusuf, "Insect pests infesting stored pearl millet *Pennisetum glaucum* (L.) R. Br. in Northeastern Nigeria and their damage potential," *Cereal Research Communications*, vol. 28, no. 1-2, pp. 181–186, 2000.

[36] S. A. Babarinde, G. O. Babarinde, and O. Olasesan, "Physical and biophysical deterioration of stored plantain chips (*Musa Sapientum* L.) due to infestation of *Tribolium castaneum* herbst (Coleoptera: Tenebrionidae)," *Journal of Plant Protection Research*, vol. 50, no. 3, pp. 303–306, 2010.

[37] L. Soldati, G. J. Kergoat, and F. L. Condamine, "Preliminary report on the Tenebrionidae (Insecta, Coleoptera) collected during the SANTO 2006 expedition to Vanuatu, with description of a new species of the genus *Uloma* Dejean, 1821," *Zoosystema*, vol. 34, no. 2, pp. 305–317, 2006.

[38] P. Vernier, R. A. Dossou, and P. Letourmy, "La fabrication d'igname à partir de *Dioscorea alata*: influence de la variété et du type de chips sur le séchage, la conservation et les qualités organoleptiques," *African Journal of Root and Tuber Crops*, vol. 3, pp. 62–67, 1999.

[39] O. C. Eneh, "Enhancing Africa's environmental management: integrated pest management for minimizing of agricultural pesticides pollution," *Research Journal of Environmental Sciences*, vol. 5, no. 6, pp. 521–529, 2011.

[40] G. W. Norton, E. G. Rajotte, and V. Gapud, "Participatory research in integrated pest management: lessons from the IPM CRSP," *Agriculture and Human Values*, vol. 16, no. 4, pp. 431–439, 1999.

[41] J. Brower and M. A. Mullen, "Effects of *Xylocoris flavipes* (Hemiptera: Anthocoridae) releases on moth populations in experimental peanut storages," *Journal of Entomological Science*, vol. 25, no. 2, pp. 268–276, 1990.

[42] A. Russo, G. E. Cocuzza, and M. C. Vasta, "Life tables of *Xylocoris flavipes* (Hemiptera: Anthocoridae) feeding on *Tribolium castaneum* (Coleoptera: Tenebrionidae)," *Journal of Stored Products Research*, vol. 40, no. 1, pp. 103–112, 2004.

[43] R. T. Arbogast and J. E. Throne, "Insect infestation of farm-stored maize in South Carolina: towards characterization of a habitat," *Journal of Stored Products Research*, vol. 33, no. 3, pp. 187–198, 1997.

[44] J. Helbig, "Efficacy of *Xylocoris flavipes* (Reuter) (Het., Anthocoridae) to suppress *Prostephanus truncatus* (Horn) (Col., Bostrichidae) in traditional maize stores in Southern Togo," *Journal of Applied Entomology*, vol. 123, no. 8, pp. 503–508, 1999.

[45] T. Imamura, M. Murata, and A. Miyanoshita, "Review biological aspects and predatory abilities of hemipterans attacking stored-product insects," *Japan Agricultural Research Quarterly*, vol. 42, no. 1, pp. 1–6, 2008.

[46] K. T. Awadallah, A. I. Afifi, and I. I. El-Sebaey, "The biology of the reduviid, *Allaeocranum biannulipes* (Mon. et Sign.), a predator of stored product insect pests," *Bulletin of the Entomological Society of Egypt*, vol. 69, pp. 169–181, 1990.

[47] D. P. Rees, "The effect of *Teretriosoma nigrescens* Lewis (Coleoptera: Histeridae) on three species of storage Bostrichidae infesting shelled maize," *Journal of Stored Products Research*, vol. 27, no. 1, pp. 83–86, 1991.

[48] M. Camara, C. Borgemeister, R. H. Markham, and H. M. Poehling, "Electrophoretic analysis of the prey spectrum of *Teretrius nigrescens* (Lewis) (Col., Histeridae), a predator of *Prostephanus truncatus* (Horn) (Col., Bostrichidae), in Mexico, Honduras, and Benin," *Journal of Applied Entomology*, vol. 127, no. 6, pp. 360–368, 2003.

[49] F. Dumaine, D. Dufour, C. Mestres et al., "Effect of yam *Dioscorea cayenenis-rotundata* post-harvests treatments on yam chips quality," in *Potential of Root Crops for Food and Industrial Resources. Twelfth Symposium of the International Society for Tropical Root Crops (ISTRC), September 10–16, 2000, Tsukuba, Japan*, M. Nakatani and K. Komaki, Eds., vol. 12 of *Symposium on Potential of Root Crops for Food and Industrial Resources*, p. 394, Cultio Corporation, Tsukuba, Japon, 2002.

[50] D. A. Akinpelu and E. M. Obuotor, "Antibacterial activity of *Piliostigma thonningii* stem bark," *Fitoterapia*, vol. 71, no. 4, pp. 442–443, 2000.

[51] S. A. M. Abdelgaleil, F. Hashinaga, and M. Nakatani, "Antifungal activity of limonoids from *Khaya ivorensis*," *Pest Management Science*, vol. 61, no. 2, pp. 186–190, 2005.

[52] M. Nakatani, S. A. M. Abdelgaleil, S. M. I. Kassem et al., "Three new modified limonoids from *Khaya senegalensis*," *Journal of Natural Products*, vol. 65, no. 8, pp. 1219–1221, 2002.

[53] D. Kubmarawa, M. E. Khan, A. M. Punah, and M. Hassan, "Phytochemical screening and antimicrobial efficacy of extracts from *Khaya senegalensis* against human pathogenic bacteria," *African Journal of Biotechnology*, vol. 7, no. 24, pp. 4563–4566, 2008.

[54] M. Sale, N. De, J. H. Doughari, and M. S. Pukuma, "In vitro assessment of antibacterial activity of bark extracts of *Khaya senegalensis*," *African Journal of Biotechnology*, vol. 7, no. 19, pp. 3443–3446, 2008.

[55] S. A. Mitchell and M. H. Ahmad, "A review of medicinal plant research at the University of the West Indies, Jamaica, 1948–2001," *West Indian Medical Journal*, vol. 55, no. 4, pp. 243–269, 2006.

[56] M. R. M. Ekué, B. Sinsin, O. Eyog-Matig, and R. Finkeldey, "Uses, traditional management, perception of variation and preferences in ackee (*Blighia sapida* K.D. Koenig) fruit traits in Benin: implications for domestication and conservation," *Journal of Ethnobiology and Ethnomedicine*, vol. 6, article 12, 14 pages, 2010.

[57] A. Khan, F. A. Gumbs, and A. Persad, "Pesticidal bioactivity of ackee (*Blighia sapida* Koenig) against three stored-product insect pests," *Tropical Agriculture*, vol. 79, no. 4, pp. 217–223, 2002.

[58] A. Khan and F. A. Gumbs, "Repellent effect of ackee (*Blighia sapida* Koenig) component fruit parts against stored product insect pests," *Tropical Agriculture*, vol. 80, no. 1, pp. 19–27, 2003.

[59] M. Tamiru, H. C. Becker, and B. L. Maass, "Diversity, distribution and management of yam landraces (*Dioscorea* spp.) in Southern Ethiopia," *Genetic Resources and Crop Evolution*, vol. 55, no. 1, pp. 115–131, 2008.

[60] G. R. Kombo, A. Dansi, L. Y. Loko et al., "Diversity of cassava (*Manihot esculenta* Crantz) cultivars and its management in the department of Bouenza in the Republic of Congo," *Genetic Resources and Crop Evolution*, vol. 59, no. 8, pp. 1789–1803, 2012.

[61] L. W. Kitch, O. Boukar, C. Endondo, and L. L. Murdock, "Farmer acceptability criteria in breeding cowpea," *Experimental Agriculture*, vol. 34, no. 4, pp. 475–486, 1998.

[62] C. S. Gold, A. Kiggundu, A. M. K. Abera, and D. Karamura, "Selection criteria of Musa cultivars through a farmer participatory appraisal survey in Uganda," *Experimental Agriculture*, vol. 38, no. 1, pp. 29–38, 2002.

[63] G. Abebe, T. Assefa, H. Harrun, T. Mesfine, and A. R. M. Al-Tawaha, "Participatory selection of drought tolerant maize varieties using mother and baby methodology: a case study in the semi-arid zones of the Central Rift Valley of Ethiopia," *World Journal of Agricultural Sciences*, vol. 1, no. 1, pp. 22–27, 2005.

[64] G. Belay, H. Tefera, B. Tadesse, G. Metaferia, D. Jarra, and T. Tadesse, "Participatory variety selection in the Ethiopian cereal tef (*Eragrostis tef*)," *Experimental Agriculture*, vol. 42, no. 1, pp. 91–101, 2006.

[65] A. Teshome, D. Patterson, Z. Asfew, J. K. Torrance, and J. T. Arnason, "Changes of *Sorghum bicolor* landrace diversity and farmers' selection criteria over space and time, Ethiopia," *Genetic Resources and Crop Evolution*, vol. 54, no. 6, pp. 1219–1233, 2007.

16

Diversity of Catechin in Northeast Indian Tea Cultivars

Santanu Sabhapondit,[1] Tanmoy Karak,[2] Lakshi Prasad Bhuyan,[1] Bhabesh Chandra Goswami,[3] and Mridul Hazarika[1]

[1] Department of Biochemistry, Tocklai Experimental Station, Tea Research Association, Assam, Jorhat 785008, India
[2] Department of Soil, Tocklai Experimental Station, Tea Research Association, Assam, Jorhat 785008, India
[3] Department of Chemistry, Gauhati University, Assam, Guwahati 781014, India

Correspondence should be addressed to Santanu Sabhapondit, santanusabhapondit@yahoo.com

Academic Editor: Ornella Abollino

Tea (*Camellia sinensis* L.) leaf contains a large amount of catechins (a group of very active flavonoids) which contribute to major quality attributes of black tea. Based on morphological characters tea plants were classified as Assam, China, and Cambod varieties. The present study is an attempt for biochemical fingerprinting of the tea varieties based on catechin composition in green leaf of cultivars grown in Northeast India. Assam variety cultivars contained the highest level of catechins followed by Cambod and China. The average catechin contents were 231 ± 7 mg g^{-1}, 202 ± 5 mg g^{-1}, and 157 ± 4 mg g^{-1} of dry weight of green leaf for Assam, Cambod, and China cultivars, respectively. Among the individual catechins the variations in epigallocatechin gallate (EGCG) and epigallocatechin (EGC) were the most prominent among the varieties. High EGC content was found to be a characteristic of Assam variety which was further corroborated through multivariate analysis.

1. Introduction

Present market is a selective one, and producers with high-quality tea are likely to survive. Quality of made tea of the plains of Northeast (NE) India depends on the quality of raw materials determined primarily by the polyphenolic constituents. It is widely accepted that Crush, Tear, and Curl (CTC) black tea quality attributes depend on flavonol composition (more precisely catechins). Epigallocatechin gallate (EGCG) is an important biochemical marker of Northeast Indian tea as it contributes 50% of total catechins [1]. The state of Assam (26°4′N–27°30′N and 89°58′E–95°41′E) in India is one of the major tea producing areas of the world. Tea in NE India is processed largely from the leaf of *Camellia assamica*, a source of a wide range of the catechins. Tea leaves contain about 180–360 mg g^{-1} of polyphenols, among which 70–80% are flavanols [2]. Total polyphenols including composition of catechins as well as their oxidation products were identified as being related to tea quality [3–7]. The variation in catechin composition is reflected in the variation in theaflavins (TFs) composition of black tea. It is well established that the formation of dihydroquercetin and dihydromyricetin, which are the precursors of dihydroxy catechins (epicatechin, (EC), epicatechin gallate (ECG)) and trihydroxy catechins (EGC and EGCG), respectively, is under genetic control [8–11].

Taxonomically tea is known as *Camellia sinensis* and belongs to the family Theaceae. Commercial tea cultivars are recognized under three different taxa, namely, *C. sinensis*, *C. assamica*, and *C. assamica ssp. lasiocalyx* [12]. However, tea is highly heterogeneous [9], and all the above taxa freely inter-breed, resulting in a cline extending from extreme China types to those of Assam origin [13]. Hybridization has been so extensive that it is often debated if archetypal *C. sinensis*, *C. assamica*, or *C. assamica ssp. lasiocalyx* still exist [14].

Based on morphological characteristics, tea is grouped into Assam, China, and Cambod varieties (Figure 1). The classification has been generally followed in Indian subcontinent possibly because of more varied and heterogeneous tea populations in the region [13]. The genetic differences between the hybrids are well reflected in biochemical composition of leaves. However, biochemical composition

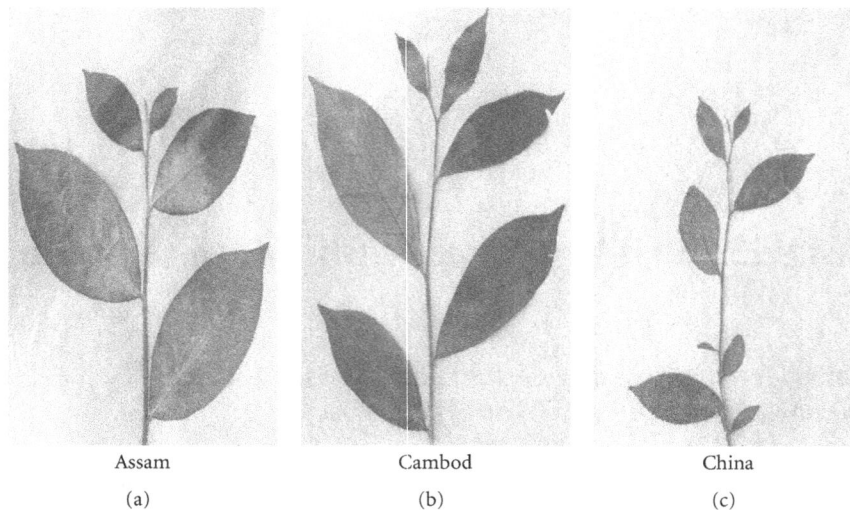

Assam   Cambod   China
(a)   (b)   (c)

FIGURE 1: The typical shoots of three varieties (note that the photographs were taken from the herbarium of Tocklai Experimental Station).

as varied between varieties is yet to be fully utilized in tea taxonomy [15]. Presence or absence of certain phenolic substances in the tea shoot has also been used as an aid in establishing interrelationship between taxa [16]. It has been reported that Assam type cultivars contain higher amount of polyphenols [2]. China variety cultivars generally possess quercetin and kaemferol-3-glucosides but these are totally absent or present only in traces in Assam variety [17, 18].

Tocklai Experimental Station, Jorhat, Assam, has released 153 germplasms to the tea industry of NE India to be grown in plains. Over 60% of 0.3 million ha of tea growing area of NE Indian plains is covered with these tea varieties. Regional variation of quality within the tea growing region (Figure 2) can be attributed to genetic diversity and its interaction with the environment.

Widespread cultivation of clonal tea for high yield and uniform quality may diminish the genetic diversity. Conservation of germplasm resources is necessary for sustainability of the tea industry. Tocklai Experimental Station has a field gene bank with over 2000 germplasms which is one of the primary centres of dispersal in the world. In order to ascertain diversity careful study of secondary metabolites, especially those which are major contributors to quality, is essential. Total catechin content could be used to indicate the quality potential of tea, with high content being related to high quality [4]. Earlier studies showed that tannin content, which is a measure of total catechin contents, could be used in the determination of genetic diversity in tea [19, 20]. However, these methods did not take into account the individual catechins present in tea leaf. Since the formation of black tea quality attributes is influenced by various catechins, characterization of cultivars based on various forms of catechins is essential to identify their quality potential [21].

The oxidative and hydrolytic enzymes endogenous to the shoots are crucial in triggering of various characteristic quality attributes of black tea. Out of the various stages of black tea processing, the fermentation stage is the most

crucial. The mechanical maceration of green tea shoots triggers the enzyme catalyzed oxidative reactions involving catechins as substrates. Upon disruption of the intercellular compartments, catechins present in the cell vacuole undergo in vivo oxidative and hydrolytic processes in presence of mild aeration. The desirable colour and briskness of made tea is dependent on the oxidative polymerization of catechins to TFs and thearubigins (TRs) by the enzymes polyphenol oxidase (PPO) and peroxidase (POD) [22].

The present study was undertaken to assess the variation of catechin (viz. EC, ECG, EGC, +C and EGCG) concentrations in extreme and cultivated varieties of Assam, China, and Cambod. The study also took into account the relative expression of individual catechins in cultivars grown in Northeast India. An understanding of catechin profile in different cultivars of tea may provide useful information on plant diversity as well as understanding their role as precursor of quality since type and quantity of catechin significantly influence the formation of two important quality attributes of tea such as theaflavins and thearubigins. This may also support future selection process for improvement of crop quality.

2. Materials and Methods

2.1. Plant Materials. Tea shoots comprising apical bud and subtending two leaves were harvested from the experimental garden of Tocklai Experimental Station, Tea Research Association, Jorhat, Assam, India ($94°12'E$ and $26°47'N$). A regular 7-day plucking during tea harvesting period was maintained. 7-day plucking interval is a common agricultural practice in tea growing areas of the NE India as it makes the young shoots produce high quality tea [23]. All the sampling plots received identical agricultural practices (fertilization at $120\,kg\,N\,ha^{-1}$, $110\,kg\,K\,ha^{-1}$ as K_2O and $30\,kg\,P\,ha^{-1}$ as P_2O_5 was applied per year, pruning and

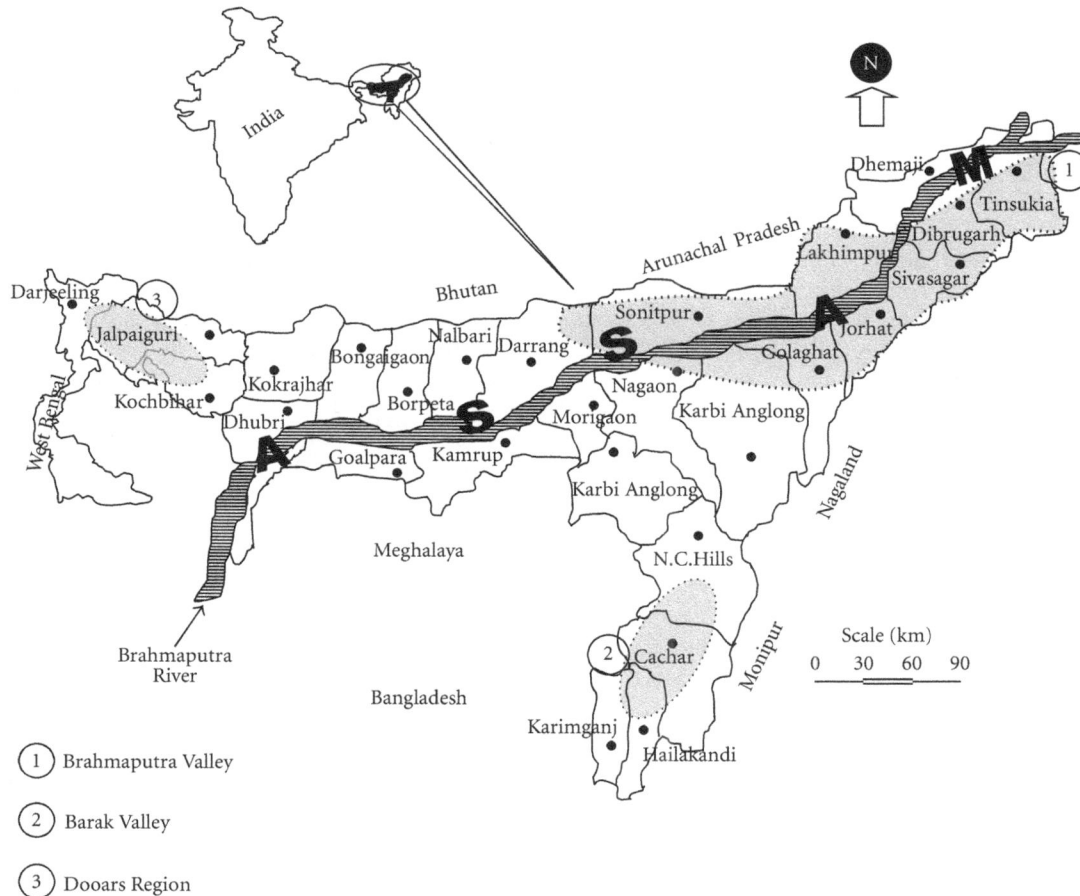

FIGURE 2: Map showing major tea growing areas of Northeast India.

plucking were maintained) where shade applied over the tea bushes contributed to 30–40% light interception. The reference samples represented pure varieties, namely, Assam (Betjan), China (Vimtal), and Cambod (extreme Cambod). Except reference samples, other cultivars representing the three varieties were as follows.

Assam Variety. TV2, TV12, TV13, TV17, TV21, S_3A_1, S_3A_3, Tingamira, TA 17, and T_3E_3 (where TV stands for Tocklai Vegetative and TA stands for Tin Ali).

China Variety. TV7, 14/13/3, 14/100/10, 14/100/16, 14/100/6, 317/1, 317/2, 317/3, 317/4, and P126 (P stands for Panitola).

Cambod Variety. TV9, TV18, TV22, TV23, TV25, TV26, and TV30.

The entire harvesting period for the sampling was from March to November for the years 2009 and 2010. Leaf samples were collected from the plots receiving similar agronomical practices. Samples were analyzed fortnightly. Soil samples collected from experimental plots were analyzed following the standard procedure described by Jackson [24]. Average soil status of the experimental plots was as follows:

well drained sandy loam soil, sand: 57.7 ± 2.1%, silt: 35.5 ± 1.4%, clay: 6.7 ± 0.7%, pH: 4.5 ± 0.002, organic carbon content: 8.0 ± 0.11 mg g^{-1}, total nitrogen: 0.8 ± 0.0001 mg g^{-1}, available P_2O_5: 0.01 ± 0.001 mg g^{-1}, and available K_2O: 0.08 ± 0.0001 mg g^{-1}.

2.2. Estimation of Catechins. About 100 g fresh tea leaf of each sample was deactivated and dried (at 90°C and dryness around 95%) in a microwave convection domestic oven (Model no. Onida PC21, India). Microwave drying of the samples did not affect catechin composition of green leaf tea samples (unpublished data of Biochemistry Department, Tocklai Experimental Station). The dried samples were ground finely for analysis. 0.2 g of sample was extracted with 5 mL 70% methanol in a water bath at 70°C over 10 min with intermittent shaking in a vortex mixture. The extract was then cooled and centrifuged at 4000 rpm (Rotanta 460R, UK) for 10 min. The supernatant was decanted into a 10 mL volumetric flask. The extraction was repeated twice and volume was made up with the solvent. 1 mL of the extract was diluted to 5 times with stabilizing agent. The stabilizing agent was prepared from EDTA (500 μg mL^{-1}), ascorbic acid (500 μg mL^{-1}), and acetonitrile (25% v/v) in water. Catechins were quantitatively estimated using waters high-performance liquid chromatography (HPLC) system with Luna 5 μ phenylhexyl phenomenax column (4.5 mm × 250 mm) and UV-Vis detector (Waters 2487, USA) set at 278 nm according to the method of International Standard Organisation [25]. During HPLC analysis, 10 μL of the diluted extract was injected into the column through Rheodyne injector. In brief, the elution made was initial 10 min with 100% mobile phase A followed by over 15 min with a linear gradient to 68% mobile phase A and 32% mobile phase B and held at this composition for another 10 min with flow rate 1 mL per min. The mobile phase A consists of 2% acetic acid and 9% acetonitrile and mobile phase B 80% acetonitrile. The chromatographic peaks were identified and estimated by external standard method from response factors (concentration of standards/peak area of standards) determined from different catechin standards procured from Sigma Aldrich, USA (ISO-14502 2005). The solvents used for extraction and analyses were of HPLC grade (E. mark, Mumbai, India).

2.3. Statistical Analysis. Raw data of various catechins of analysed tea samples were arranged in a data table where each row referred to an individual, and columns were associated to different variables.

The data were also log-transformed so as to more closely correspond to normal distribution. Further, all the variables were standardized by calculating their standard scores (*z*-scores) as follows:

$$z_i = \frac{x_i - \overline{x}}{s}, \tag{1}$$

where z_i is the standard score of the sample i; x_i is the value of sample i, $\overline{x}$ is the mean and s is the standard deviation.

Standardization scales the log-transformed data to a range of approximately ±3 standard deviations, centered about a mean of zero. In this way, each variable has equal weight in the statistical analyses. Besides normalizing and reducing outliers, these transformations also tend to homogenize the variance of the distribution [26–28]. Standardization also tends to increase the influence of variables whose variance is small and reduce the influence of variables whose variance is large. Furthermore, the standardization procedure eliminates the influence of different units of measurement and renders the data dimensionless.

The data were used for hierarchical agglomerative cluster analysis (HCA) described by Singh et al. [29 and principal component analysis (PCA) described by Kano et al. [30]. All these statistical analyses were performed using SPSS version 13 (SPSS Inc., Chicago, USA) [31].

3. Results and Discussion

Biochemical parameters of green leaf influencing black tea quality of the plains of NE India consist of catechins which are converted to TFs and TRs, the critical parameters of quality of CTC tea [32, 33]. Notwithstanding total polyphenols correlate with black tea quality, some polyphenols do not contribute to the formation of any black tea quality

parameter [34]. Only flavan-3-ols are critical for black tea quality [7]. The average catechin compositions of green tea leaves of the cultivars of three varieties is presented in Table 1. Large variations in the catechin compositions were observed among the cultivars reflecting genetic variability [6].

It was observed from this study that the total catechin and some individual catechins could be used as markers to differentiate between the three major varieties. The clear differentiation of China variety from Assam and Cambod could be established using the catechin as marker (see below). Similar observations were reported in Japanese tea [19]. Total green leaf catechin concentration and the ratio of dihydroxy to trihydroxy catechins were used to establish genetic diversity in the tea germplasms of Kenya [6]. Distribution of various catechins in all the three varieties showed that trihydroxy catechins accounted for 71–76% followed by dihydroxy for 22–27%. It is worth mentioning that EGCG which alone contributed 52–58% of total catechins was responsible for higher values of trihydroxy catechins. EGCG accounted for around 55% of total catechins in cultivars grown in Assam which was higher than the Central and Southern African tea leaf and much higher than Kenyan tea where contribution of EGCG was around 25% [5, 7].

Total catechin contents in green leaf were 231 ± 7.40 mg g^{-1}, 202 ± 4.58 mg g^{-1}, and 157 ± 3.82 mg g^{-1} for Assam, Cambod and China varieties, respectively. Large variations in individual catechins and total catechins among the varieties were observed. Assam variety cultivars contained the highest catechins followed by Cambod, and China (Table 1). The average EGCG contents of the varieties were 121.7 ± 2.4 mg g^{-1} for Assam, 112.6 ± 2.9 mg g^{-1} for Cambod and 86.2 ± 1.3 mg g^{-1} for China. Out of the eleven Assam cultivars studied, EGCG content of the cultivar S_3A_3 was found to be the highest. As the results indicated, the catechin content in China variety was substantially lower than the other two varieties.

The second largest contributor to total catechin content was EGC for Assam variety while it was ECG for Cambod and China variety. The variation in EGC content was more prominent between the varieties. The average EGC contents were 51.0 ± 1.0 mg g^{-1} for Assam, 36.1 ± 1.3 mg g^{-1} for Cambod, and 25.7 ± 0.8 mg g^{-1} for China variety. The average ECG content was found lower than EGC in Assam cultivars while in Cambod, and China cultivars it was higher. The average ECG contents in Assam, Cambod and China varieties were 38.6 ± 1.0 mg g^{-1}, 37.5 ± 1.2 mg g^{-1}, and 30.4 ± 1.2 mg g^{-1} respectively. Therefore, high EGC was a characteristic precursor of Assam variety. The dihydroxy-to-trihydroxy-catechin ratio (CATRAT) among the varieties was between 0.3 and 0.7. The highest CATRAT was found in TV7 of China variety.

3.1. Cluster Analysis and Principal Components Analysis. Hierarchical agglomerative cluster analysis (HCA) in the form of dendrogram and principal components analysis (PCA) were used to explore structure and relationships in multivariate data [27, 28]. The rationale of cluster analysis was to partition a set of objects into two or more groups

TABLE 1: Catechin profile of different green tea leaves (all units are in mg g^{-1} except catechin ratio, data represent the mean of three replicates ± standard error).

Cultivar	EGC	+C	EC	EGCG	ECG	Total Catechin (CAT)	Dihydroxy Catechin (EC+ECG)	Trihydroxy Catechin (EGC+EGCG)	CATRAT*
Assam varieties									
TV2	53.3 ± 0.5	5.4 ± 0.4	21.2 ± 0.6	114.7 ± 0.9	34.2 ± 0.4	228.1 ± 1.0	55.4 ± 0.5	167.9 ± 1.0	0.3 ± 0.1
TV12	43.2 ± 0.4	6.0 ± 0.2	12.0 ± 0.1	119.8 ± 0.8	36.6 ± 0.8	217.6 ± 0.8	48.6 ± 0.7	163.0 ± 0.5	0.3 ± 0.0
TV13	55.2 ± 0.7	4.5 ± 0.7	12.8 ± 0.3	140.8 ± 1.7	43.0 ± 0.7	256.6 ± 1.5	54.8 ± 0.9	195.9 ± 1.7	0.3 ± 0.0
TV17	41.2 ± 1.6	9.5 ± 0.4	12.5 ± 0.9	129.4 ± 1.1	62.6 ± 1.7	256.0 ± 0.6	75.2 ± 1.6	170.6 ± 1.2	0.5 ± 0.2
TV21	58.3 ± 1.5	4.6 ± 0.5	12.3 ± 0.5	111.2 ± 0.4	34.3 ± 0.3	220.7 ± 0.9	46.7 ± 0.4	169.5 ± 1.1	0.3 ± 0.1
S$_3$A$_1$	62.3 ± 0.6	3.5 ± 0.2	14.6 ± 0.2	142.5 ± 0.7	46.2 ± 0.9	269.2 ± 0.7	61.3 ± 0.7	212.8 ± 1.5	0.3 ± 0.0
S$_3$A$_3$	63.1 ± 0.8	7.7 ± 0.1	12.5 ± 0.0	147.0 ± 0.6	43.6 ± 0.9	274.0 ± 0.4	55.4 ± 0.8	210.5 ± 0.2	0.3 ± 0.1
Tingamira	48.4 ± 0.4	5.3 ± 0.5	19.8 ± 0.6	109.7 ± 0.6	35.4 ± 0.3	218.5 ± 0.6	55.2 ± 0.4	158.0 ± 0.5	0.4 ± 0.0
TA17	49.9 ± 0.8	8.6 ± 0.2	10.0 ± 0.1	139.7 ± 0.8	37.9 ± 0.5	246.2 ± 1.0	48.0 ± 0.5	189.8 ± 1.1	0.3 ± 0.0
T$_3$E$_3$	50.8 ± 1.0	4.1 ± 0.6	20.0 ± 0.6	113.1 ± 0.8	35.0 ± 0.6	222.3 ± 0.9	54.8 ± 0.6	163.9 ± 0.8	0.3 ± 0.1
Betjan	48.4 ± 0.5	4.6 ± 0.4	20.2 ± 0.5	108.0 ± 0.9	34.4 ± 0.7	215.2 ± 0.9	54.6 ± 0.7	156.4 ± 0.9	0.4 ± 0.1
Cambod varieties									
TV9	46.6 ± 0.4	4.5 ± 0.8	11.4 ± 0.7	122.6 ± 1.2	35.1 ± 0.7	220.3 ± 1.0	46.5 ± 0.8	169.2 ± 1.1	0.3 ± 0.1
TV18	31.0 ± 0.2	4.3 ± 0.1	5.7 ± 0.1	163.6 ± 0.2	60.8 ± 0.4	265.4 ± 0.3	65.9 ± 0.4	195.5 ± 0.2	0.3 ± 0.0
TV22	46.9 ± 0.1	7.2 ± 0.1	14.1 ± 0.7	131.0 ± 1.0	51.7 ± 0.1	250.9 ± 0.6	65.8 ± 0.2	177.9 ± 0.9	0.4 ± 0.1
TV23	39.1 ± 1.0	4.9 ± 0.3	19.8 ± 0.9	116.1 ± 0.5	41.1 ± 0.8	220.6 ± 0.6	60.0 ± 1.1	155.1 ± 0.2	0.4 ± 0.1
TV25	52.3 ± 0.6	8.1 ± 0.5	17.0 ± 0.5	122.0 ± 0.3	44.2 ± 0.1	243.7 ± 0.3	61.2 ± 0.2	174.4 ± 0.1	0.4 ± 0.0
TV26	32.1 ± 1.3	3.6 ± 0.6	12.3 ± 0.8	120.1 ± 0.5	41.7 ± 0.2	209.8 ± 0.4	54.0 ± 0.3	152.2 ± 0.4	0.4 ± 0.0
TV30	30.2 ± 1.7	5.1 ± 0.6	12.6 ± 0.8	91.4 ± 1.4	31.9 ± 0.7	175.0 ± 1.5	44.5 ± 0.8	125.4 ± 1.7	0.4 ± 0.1
Ex Cambod#	38.9 ± 2.7	1.3 ± 0.2	17.5 ± 1.1	109.0 ± 0.6	28.2 ± 1.0	204.8 ± 1.5	45.6 ± 1.1	157.9 ± 1.8	0.3 ± 0.1
China varieties									
TV7	18.2 ± 0.4	8.2 ± 0.7	17.2 ± 0.9	76.7 ± 1.3	45.5 ± 0.8	168.1 ± 1.5	60.7 ± 0.6	98.2 ± 1.8	0.7 ± 0.1
14/13/3	34.7 ± 0.4	2.4 ± 0.7	15.1 ± 1.4	85.0 ± 0.7	24.7 ± 0.6	157.4 ± 0.9	39.8 ± 1.2	119.7 ± 0.6	0.3 ± 0.1
14/100/10	16.5 ± 0.3	3.6 ± 0.7	6.6 ± 0.2	90.0 ± 0.3	25.4 ± 0.5	142.1 ± 0.2	32.0 ± 0.5	106.5 ± 0.4	0.3 ± 0.1
14/100/16	23.4 ± 0.4	2.9 ± 0.1	6.9 ± 0.1	77.6 ± 0.3	31.6 ± 0.4	142.4 ± 0.6	38.5 ± 0.4	101.1 ± 0.5	0.4 ± 0.0
14/100/6	24.3 ± 0.2	5.8 ± 0.1	7.0 ± 0.1	85.8 ± 0.3	25.9 ± 0.2	148.7 ± 0.2	32.8 ± 0.2	110.1 ± 0.3	0.3 ± 0.1
317/7	25.7 ± 0.2	6.1 ± 0.1	7.9 ± 0.3	77.7 ± 0.2	24.4 ± 0.2	141.7 ± 0.1	32.2 ± 0.2	103.5 ± 0.2	0.3 ± 0.0
317/2	31.2 ± 0.1	3.9 ± 0.1	15.4 ± 0.1	89.0 ± 0.1	27.1 ± 0.1	166.7 ± 0.1	42.5 ± 0.1	120.2 ± 0.1	0.4 ± 0.0
317/3	31.2 ± 0.2	2.0 ± 0.3	14.2 ± 0.3	86.5 ± 0.1	20.2 ± 0.1	154.1 ± 0.2	34.3 ± 0.3	117.7 ± 0.1	0.3 ± 0.0
317/4	33.4 ± 0.1	2.1 ± 0.1	7.5 ± 0.3	85.9 ± 0.1	25.2 ± 0.2	154.1 ± 0.2	32.67 ± 0.2	119.3 ± 0.1	0.3 ± 0.0
P126	34.0 ± 0.2	1.8 ± 0.2	15.4 ± 0.2	81.3 ± 0.1	21.2 ± 0.0	153.3 ± 0.1	36.5 ± 0.1	115.4 ± 0.1	0.3 ± 0.0
Vimtal	30.2 ± 0.2	3.0 ± 0.1	15.4 ± 0.1	88.0 ± 0.1	26.8 ± 0.1	157.6 ± 1.1	42.1 ± 0.1	118.6 ± 0.2	0.4 ± 0.0

*CATRAT = (EC+ECG)/(EGC+EGCG); #Ex Cambod: Extreme Cambod.

based upon the similarity of the objects in many disciplines with respect to a chosen set of characteristics so that similar objects were in the same class [35]. In the cluster analysis, emphasis was to differentiate biochemical parameters, based upon multiple tea samples and quality parameters, and it was done through HCA. Therefore, HCA was applied to the biochemical data sets with a view to grouping the similar spatial variabilities spread over the variety of tea samples and in the resultant dendrogram. This method used the analysis of variable approach to evaluate the distances between clusters, attempting to minimize the sum of squares of any two clusters that could be formed at each step. It yielded a dendrogram (Figures 3(a), 3(b), and 3(c)), grouping all

variables of the samples into two statistically significant clusters. For Assam tea, three clusters were constructed. One cluster included (+) catechin (+C), dihydroxy-to-trihydroxy-catechin ratio (CATRAT) and EC, another one included EGC, dihydroxy catechin (EC + ECG), and ECG. These two clusters were interrelated with another cluster having EGCG, trihydroxy catechin (EGC + EGCG), and total catechin (CAT) (Figure 3(a)). The similar pattern of dendrogram was also observed for Cambod (Figure 3(b)) and China varieties (Figure 3(c)). Therefore, this indicated that all parameters were likely having direct influence on the quality of tea leaf irrespective of their varieties.

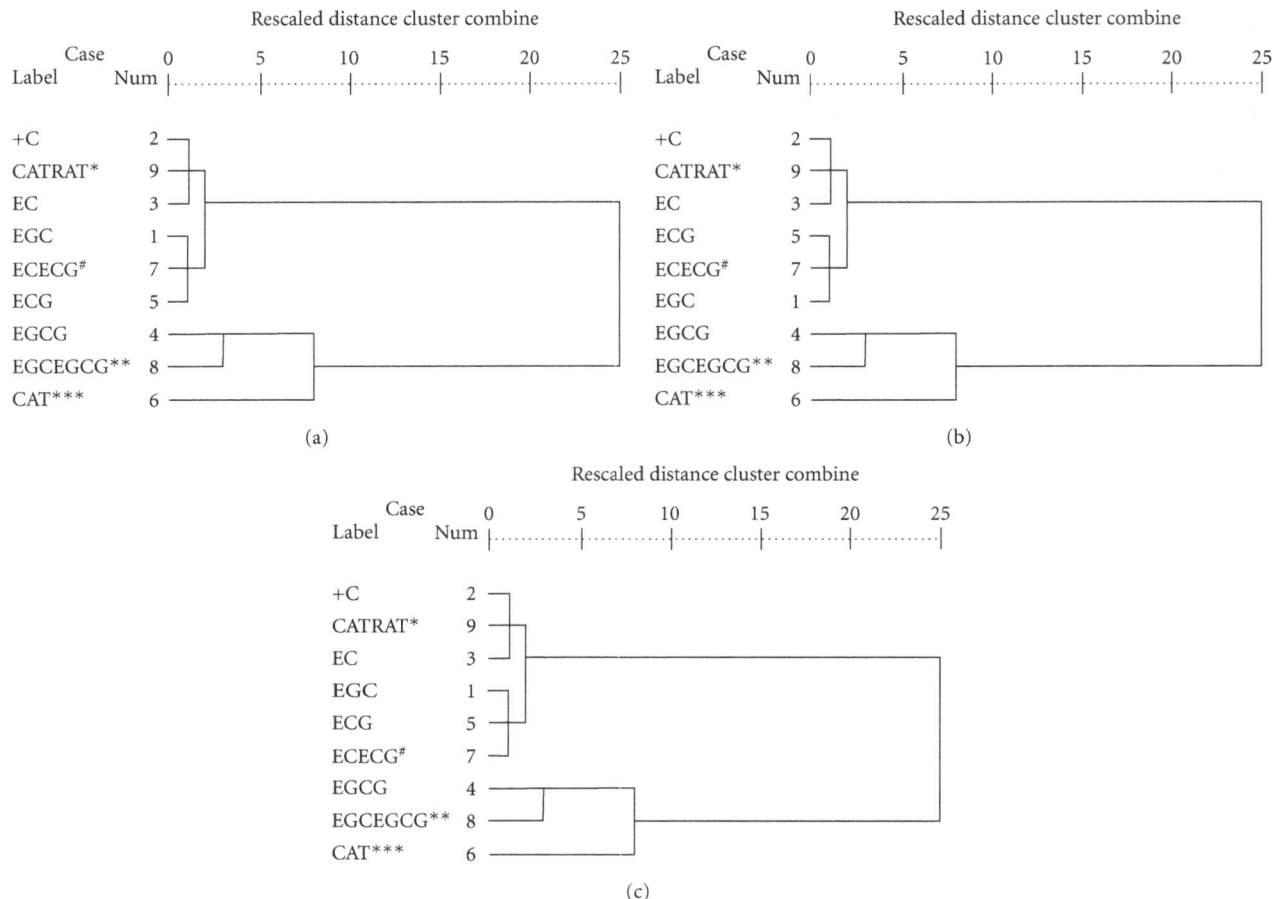

FIGURE 3: Dendrogram for cluster analysis. The dissimilarity defined by Euclidean distance and the combination of clusters is based on Ward method. (a) Assam, (b) Cambod, and (c) China tea (*CATRAT: catechin ratio; #ECECG: dihydroxy catechin; **EGCEGCG: trihydroxy catechin, and ***CAT: total catechin).

From HCA, we could not clearly distinguish the relations among the different varieties of tea samples. Therefore, all the parameters were transformed into three main comprehensive matrices referring to PCA technique. On plotting the first two principal components (Varimax 1 and Varimax 2), they showed two clusters for Assam tea, one cluster for Cambod tea, and three clusters for China tea in PCA (Figures 4(a), 4(b), and 4(c)). Principal component analysis (PCA) is one of the best statistical techniques for extracting linear relationships among a set of variables. Principal components are the linear combinations of original variables and are the eigenvectors. The Varimax rotation distributes the PC loadings such that their dispersion is minimized by maximizing the number of large and small coefficients. The Cornbach alpha and Kaiser-Meyer-Olkin (KMO) sample adequacy showed the appropriate application of PCA in the present dataset. Principal component 1 (PC1) had higher loadings for the variables like ECG and dihydroxy catechin (EC + ECG) with +C for Assam tea (Figure 4(a)). PC1 accounted for 41.8% of the total variance and could be thus interpreted as a catechin component. PC2 contained 33.6% of the variance and had a higher loading for total catechin (CAT), EGCG, and trihydroxy catechin (EGC + EGCG). This component can be explained taking into account that high levels of total catechin contributed to better quality of Assam tea. Figure 4(b) reflects the PCA of Cambod tea. Here only one PC (PC1) was obtained containing ECG, EGCG, dihydroxy catechin (EC + ECG), and trihydroxy catechin (EGC + EGCG) with total catechin (CAT). PC1 contained 83% of the variance. Therefore, comparing Figures 4(a) and 4(b) it can be concluded that the pattern of catechins in Cambod tea differed from the one present in Assam tea. China tea gave three principal components. PC1 explained 44.43% of the total variance, whereas PC2 and PC3 expressed 33.37% and 9.70%, respectively, of the variance. PC1 can be interpreted as a major quality component of China tea where the contributing factors were EGC, EGCG, and trihydroxy catechin (EGC + EGCG), as shown in Figure 4(c).

4. Conclusion

Differential display of catechins in cultivars forms a basis for future elucidation of catechin metabolism in tea. Profiling of individual and total catechins was found to be a useful

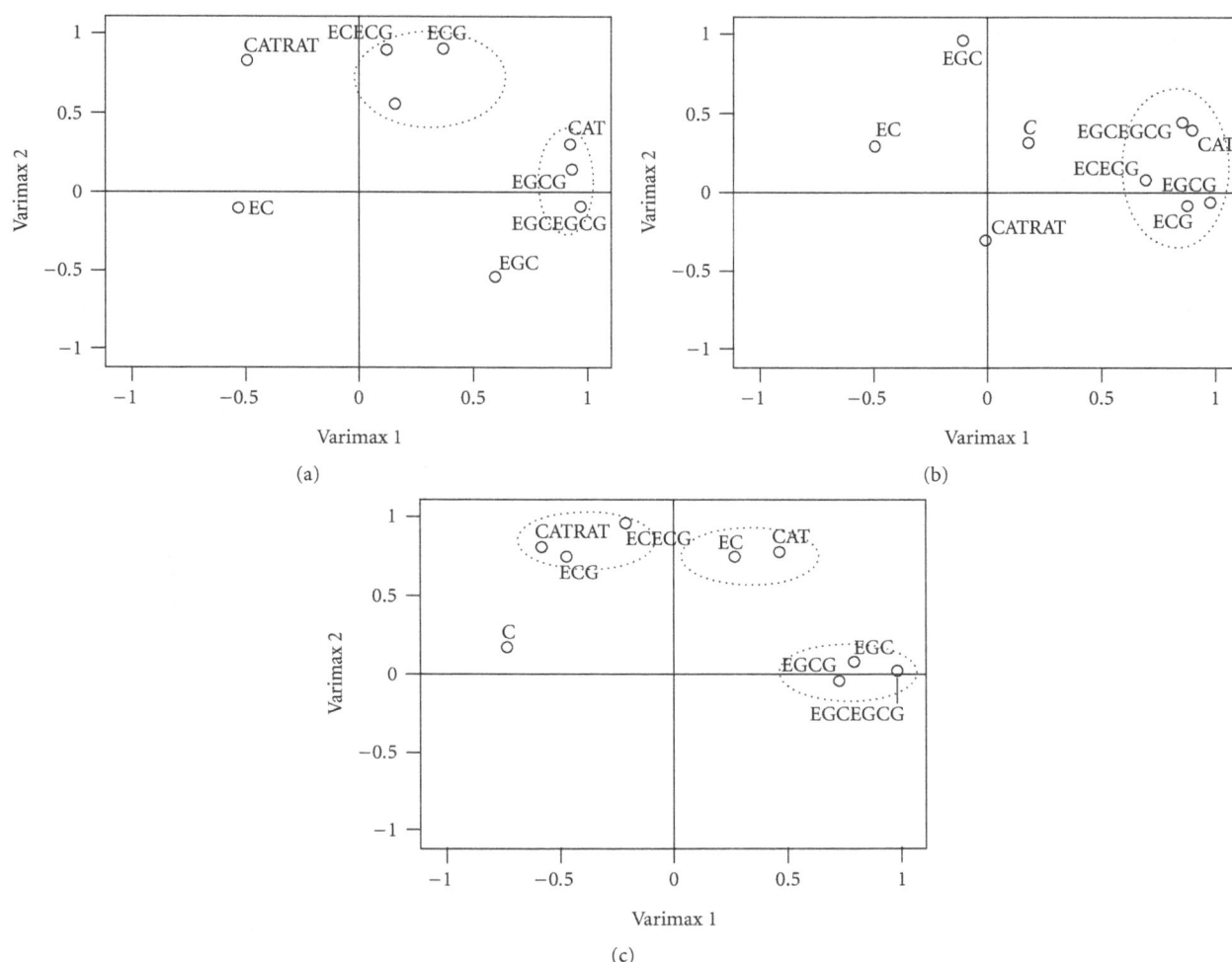

FIGURE 4: Factor loading pattern of the analyzed variables of the three varieties obtained by PCA followed by Varimax rotation. (a) Assam, (b) Cambod, and (c) China tea (note that in these figures "C" indicates "+C").

technique to determine genetic diversity in tea germplasms. Among the three pure varieties China variety cultivars contained lower catechins. PCA showed different groupings of catechins for Assam, Cambod, and China teas, and such groupings might be used to differentiate between such varieties.

Acknowledgment

The authors gratefully acknowledge the helpful comments and suggestions made by three anonymous reviewers of this paper.

References

[1] S. Gupta, B. Saha, and A. K. Giri, "Comparative antimutagenic and anticlastogenic effects of green tea and black tea: a review," *Mutation Research*, vol. 512, no. 1, pp. 37–65, 2002.

[2] L. P. Bhuyan, A. Hussain, P. Tamuly, R. C. Gogoi, P. K. Bordoloi, and M. Hazarika, "Chemical characterisation of CTC black tea of northeast India: correlation of quality parameters with tea tasters' evaluation," *Journal of the Science of Food and Agriculture*, vol. 89, no. 9, pp. 1498–1507, 2009.

[3] M. Obanda, P. O. Owuor, and C. K. Njuguna, "The impact of clonal variation of total polyphenols content and polyphenol oxidase activity of fresh tea shoots on plain black tea quality parameters," *Tea*, vol. 13, pp. 129–133, 1992.

[4] M. Obanda, P. O. Owuor, and S. J. Taylor, "Flavanol composition and caffeine content of green leaf as quality potential indicators of Kenyan black teas," *Journal of the Science of Food and Agriculture*, vol. 74, no. 2, pp. 209–215, 1997.

[5] L. P. Wright, N. I. K. Mphangwe, H. E. Nyirenda, and Z. Apostolides, "Analysis of caffeine and flavan-3-ol composition in the fresh leaf of *Camellia sinesis* for predicting the quality of the black tea produced in Central and Southern Africa," *Journal of the Science of Food and Agriculture*, vol. 80, no. 13, pp. 1823–1830, 2000.

[6] G. N. Magoma, F. N. Wachira, M. Obanda, M. Imbuga, and S. G. Agong, "The use of catechins as biochemical markers in diversity studies of tea (*Camellia sinensis*)," *Genetic Resources and Crop Evolution*, vol. 47, no. 2, pp. 107–114, 2000.

[7] P. O. Owuor and M. Obanda, "The use of green tea (*Camellia sinensis*) leaf flavan-3-ol composition in predicting plain black

tea quality potential," *Food Chemistry*, vol. 100, no. 3, pp. 873–884, 2007.

[8] A. M. Gerats and C. Martin, "Flavanoid synthesis in Petunia Hybrida; genetics and molecular biology of flower colour," in *Phenolic Metabolism in Plants*, H. A. Stafford and R. K. Ibrahim, Eds., pp. 167–175, Plenum Press, New York, NY, USA, 1992.

[9] A. Gulati, S. Rajkumar, S. Karthigeyan et al., "Catechin and catechin fractions as biochemical markers to study the diversity of Indian tea (*Camellia sinensis* (L.) O. Kuntze) germplasm," *Chemistry and Biodiversity*, vol. 6, no. 7, pp. 1042–1052, 2009.

[10] M. A. S. Marles, H. Ray, and M. Y. Gruber, "New perspectives on proanthocyanidin biochemistry and molecular regulation," *Phytochemistry*, vol. 64, no. 2, pp. 367–383, 2003.

[11] G. J. Tanner, K. T. Francki, S. Abrahams, J. M. Watson, P. J. Larkin, and A. R. Ashton, "Proanthocyanidin biosynthesis in plants. Purification of legume leucoanthocyanidin reductase and molecular cloning of its cDNA," *Journal of Biological Chemistry*, vol. 278, no. 34, pp. 31647–31656, 2003.

[12] P. K. Baruah, "Classification of tea plants: species hybrids," *Two and a Bud*, vol. 13, no. 1, pp. 14–16, 1965.

[13] W. Wight, "Tea classification revised," *Current Science*, vol. 31, pp. 298–299, 1962.

[14] H. P. Bezbaruah, "The tea varieties in cultivation—an appraisal," *Two and a Bud*, vol. 23, pp. 13–19, 1976.

[15] G. W. Sanderson and P. Kanapathipillai, "Further study on the effect of climate and clone on the chemical composition of fresh tea flush," *Tea Quarterly*, vol. 35, no. 4, pp. 222–229, 1964.

[16] E. A. H. Roberts, W. Wight, and D. J. Wood, "Paper chromatography as an aid to the identification of *Thea Camellias*," *New Phytologist*, vol. 57, pp. 211–225, 1958.

[17] M. Hazarika and P. K. Mahanta, "Compositional changes in chlorophylls and carotenoids during the four flushes of tea in north east India," *Journal of the Science of Food and Agriculture*, vol. 35, pp. 298–303, 1984.

[18] B. Banerjee, "Botanical classification of tea," in *Tea Cultivation to Consumption*, K. C. Wilson and M. N. Clifford, Eds., pp. 25–51, Chapman and Hall, London, UK, 1992.

[19] Y. Takeda, "Differences in caffeine and tannin contents between tea cultivars, and application to tea breeding," *Japan Agricultural Research Quarterly*, vol. 28, no. 2, pp. 117–123, 1994.

[20] K. Wei, L. Wang, J. Zhou et al., "Catechin contents in tea (*Camellia sinensis*) as affected by cultivar and environment and their relation to chlorophyll contents," *Food Chemistry*, vol. 125, no. 1, pp. 44–48, 2011.

[21] J. Thomas, R. Raj Kumar, and A. K. A. Mandal, "Diversity among various forms of flavanols in selected UPASI tea germplasms," *Journal of Plantation Crops*, vol. 36, no. 3, pp. 171–174, 2008.

[22] A. Robertson, "The chemistry and biochemistry of black tea production, the non-volatiles," in *Tea Cultivation to Consumption*, K. C. Wilson and M. N. Clifford, Eds., pp. 603–647, Chapman and Hall, London, UK, 1992.

[23] S. Baruah, M. Hazarika, P. K. Mahanta, H. Horita, and T. Murai, "Effect of plucking intervals on the chemical constituents of CTC black teas," *Agricultural and Biological Chemistry*, vol. 50, pp. 1039–1041, 1986.

[24] M. L. Jackson, *Soil Chemical Analysis*, Prentice Hall of India, New Delhi, India, 1973.

[25] ISO 14502-2, "Determination of substances characteristics of green and black tea-Part2: content of catechins in green tea-Method using high-performance liquid chromatography," *International Standard Organisation*, 2005.

[26] R. J. Rummel, *Applied Factor Analysis*, Northwestern University Press, Evanston, Ill, USA, 1970.

[27] O. Abollino, M. Malandrino, A. Giacomino, and E. Mentasti, "The role of chemometrics in single and sequential extraction assays: a review. Part I. Extraction procedures, uni- and bivariate techniques and multivariate variable reduction techniques for pattern recognition," *Analytica Chimica Acta*, vol. 688, no. 2, pp. 104–121, 2011.

[28] A. Giacomino, O. Abollino, M. Malandrino, and E. Mentasti, "The role of chemometrics in single and sequential extraction assays: a Review. Part II. Cluster analysis, multiple linear regression, mixture resolution, experimental design and other techniques," *Analytica Chimica Acta*, vol. 688, no. 2, pp. 122–139, 2011.

[29] K. P. Singh, A. Malik, D. Mohan, and S. Sinha, "Multivariate statistical techniques for the evaluation of spatial and temporal variations in water quality of Gomti River (India) -a case study," *Water Research*, vol. 38, no. 18, pp. 3980–3992, 2004.

[30] M. Kano, S. Hasebe, I. Hashimoto, and H. Ohno, "A new multivariate statistical process monitoring method using principal component analysis," *Computers and Chemical Engineering*, vol. 25, no. 7-8, pp. 1103–1113, 2001.

[31] M. J. Norusis, *SPSS@10.0. Guide to Data Analysis*, Prentice-Hal, Chicago, Ill, USA, 2000.

[32] P. J. Hilton, R. Palmer-Jones, and R. T. Ellis, "Effects of season and nitrogen fertiliser upon the flavanol composition and tea making quality of fresh shoots of tea (*Camellia sinensis* L.) in central Africa," *Journal of the Science of Food and Agriculture*, vol. 24, pp. 819–826, 1973.

[33] D. J. Millin, "Factors affecting the quality of tea," in *Quality Control in the Food Industry*, S. M. Herschdoerfer, Ed., vol. 4, pp. 127–160, Academic Press, London, UK, 1987.

[34] E. A. H. Roberts, "Economic importance of flavonoid substances: tea fermentation," in *The Chemistry of Flavonoid Compounds*, T. A. Geisman, Ed., pp. 468–510, Pergamon Press, New York, NY, USA, 1962.

[35] J. F. Hair, R. E. Anderson, R. L. Tatham, and W. C. Black, *Multivariate Data Analysis with Readings*, Prentice-Hall International, New Jersey, NJ, USA, 1995.

17

Influence of Drought and Sowing Time on Protein Composition, Antinutrients, and Mineral Contents of Wheat

Sondeep Singh, Anil K. Gupta, and Narinder Kaur

Department of Biochemistry, Punjab Agricultural University, Ludhiana 141004, India

Correspondence should be addressed to Anil K. Gupta, anilkgupta@sify.com

Academic Editor: Enrico Porceddu

The present study in a two-year experiment investigated the influence of drought and sowing time on protein composition, antinutrients, and mineral contents of wheat whole meal of two genotypes differing in their water requirements. Different thermal conditions prevailing during the grain filling period under different sowing time generated a large effect on the amount of total soluble proteins. Late sown conditions offered higher protein content accompanied by increased albumin-globulin but decreased glutenin content. Fe content was increased to 20–23%; however, tannin decreased to 18–35% under early sown rain-fed conditions as compared to irrigated timely sown conditions in both the genotypes. Activity of trypsin inhibitor was decreased under rain-fed conditions in both genotypes. This study inferred that variable sowing times and irrigation practices can be used for inducing variation in different wheat whole meal quality characteristics. Lower temperature prevailing under early sown rain-fed conditions; resulted in higher protein content. Higher Fe and lower tannin contents were reported under early sown rain-fed conditions however, late sown conditions offered an increase in phytic acid accompanied by decreased micronutrients and glutenin contents.

1. Introduction

Wheat (*Triticum aestivum* L.), the most important cereal crop along with rice and corn, feeds most of the world's population. It is contributing to 28% of the world edible dry matter and up to 60% of the daily calorie intake in several developing countries. Wheat consumption is increasing worldwide as a result of higher income levels, urbanization, and substitution with other cereals. Therefore, the nutritional quality of the wheat whole meal has a significant impact on human health and well-being especially in the developing world [1].

Wheat endosperm acts as a starch storage region of the kernel while germ is a concentrated source of minerals and protein. Globally, wheat is the leading source of vegetable protein in human food having higher protein content than the other major cereals. Amount as well as composition of grain proteins determines the protein quality and hence end-use of wheat. The gluten proteins, the gliadins, and glutenins constitute up to 80–85% of total flour protein, and they confer properties of elasticity and extensibility that are essential for functionality of wheat whole meals [2]. Apart from the gluten proteins, albumins and globulins constituting 10–22% of total flour protein [3] are important from nutritional point because of high amounts of essential amino acids.

Mineral nutrients play a fundamental role in the biochemical and physiological functions of biological systems. Micronutrient malnutrition particularly of Fe and Zn is a growing concern worldwide. It is mainly due to extensive consumption of staple cereals which have low amount and availability of micronutrients [4]. Modern wheat (*T. aestivum*) cultivars are poor in Fe and Zn contents as compared to wild and primitive wheat such as einkorn wheat (*Triticum monococum*), emmer wheat (*Triticum dicocum*), and wild emmer wheat (*Triticum dicoccoides*) [5]. One sustainable agricultural approach to reducing micronutrient malnutrition globally is to enrich staple food crops with micronutrients or decreasing antinutrient substances that inhibit micronutrient bioavailability. So the nutritional quality of wheat whole meal is further dependent on the status of antinutrients such as phytates, tannins, and protease inhibitors in grains.

Phytate or phytic acid accounts for as much as 85% of the total phosphorus content of cereal grains and significantly influences the functional and nutritional properties of foods. Due to its strong binding capacity, phytic acid readily forms insoluble complexes with multivalent cations and proteins at physiological pH and hence renders several minerals biologically unavailable to animals and humans [6]. Low phytic acid-containing cereals, therefore, are desirable from nutritional point. Tannins are polyhydric phenols which form insoluble complexes with proteins, carbohydrates, and lipids leading to reduction in digestibility of these nutrients. Other effects that have been attributed to tannins include damage to the intestinal tract and interference with the absorption of iron and a possible carcinogenic effect [7]. Similarly protease inhibitors are known to cause growth inhibition by interfering with digestion causing pancreatic hypertrophy and metabolic disturbance of sulfur amino acid utilization [8].

During the past several decades, the primary objective of plant breeding programs has been to increase yield, a quest that will remain a principal concern in providing the calorie intake required for the growing world population. However, equally important but largely overlooked in breeding programs is the nutritional quality of wheat whole meal particularly the protein quality and micronutrients concentrations and their bioavailability [5].

The wheat quality characteristics are usually influenced by genotype, environmental factors, and interactions between genotype and environment. Adverse environmental conditions such as extreme temperature and drought during the anthesis and grain filling period have been identified as major constraints to wheat protein content and composition [9, 10]. Most of the earlier studies were focused on the effect of either heat or drought stress during grain filling on composition of proteins and starch characteristics. Few studies have examined the combined effect of high temperature and drought on protein and gluten contents with limited emphasis on nutritional quality. However, little is known concerning the combined effect of sowing time and drought stress on nutritional quality of wheat whole meal. In our earlier work we analysed the effect of rain-fed and different sowing times on monomeric/polymeric proteins and starch pasting characteristics of wheat whole meal [11]. The study clearly indicated the significant variation in the content and composition of storage proteins under different sowing conditions. Rain-fed conditions resulted in generation of new correlations among various protein fractions and starch pasting characteristics, which were far from those observed under irrigated conditions. The study suggested that utilization of these new correlation trends observed between different protein components under rain-fed conditions could be useful for enhancing protein content without affecting protein quality of wheat grains. However, the work was confined up to the level of monomeric and polymeric protein fractions and starch pasting characteristics. In the present study, therefore, we are attempting to further fractionate the protein to albumin-globulins and gliadin components and to observe the variation in the minerals and antinutrients, to better understand the effect of varying sowing conditions on wheat whole-meal quality.

2. Materials and Methods

2.1. Plant Material and Experimental Conditions. Two wheat (T. aestivum L.) genotypes, PBW 343 (high yielding, drought susceptible) and C 306 (drought tolerant), were grown in fields of the Punjab Agricultural University, Ludhiana, (30°54′N, 75°48′E, elevation 247 m above mean sea level), India. The experimental soil was loamy sand with pH about 7.8–8.0. The seeds were sown in $2 \, m^2$ plots with a row space of 20 cm. The sowing dates were early sown (ES) (15 October), timely sown (TS) (15 November), and late sown (LS) (15 December) in 2007 (year I) and 2008 (year II). The experimental design was a randomized complete block with three replications. Under the irrigated treatment, plants were watered throughout the period from sowing to maturity according to the recommended agronomic practices [12]. All irrigations were withheld from the plants subjected to the drought treatment except the presowing irrigation for field preparation. Therefore, the drought-treated plants received water only available through rainfall. The weather data of total rainfall and evapotranspiration rate for the crop season in both years was collected from the field meteorological observatory. Total rainfall as an average of two years received by the crop from sowing to maturity was 94, 81, and 82 mm, while during grain filling period was 23, 55, and 45 mm for early, timely and late sown crops, respectively. During the crop season rainfall was scanty while evapotranspiration was higher, which helped in the development of severe drought stress.

2.2. Grain Plumpness. Wheat grains were separated according to the grain diameters (>2.8 mm and <2.5 mm) using grain sorter (Sortimat, Germany) fitted with respective sized meshes. The grains of different diameters were collected and then weighed on an electronic balance and expressed in percentage.

2.3. Test Weight. Test weight was determined using the apparatus developed by the Directorate of Wheat Research, Karnal, India, which employs a standard container of 100 mL capacity. The container is filled with the sample of wheat grains by removing all shrunken and broken kernels and other foreign material. The grains were weighed and the test weight expressed in kg/hectoliter (hl) [13].

2.4. Milling and Protein Fractionation. Grains from each treatment were finally milled using Cyclotec 1093 sample mill (Foss, Tecator, Sweden) to obtain wheat whole meal in both the years. Protein fractionation was carried out from wheat whole meal according to Triboi et al. [10] with a little modification. Briefly, the protein fractions albumin-globulin, amphiphil, gliadin, and glutenin were sequentially extracted from 800 mg of flour. During each extraction step, the samples were continuously shaken in a volumetric flask (100 mL) placed in a temperature controlled shaker (orbital shaking incubator CIS-24, Remi, India) for 60 min. Soluble and insoluble fractions were separated by centrifugation at $8,000 \times g$ (Sigma, USA) for 30 min at the extraction

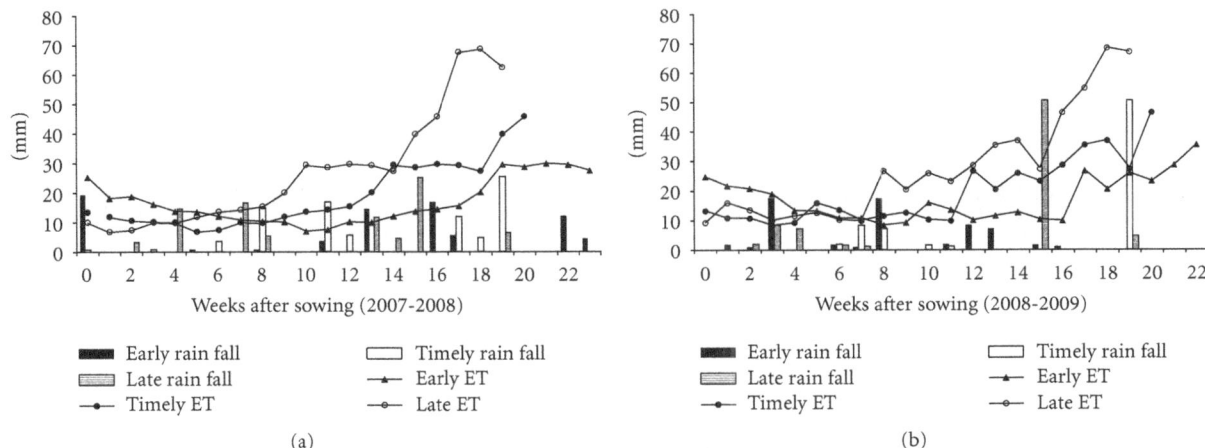

FIGURE 1: Patterns of total rain fall and evapotranspiration during the wheat growing season for the year 2007-2008 (year I) and 2008-2009 (year II). ET, evapotranspiration.

temperature. Albumins-globulins were extracted at 4°C with 25 mL 0.05 M NaCl, 0.05 M sodium phosphate buffer pH 7.8. After removal of amphiphilic proteins by extraction at 4°C from the previous pellet with 25 mL 2% (v/v) Triton X-114, 0.1 M NaCl, 0.05 M sodium phosphate buffer pH 7.8, gliadins were extracted at 20°C from the previous pellet with 25 mL 70% (v/v) ethanol. Glutenins were extracted at 20°C from the previous pellet with 25 mL 2% (w/v) SDS (sodium dodecyl sulfate), 2% (v/v) β-mercaptoethanol, 0.05 M tetraborate buffer pH 8.5. After centrifugation, the glutenins were recovered in the supernatant.

2.5. Antinutrients. Antinutrients like phytic acid was extracted from wheat whole meal with 1.2% HCl and precipitated with 0.4% $FeCl_3$, and inorganic phosphorus was estimated as described previously [14]. Trypsin inhibitor was isolated and quantified by inhibiting the bovine trypsin by using N-α-benzoyl-DL-arginine-p-nitroanilide as a substrate [15]. One trypsin inhibitor unit is defined as the quantity of inhibitor which inhibits 50% of trypsin inhibitor activity. Tannins were extracted and estimated as described earlier [16]. Briefly, tannins were extracted twice from wheat whole meal with boiling water. All the fractions were filtered through the Whatman no. 1 filter paper and pooled. The intensity of blue colour developed after addition of Folin-Denis' reagent and sodium carbonate solutions was measured at 700 nm.

2.6. Minerals. Wheat whole meal was analysed for Fe and Zn contents using atomic absorption spectrophotometer (AAS 240, Varian). Flour samples were digested in a mixture of nitric acid:perchloric acid (10 : 1 v/v); volume was made with quartz glass distilled water and analysed by AAS 240.

2.7. Statistical Analysis. The results are expressed as mean ± standard deviation of three replicates obtained from three different sowing plots in the fields. A completely randomized design was used for analysis of variance (ANOVA) using CPCS 1 software package. Correlation analysis was carried

out using MS Excel 2003. Statistical significance was set at $P \leq 0.05$, except where specified.

3. Results

The different sowing times selected in this study resulted in exposure of plants to varied temperatures (Figure 1) and rainfall (Figure 2) events before and during grain filling periods. The mean maximum temperatures during grain filling periods under early sown (ES), timely sown (TS), and late sown (LS) conditions as an average of both the years were 22.4°C, 26.3°C, and 29.4°C, respectively. Similarly, mean minimum temperatures were 8.8°C, 11.8°C, and 14.2°C in ES, TS, and LS conditions, respectively. Diverse temperature range and amount of rainfall in different sowing times resulted in significant differences in test weight and grain plumpness (Table 1) for both genotypes.

3.1. Flour Protein Fractions. Amount of total soluble protein (TSP) was higher in genotype C 306 as compared to PBW 343 under all treatments (Table 2). Rain-fed (RF) conditions resulted in decrease in TSP under timely sown (TS) and late sown (LS) crop whereas significant increase was observed under early sown (ES) conditions. Crop in irrigated ES conditions accumulated lesser TSP; however, a significant increase in TSP accumulation was observed for LS conditions in both genotypes. This resulted in higher grain protein percentage due to delayed sowing (Table 2).

Different protein fractions were also influenced by sowing times and RF conditions in both genotypes (Table 2). Albumin-globulin and gliadin proteins ranged from 9 to 13% and 21 to 32% of total protein, respectively, under irrigated conditions (Table 2). When compared to irrigated TS conditions, albumin-globulin proteins were enhanced under irrigated LS as well as under RF conditions in both genotypes. Both in irrigated + TS and RF + TS conditions, the gliadin content was significantly low in C 306 as compared to PBW 343 (Table 2). However, under RF + ES

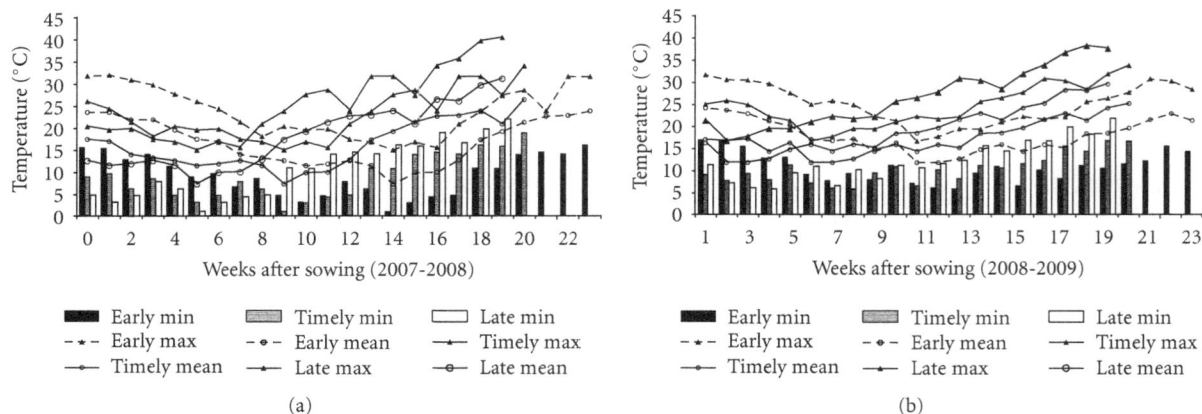

FIGURE 2: Patterns of weekly temperatures during the wheat growing season for the year 2007-2008 (year I) and 2008-2009 (year II).

TABLE 1: Effect of sowing time on test weight and grain plumpness of drought-susceptible (PBW 343) and drought-tolerant (C 306) wheat genotypes grown under irrigated and rain-fed conditions. Values are mean ± standard deviation of three plots; CD, critical difference; hl, hectoliter; NS, not significant.

Year	Sowing time	Genotype	Test weight (Kg/hl) Treatment		Grain plumpness > 2.8 mm(%) Treatment		Grain plumpness < 2.5 mm (%) Treatment	
			Irrigated	Rain fed	Irrigated	Rain fed	Irrigated	Rain fed
Year I (2008)	Early sown	PBW 343	71.82 ± 0.40	66.50 ± 3.98	26.71 ± 1.97	39.12 ± 0.75	21.21 ± 0.59	17.34 ± 0.27
		C 306	77.81 ± 0.96	71.48 ± 0.51	53.73 ± 2.70	23.48 ± 1.25	12.16 ± 0.54	29.24 ± 1.23
	Timely sown	PBW 343	72.74 ± 0.04	68.89 ± 0.45	24.70 ± 0.36	7.79 ± 1.87	20.25 ± 0.71	40.32 ± 1.95
		C 306	75.17 ± 2.28	76.94 ± 0.30	56.18 ± 6.40	41.65 ± 1.00	13.22 ± 5.29	19.26 ± 0.28
	Late sown	PBW 343	70.18 ± 0.14	71.78 ± 0.08	9.16 ± 0.81	14.52 ± 0.54	38.61 ± 1.44	31.74 ± 1.59
		C 306	76.87 ± 0.64	74.04 ± 0.09	63.17 ± 1.12	39.94 ± 0.06	9.83 ± 0.85	19.42 ± 0.15
Year II (2009)	Early sown	PBW 343	74.08 ± 0.60	72.04 ± 0.14	66.87 ± 17.77	36.42 ± 0.53	7.02 ± 4.21	21.85 ± 0.57
		C 306	78.95 ± 1.33	77.70 ± 0.87	65.55 ± 3.00	42.69 ± 8.24	8.54 ± 0.77	18.63 ± 3.10
	Timely sown	PBW 343	75.19 ± 1.40	72.45 ± 1.21	58.82 ± 16.80	33.40 ± 3.59	7.82 ± 4.90	22.56 ± 2.02
		C 306	78.17 ± 2.28	77.26 ± 0.45	63.84 ± 10.43	43.25 ± 7.60	9.22 ± 4.09	19.39 ± 2.97
	Late sown	PBW 343	68.09 ± 2.23	72.66 ± 1.02	13.49 ± 5.52	13.95 ± 0.81	32.42 ± 6.81	34.03 ± 0.96
		C 306	76.86 ± 1.15	75.90 ± 1.67	37.00 ± 6.78	31.84 ± 7.18	20.09 ± 3.28	24.53 ± 2.55

		Test weight		Grain plumpness > 2.8 mm		Grain plumpness < 2.5 mm	
		Year I	Year II	Year I	Year II	Year I	Year II
CD $P \leq 0.05$	Sowing time	1.12	1.10	1.55	3.09	1.55	3.09
	Genotype	0.91	0.90	1.27	2.52	1.27	2.52
	Treatment	0.91	NS	1.27	2.52	1.27	2.52

conditions, relative content of gliadin decreased in both genotypes as compared to RF + TS conditions. This led to lower gliadin percentage in C 306 as compared to PBW 343 under RF + ES conditions (Table 2). In PBW 343, LS conditions resulted in lower accumulation of gliadin both under irrigated and rain-fed conditions, and such an effect was also observed in crop of C 306. Rain-fed conditions in general resulted in lowering of glutenin in ES and TS crop. Grains of C 306 obtained from ES and TS crops had relatively higher percentage of glutenin as compared to the respective grains of PBW 343 (Table 2). However, this trend is reversed in LS crop (Table 2).

3.2. Mineral Contents. In the present study, different sowing times and irrigation/RF conditions resulted in significant variation in grain micronutrients particularly in Fe contents. In general, grain Fe and Zn contents ranged from 26 to 41 and 19 to 24 mg/Kg, respectively, under irrigated conditions (Table 3). Fe content was significantly higher in PBW 343 as compared to C 306 under irrigated ES and TS conditions as well as under RF conditions. However, under irrigated LS conditions, C 306 registered higher Fe content (Table 3). Effect of sowing time on Fe content was variable among the genotypes and across the years. On comparing the average of both years, 22% increase in Fe content was recorded under

TABLE 2: Effect of sowing time on albumin-globulin, gliadin, and glutenin proteins in mature grains of drought-susceptible (PBW 343) and drought-tolerant (C 306) wheat genotypes grown under irrigated and rain-fed conditions. Values are mean ± SD of three replicates; TSP, total soluble proteins; NS, not significant.

Year	Sowing time	Genotype	TSP (%) Treatment		Amphiphil (% of total protein) Treatment		Albumin-globulin (% of total protein) Treatment		Gliadin (% of total protein) Treatment		Glutenin (% of total protein) Treatment	
			Irrigated	Rain fed	Irrigated	Rain fed	Irrigated	Rain fed	Irrigated	Rain fed	Irrigated	Rain fed
Year I (2008)	Early sown	PBW 343	9.32 ± 0.79	11.62 ± 0.59	5.06 ± 0.49	6.13 ± 0.15	11.42 ± 0.41	13.46 ± 0.15	31.62 ± 1.08	27.54 ± 0.65	39.31±1.36	34.36±0.74
		C 306	10.17 ± 0.75	14.02 ± 0.61	3.43 ± 0.31	6.74 ± 0.12	12.25 ± 0.25	13.90 ± 0.12	30.40 ± 0.55	26.12 ± 0.94	44.25±1.34	40.18±0.66
	Timely sown	PBW 343	10.51 ± 0.13	9.40 ± 0.51	4.60 ± 0.39	5.75 ± 0.33	10.79 ± 0.21	13.97 ± 0.50	31.26 ± 0.70	36.56 ± 0.95	44.56±1.46	41.73±0.97
		C 306	12.48 ± 0.21	11.62 ± 0.29	5.37 ± 0.10	7.12 ± 0.19	11.68 ± 0.34	13.53 ± 0.29	20.59 ± 1.50	17.87 ± 0.69	48.37±2.14	44.37±1.06
	Late sown	PBW 343	12.31 ± 0.17	11.60 ± 0.49	5.76 ± 0.18	6.28 ± 0.17	13.01 ± 0.29	13.46 ± 0.31	22.33 ± 0.56	27.14 ± 0.38	46.69±1.24	44.08±1.06
		C 306	13.76 ± 0.59	11.97 ± 0.24	5.25 ± 0.12	7.27 ± 0.16	11.94 ± 0.16	13.35 ± 0.35	29.95 ± 1.29	34.42 ± 0.51	40.91±0.56	41.93±0.88
Year II (2009)	Early sown	PBW 343	10.09 ± 0.45	12.22 ± 0.15	6.16 ± 0.19	6.85 ± 0.17	9.17 ± 0.51	10.42 ± 0.21	29.51 ± 0.88	26.89 ± 0.45	41.08±1.32	35.07±0.96
		C 306	10.26 ± 0.51	12.74 ± 0.14	4.15 ± 0.21	5.91 ± 0.12	10.12 ± 0.24	11.30 ± 0.22	28.05 ± 0.68	23.25 ± 0.46	45.47±1.80	41.44±1.36
	Timely sown	PBW 343	10.94 ± 0.81	10.17 ± 0.56	3.17 ± 0.10	4.83 ± 0.19	9.17 ± 0.15	11.67 ± 0.24	28.26 ± 0.55	35.26 ± 0.69	47.47±0.88	40.11±1.73
		C 306	12.82 ± 0.16	10.60 ± 0.17	4.60 ± 0.23	6.15 ± 0.18	11.15 ± 0.19	12.86 ± 0.37	24.40 ± 0.56	19.13 ± 0.33	51.09±1.58	43.52±0.72
	Late sown	PBW 343	11.54 ± 0.13	10.43 ± 0.34	5.81 ± 0.11	6.90 ± 0.15	12.44 ± 0.33	12.80 ± 0.65	23.67 ± 0.50	30.14 ± 0.70	48.55±1.57	45.96±2.23
		C 306	13.16 ± 0.77	11.03 ± 0.82	5.14 ± 0.14	8.84 ± 0.18	10.30 ± 0.35	11.62 ± 0.29	27.45 ± 0.48	30.12 ± 0.64	42.24±1.04	44.21±1.10

CD $P \leq 0.05$	TSP		Amphiphil		Albumin-globulin		Gliadin		Glutenin	
	Year I	Year II	Year I	Year II	Year I	Year II	Year I	Year II	Year I	Year II
Sowing time	0.15	0.19	0.36	0.24	NS	0.48	1.14	NS	1.68	2.00
Genotype	0.12	0.15	NS	NS	NS	NS	0.93	1.19	1.37	NS
Treatment	0.12	0.15	0.29	0.19	0.34	0.39	NS	NS	1.37	1.63

TABLE 3: Effect of sowing time on iron (mg/Kg) and zinc (mg/Kg) contents in mature grains of drought-susceptible (PBW 343) and drought-tolerant (C 306) wheat genotypes grown under irrigated and rain-fed conditions. Values are mean ± standard deviation of three replicates; CD, critical difference; NS, not significant.

Year	Sowing Time	Genotype	Iron (mg/Kg) Treatment		Zinc (mg/Kg) Treatment	
			Irrigated	Rain fed	Irrigated	Rain fed
Year I (2008)	Early sown	PBW 343	40.85 ± 0.58	43.85 ± 0.43	23.06 ± 0.27	23.42 ± 0.07
		C 306	35.50 ± 0.98	42.95 ± 1.78	19.28 ± 2.52	23.04 ± 0.60
	Timely sown	PBW 343	27.35 ± 0.37	40.05 ± 0.18	22.36 ± 0.42	18.14 ± 3.88
		C 306	26.35 ± 0.54	37.15 ± 1.44	22.61 ± 0.25	22.49 ± 0.32
	Late sown	PBW 343	33.55 ± 1.36	35.70 ± 0.73	22.53 ± 0.34	18.10 ± 3.04
		C 306	35.05 ± 1.32	30.20 ± 1.31	19.79 ± 2.14	18.16 ± 5.45
Year II (2009)	Early sown	PBW 343	37.80 ± 1.75	42.05 ± 1.87	22.82 ± 0.12	23.89 ± 0.10
		C 306	26.35 ± 1.60	34.20 ± 2.00	22.57 ± 0.18	23.64 ± 0.20
	Timely sown	PBW 343	36.95 ± 2.07	41.70 ± 0.67	23.01 ± 0.48	23.67 ± 0.18
		C 306	36.25 ± 0.80	34.80 ± 2.02	23.48 ± 0.19	22.30 ± 0.26
	Late sown	PBW 343	29.10 ± 0.87	30.15 ± 1.33	23.56 ± 0.14	23.26 ± 0.09
		C 306	31.20 ± 2.78	29.32 ± 3.63	23.70 ± 0.14	23.00 ± 0.20

		Iron		Zinc	
		Year I	Year II	Year I	Year II
CD P ≤ 0.05	Sowing time	1.24	2.34	NS	NS
	Genotype	1.01	1.91	NS	0.21
	Treatment	1.01	1.91	NS	NS

irrigated ES conditions in PBW 343 as compared to irrigated TS conditions. Under RF + ES conditions, both genotypes registered 5–7% increase in Fe content as compared to RF + TS conditions, and 20–23% increase as compared to irrigated TS conditions (Table 3). In contrast, LS+RF conditions resulted in 20–24% decrease in Fe content in both the genotypes as compared to TS+RF conditions. Further, an increase (5–21%) in Fe content was observed in both the genotypes in response to RF conditions except in C 306 under LS conditions. Effect of sowing time and RF conditions on Zn contents was not significant in both years. In general, Zn content increased in response to RF conditions under ES conditions but decreased under TS and LS conditions in both genotypes (Table 3).

Cultivar C 306 had lower tannin content as compared to PBW 343 under irrigated as well as RF conditions (Table 4). Trypsin inhibitor activity was more in C 306 as compared to PBW 343 irrespective of sowing time and irrigation conditions. The ES crop appeared to have higher trypsin inhibitor activity (Table 4). In ES and TS crops, on an average PBW 343 had slightly higher phytic acid content in irrigated crop (Table 4).

4. Discussion

Proteins are the most important components of wheat grains governing end-use quality. Both amount and composition of proteins determine the protein quality and hence end-use quality of wheat. Environmental conditions during grain filling influence the accumulation of protein in the developing wheat kernel and can alter the functional properties of the resulting flour. Variations in both protein content and composition significantly modify flour quality for different end products. Although grain protein composition depends primarily on genotype, it is significantly affected by environmental factors and their interactions [17]. Increase in flour protein under water deficit conditions has been reported mainly due to higher rates of accumulation of grain nitrogen and lower rates of accumulation of carbohydrates. Irrigation, on the other hand, may decrease flour protein content by dilution of nitrogen with carbohydrates [18]. Changing the sowing time generated a large effect on the amount of TSP (Table 2), probably driven by the different thermal conditions prevailing during the grain filling period (Figure 1). This was particularly evident on comparing the early and late sowings. Similar results have been reported earlier [19]. Although albumin-globulins do not have significant effect on dough quality, but they are important from nutritional point of view due to high content of essential amino acids in them. Hurkman et al. have reported accumulation of albumin-globulins in response to high temperature in wheat grains [20].

It appears that optimum temperature of gliadin synthesis in grains is genotype dependent. The differential effect of temperature leading to relatively higher production of gliadins resulting in reduced dough strength has been reported earlier [19]. The results indicate that C 306 and PBW 343 have different optima temperature for synthesis of glutenin. An increased grain protein and gluten content in

TABLE 4: Effect of sowing time on tannin (μg g^{-1} DW), phytic acid (mg g^{-1} DW), and trypsin inhibitor (units g^{-1} DW) contents in mature grains of drought-susceptible (PBW 343) and drought tolerant (C 306) wheat genotypes grown under irrigated and rain-fed conditions. Values are mean $\pm$ standard deviation of three replicates; CD, critical difference; NS, not significant; One inhibitor unit is defined as the quantity of inhibitor which inhibits 50% of trypsin activity.

Year	Sowing time	Genotype	Tannin (μg g^{-1} DW)		Phytic acid (mg g^{-1} DW)		Trypsin inhibitors (units g^{-1} DW)	
			Treatment		Treatment		Treatment	
			Irrigated	Rain fed	Irrigated	Rain fed	Irrigated	Rain fed
Year I (2008)	Early sown	PBW 343	1608 $\pm$ 63.68	1382 $\pm$ 50.69	8.49 $\pm$ 0.50	6.00 $\pm$ 0.42	31.94 $\pm$ 0.98	33.80 $\pm$ 2.27
		C 306	1083 $\pm$ 48.03	1010 $\pm$ 18.03	7.85 $\pm$ 0.70	8.84 $\pm$ 0.68	45.83 $\pm$ 1.96	42.13 $\pm$ 1.13
	Timely sown	PBW 343	1840 $\pm$ 24.74	1689 $\pm$ 29.68	7.95 $\pm$ 0.62	9.38 $\pm$ 0.96	31.48 $\pm$ 3.97	27.31 $\pm$ 2.84
		C 306	1434 $\pm$ 16.00	1353 $\pm$ 33.79	8.00 $\pm$ 0.53	9.48 $\pm$ 1.02	40.28 $\pm$ 5.89	27.59 $\pm$ 6.57
	Late sown	PBW 343	1864 $\pm$ 20.87	1842 $\pm$ 19.00	9.58 $\pm$ 0.82	7.51 $\pm$ 0.45	29.63 $\pm$ 2.84	21.76 $\pm$ 2.27
		C 306	1767 $\pm$ 10.87	1470 $\pm$ 30.69	8.84 $\pm$ 0.40	6.62 $\pm$ 0.84	36.11 $\pm$ 5.97	32.87 $\pm$ 1.13
Year II (2009)	Early sown	PBW 343	1901 $\pm$ 24.17	1783 $\pm$ 32.50	7.70 $\pm$ 1.12	9.33 $\pm$ 1.28	34.26 $\pm$ 1.40	31.48 $\pm$ 1.57
		C 306	1426 $\pm$ 25.50	913 $\pm$ 12.50	7.56 $\pm$ 0.97	7.48 $\pm$ 0.64	46.30 $\pm$ 1.60	45.37 $\pm$ 1.52
	Timely sown	PBW 343	2020 $\pm$ 19.14	1426 $\pm$ 14.00	8.52 $\pm$ 0.36	9.26 $\pm$ 0.23	39.81 $\pm$ 1.13	31.02 $\pm$ 1.23
		C 306	1545 $\pm$ 29.48	1307 $\pm$ 19.67	9.74 $\pm$ 1.11	7.33 $\pm$ 0.59	41.20 $\pm$ 2.84	30.56 $\pm$ 0.98
	Late sown	PBW 343	1901 $\pm$ 22.03	1664 $\pm$ 27.57	10.00 $\pm$ 0.44	13.93 $\pm$ 0.28	30.56 $\pm$ 3.54	24.54 $\pm$ 1.50
		C 306	1248 $\pm$ 26.50	832 $\pm$ 20.88	9.22 $\pm$ 0.56	10.59 $\pm$ 0.90	39.81 $\pm$ 3.97	32.41 $\pm$ 1.45
			Tannin		Phytic acid		Trypsin inhibitors	
			Year I	Year II	Year I	Year II	Year I	Year II
CD $P \leq 0.05$		Sowing time	25.56	17.27	0.58	0.66	3.77	2.5
		Genotype	20.87	14.10	NS	0.54	3.08	2.05
		Treatment	20.87	14.10	0.48	0.54	3.08	2.05

response to late water stress as compared to the fully irrigated treatment in a winter bread variety has also been reported earlier [21]. Labuschagne et al. also reported different temperature requirements of polymeric glutenin accumulation in different wheat cultivars [22]. LS conditions are offering higher protein content accompanied by increased albumin-globulin; however, decrease in glutenin content made the flour technically of poor quality. After anthesis, environmental conditions primarily affect kernel size, protein concentration and composition. Temperature and rain fall during grain filling strongly affect grain protein content and gliadin to HMW-GS and LMW-GS ratios [23]. An evaluation of drought and heat effects on wheat in a Mediterranean climate showed the highest grain protein content under warm dry rain-fed conditions and the lowest in the irrigated environment [24].

The gluten proteins, glutenin and gliadin, are responsible for flexibility and extensibility of dough. The deterioration in dough quality has been attributed to decline in glutenin-to-gliadin ratio. The glutenin-to-gliadin ratio was higher in drought-tolerant genotype C 306 as compared to PBW 343 when crop was raised under ES and TS conditions. Under rain-fed conditions, this ratio increased in C 306 whereas it either remained unchanged or declined in PBW 343 on the basis of data of two years. Therefore, it can be concluded that dough quality in C 306 flour is not adversely affected when crop is grown under rain-fed conditions as ES and TS crops. However, when crop is exposed to both drought

and high temperature (Figure 1) during LS conditions, the glutenin-to-gliadin ratio declines both in C 306 and PBW 343 (Table 2).

Changes in the amount of gluten transcripts or in the temporal regulation of gluten protein genes in response to environmental conditions could lead to alterations in flour quality. Ratios of the different classes of proteins or of specific proteins within each class could change and thus affect the formation of glutenin polymers [25]. Water deficit and temperature being the important environmental conditions influencing the amount, composition, and/or polymerization of the wheat storage proteins have been reported in earlier studies [21, 26].

Among grain mineral nutrients, Zn and Fe deficiencies are the most important global challenge. According to the World Health Organization, deficiencies in Zn and Fe rank 5th and 6th, respectively, among the risk factors responsible for illnesses in developing countries. Deficiency in Zn and Fe afflicts over three billion people worldwide resulting in overall poor health, anemia, increased mortality rates, and lower worker productivity [1]. Producing micronutrient-enriched cereals and improving their bioavailability are considered promising and cost-effective approaches for diminishing malnutrition.

Gomez-Becerra et al. reported that the concentration of grain's Fe and Zn in modern wheat cultivars is on an average around 35 and 25 mg/Kg, respectively [27]. Peleg et al. have reported that water stress conditions either result in increase, decrease, or no changes in grain micronutrients in wheat [1]. In other words, no definite trend on accumulation of micronutrient under water deficit conditions can be predicted. In the present work, just by changing sowing time and irrigation conditions, a significant variation in micronutrient particularly grain Fe contents (Table 3) and protein composition was observed. This variation could be useful for selection of superior germplasm for breeding programs aimed at mineral biofortification. For example, under RF conditions, an increase in grain Fe content in both genotypes in a two-year trial were observed (Table 3). Obviously these data gave an encouraging indication for a study involving large number of wheat genotypes under restricted/controlled irrigation to increase grain Fe content. Increasing evidence suggests that enhancement of grain protein content of wheat could also greatly contribute to biofortification with micronutrients [28–30]. Distelfeld et al. suggested the possible role of proteins as potential candidates for chelating micronutrients [31]. Further, cereal grains are inherently poor in concentration of micronutrients and rich in compounds depressing their bioavailability such as phytic acid [32]. At physiological pH, phytic acid due to high negative charge density acts as a strong chelator of positively charged mineral cations such as Fe and Zn [33]. When comparing the average data of phytic acid contents for two years (Table 4), minimum phytic acid content was observed in grains of ES crop. Under irrigated conditions, PBW 343 has a higher phytic acid content in ES and TS crops. Maximum content of phytic acid was observed in LS crop of PBW 343. Although there are reports that Zn and Fe content increased with higher phytic acid and protein content [34], such a behavior was not observed with respect to phytic acid in the same genotype when sown under different climatic conditions. In LS crop, combination of higher temperature and water-deficit conditions probably caused maximum phytic acid content in the grains.

Both the genotypes registered 18–35% reduction in tannin content under ES+RF conditions when compared to TS-irrigated conditions. The inhibitory effects on absorption and utilization of minerals such as iron and zinc have also been attributed to the presence of trypsin inhibitors [33]. Effect of ES conditions was variable among the genotypes since inhibitor activity increased in C 306 but decreased in PBW 343 when compared to irrigated TS conditions (Table 4). However, RF conditions clearly resulted in significant decrease in trypsin inhibitor as well as tannin contents in both genotypes. This suggested that ES+RF conditions resulted not only in higher grain Fe content (Table 3) but also lower antinutrients (Table 4) associated with higher grain protein contents (Table 2). In contrast, in the present study LS conditions are offering an increase in phytic acid content accompanied by decreased micronutrients (Table 3) and glutenin content (Table 2), thereby decreasing the grain nutritional quality. So lower temperature of ES conditions (around 22°C during grain filling in present study) along with limited irrigation could offer a new approach for biofortification of wheat grains without any additional economic inputs.

Disclosure

This paper is a part of a doctoral thesis of the senior author.

References

[1] Z. Peleg, Y. Saranga, A. Yazici, T. Fahima, L. Ozturk, and I. Cakmak, "Grain zinc, iron and protein concentrations and zinc-efficiency in wild emmer wheat under contrasting irrigation regimes," *Plant and Soil*, vol. 306, no. 1-2, pp. 57–67, 2008.

[2] P. R. Shewry, J. A. Napier, and A. S. Tatham, "Seed storage proteins: structures and biosynthesis," *Plant Cell*, vol. 7, no. 7, pp. 945–956, 1995.

[3] H. Singh and F. MacRitchie, "Application of polymer science to properties of gluten," *Journal of Cereal Science*, vol. 33, no. 3, pp. 231–243, 2001.

[4] Z. Peleg, I. Cakmak, L. Ozturk et al., "Quantitative trait loci conferring grain mineral nutrient concentrations in durum wheat × wild emmer wheat RIL population," *Theoretical and Applied Genetics*, vol. 119, no. 2, pp. 353–369, 2009.

[5] I. Cakmak, "Enrichment of cereal grains with zinc: agronomic or genetic biofortification?" *Plant and Soil*, vol. 302, no. 1-2, pp. 1–17, 2008.

[6] B. K. Shashi, S. Sharan, S. Hittalamani, A. G. Shankar, and T. K. Nagarathna, "Micronutrient composition, antinutritional factors and bioaccessibility of iron in different finger millet (*Eleusine coracana*) genotypes," *Karnataka Journal of Agricultural Sciences*, vol. 20, pp. 583–585, 2007.

[7] A. S. Ekop, I. B. Obot, and E. N. Ikpatt, "Anti-nutritional factors and potassium bromate content in bread and flour samples in Uyo Metropolis, Nigeria," *E-Journal of Chemistry*, vol. 5, no. 4, pp. 736–741, 2008.

[8] I. E. Liener, "Implications of antinutritional components in soybean foods," *Critical Reviews in Food Science and Nutrition*, vol. 34, no. 1, pp. 31–67, 1994.

[9] D. Jiang, H. Yue, B. Wollenweber et al., "Effects of post-anthesis drought and waterlogging on accumulation of high-molecular-weight glutenin subunits and glutenin macropolymers content in wheat grain," *Journal of Agronomy and Crop Science*, vol. 195, no. 2, pp. 89–97, 2009.

[10] E. Triboï, P. Martre, and A. M. Triboï-Blondel, "Environmentally-induced changes in protein composition in developing grains of wheat are related to changes in total protein content," *Journal of Experimental Botany*, vol. 54, no. 388, pp. 1731–1742, 2003.

[11] S. Singh, A. K. Gupta, S. K. Gupta, and N. Kaur, "Effect of sowing time on protein quality and starch pasting characteristics in wheat (*Triticum aestivum* L.) genotypes grown under irrigated and rain-fed conditions," *Food Chemistry*, vol. 122, no. 3, pp. 559–565, 2010.

[12] *Package of Practices for Rabi Crops*, Punjab Agricultural University, Ludhiana, India, 2008.

[13] B. K. Mishra, "Quality needs for Indian traditional products," in *Wheat: Research needs beyond 2000 AD*, S. Nagarajan, G. Singh, and B. S. Tyagi, Eds., pp. 939–977, Narosa, New Delhi, India, 1998.

[14] A. K. Gupta, V. Kaur, and N. Kaur, "Appearance of different phosphatase forms and phosphorus partitioning in nodules of chickpea (Cicer arietinum L.) during development," *Acta Physiologiae Plantarum*, vol. 20, no. 4, pp. 369–374, 1998.

[15] N. Hajela, A. H. Pande, S. Sharma, D. N. Rao, and K. Hajela, "Studies on a doubleheaded protease inhibitor from Phaseolus

mungo," *Journal of Plant Biochemistry and Biotechnology*, vol. 8, no. 1, pp. 57–60, 1999.

[16] S. Sadasivam and A. Manickam, "Phenolics," in *Biochemical Methods for Agricultural Sciences*, pp. 187–188, Wiley Eastern Ltd, New Delhi, India, 1992.

[17] J. Zhu and K. Khan, "Effects of genotype and environment on glutenin polymers and breadmaking quality," *Cereal Chemistry*, vol. 78, no. 2, pp. 125–130, 2001.

[18] M. J. Guttieri, R. McLean, J. C. Stark, and E. Souza, "Managing irrigation and nitrogen fertility of hard spring wheats for optimum bread and noodle quality," *Crop Science*, vol. 45, no. 5, pp. 2049–2059, 2005.

[19] R. Motzo, S. Fois, and F. Giunta, "Protein content and gluten quality of durum wheat (*Triticum turgidum* subsp. durum) as affected by sowing date," *Journal of the Science of Food and Agriculture*, vol. 87, no. 8, pp. 1480–1488, 2007.

[20] W. J. Hurkman, W. H. Vensel, C. K. Tanaka, L. Whitehand, and S. B. Altenbach, "Effect of high temperature on albumin and globulin accumulation in the endosperm proteome of the developing wheat grain," *Journal of Cereal Science*, vol. 49, no. 1, pp. 12–23, 2009.

[21] A. Ozturk and F. Aydin, "Effect of water stress at various growth stages on some quality characteristics of winter wheat," *Journal of Agronomy and Crop Science*, vol. 190, no. 2, pp. 93–99, 2004.

[22] M. T. Labuschagne, O. Elago, and E. Koen, "Influence of extreme temperatures during grain filling on protein fractions, and its relationship to some quality characteristics in bread, biscuit, and durum wheat," *Cereal Chemistry*, vol. 86, no. 1, pp. 61–66, 2009.

[23] P. J. Randall and H. J. Moss, "Some effects of temperature regime during grain filling on wheat quality," *Australian Journal of Agricultural Research*, vol. 41, pp. 603–617, 1990.

[24] L. F. García Del Moral, Y. Rharrabti, V. Martos, and C. Royo, "Environmentally induced changes in amino acid composition in the grain of durum wheat grown under different water and temperature regimes in a Mediterranean environment," *Journal of Agricultural and Food Chemistry*, vol. 55, no. 20, pp. 8144–8151, 2007.

[25] S. B. Altenbach, K. M. Kothari, and D. Lieu, "Environmental conditions during wheat grain development alter temporal regulation of major gluten protein genes," *Cereal Chemistry*, vol. 79, no. 2, pp. 279–285, 2002.

[26] E. Johansson, M. L. Prieto-Linde, G. Svensson, and J. Ö. Jönsson, "Influences of cultivar, cultivation year and fertilizer rate on amount of protein groups and amount and size distribution of mono- and polymeric proteins in wheat," *Journal of Agricultural Science*, vol. 140, no. 3, pp. 275–284, 2003.

[27] H. F. Gomez-Becerra, A. Yazici, L. Ozturk et al., "Genetic variation and environmental stability of grain mineral nutrient concentrations in *Triticum dicoccoides* under five environments," *Euphytica*, vol. 171, no. 1, pp. 39–52, 2009.

[28] I. Cakmak, A. Torun, E. Millet et al., "*Triticum dicoccoides*: an important genetic resource for increasing zinc and iron concentration in modern cultivated wheat," *Soil Science and Plant Nutrition*, vol. 50, no. 7, pp. 1047–1054, 2004.

[29] A. Morgounov, H. F. Gómez-Becerra, A. Abugalieva et al., "Iron and zinc grain density in common wheat grown in Central Asia," *Euphytica*, vol. 155, no. 1-2, pp. 193–203, 2007.

[30] U. B. Kutman, B. Yildiz, L. Ozturk, and I. Cakmak, "Biofortification of durum wheat with zinc through soil and foliar applications of nitrogen," *Cereal Chemistry*, vol. 87, no. 1, pp. 1–9, 2010.

[31] A. Distelfeld, I. Cakmak, Z. Peleg et al., "The high grain protein concentration locus, *Gpc-B1*, on chromosome arm 6BS of wheat is also associated with high grain iron and zinc concentrations," *Physiologia Plantarum*, vol. 129, no. 3, pp. 635–643, 2007.

[32] V. Raboy, "Low-phytic-acid grains," *Food and Nutrition Bulletin*, vol. 21, no. 4, pp. 423–427, 2000.

[33] A. Yasmin, A. Zeb, A. W. Khalil, G. M. U. D. Paracha, and A. B. Khattak, "Effect of processing on anti-nutritional factors of red kidney bean (*Phaseolus vulgaris*) grains," *Food and Bioprocess Technology*, vol. 1, no. 4, pp. 415–419, 2008.

[34] F. J. Zhao, Y. H. Su, S. J. Dunham et al., "Variation in mineral micronutrient concentrations in grain of wheat lines of diverse origin," *Journal of Cereal Science*, vol. 49, no. 2, pp. 290–295, 2009.

18

Establishing an Efficient Way to Utilize the Drought Resistance Germplasm Population in Wheat

Jiancheng Wang,[1,2] **Yajing Guan,**[1] **Yang Wang,**[1] **Liwei Zhu,**[1]
Qitian Wang,[1] **Qijuan Hu,**[1] **and Jin Hu**[1]

[1] *Seed Science Center, College of Agriculture and Biotechnology, Zhejiang University, Hangzhou 310058, China*
[2] *Shandong Crop Germplasm Center, Shandong Academy of Agricultural Sciences, Jinan 250100, China*

Correspondence should be addressed to Jin Hu; jhu@zju.edu.cn

Academic Editors: J. Huang and Z. Wang

Drought resistance breeding provides a hopeful way to improve yield and quality of wheat in arid and semiarid regions. Constructing core collection is an efficient way to evaluate and utilize drought-resistant germplasm resources in wheat. In the present research, 1,683 wheat varieties were divided into five germplasm groups (high resistant, HR; resistant, R; moderate resistant, MR; susceptible, S; and high susceptible, HS). The least distance stepwise sampling (LDSS) method was adopted to select core accessions. Six commonly used genetic distances (Euclidean distance, Euclid; Standardized Euclidean distance, Seuclid; Mahalanobis distance, Mahal; Manhattan distance, Manhat; Cosine distance, Cosine; and Correlation distance, Correlation) were used to assess genetic distances among accessions. Unweighted pair-group average (UPGMA) method was used to perform hierarchical cluster analysis. Coincidence rate of range (CR) and variable rate of coefficient of variation (VR) were adopted to evaluate the representativeness of the core collection. A method for selecting the ideal constructing strategy was suggested in the present research. A wheat core collection for the drought resistance breeding programs was constructed by the strategy selected in the present research. The principal component analysis showed that the genetic diversity was well preserved in that core collection.

1. Introduction

Drought is probably the most important abiotic stress that limits plant growth [1]. Drought stress is one of the most serious environmental factors that can severely limit the yield and quality of agricultural crops [2]. With global climate change, the lack of water for agronomic purposes will become the major problem for crop production [3]. In agronomical point-of-view, drought stress is a situation in which lack of water exceeds the capacity of plants which leads to the growth prevention. Thus, improving the drought tolerance is a major adaptation strategy for plant production in arid and semiarid regions [4]. In drought prone environments, crop drought resistance is a major factor in the stabilization of crop performance. Drought resistance is now considered by both breeders and molecular biologists as a valid breeding target.

Wheat (*Triticum aestivum* L.) is one of the most important cereals in the world. Drought stress may reduce all yield components in wheat [5]. Drought is the major factor limiting wheat growth and productivity in many regions of the world, and the changing global climate is making the situation more serious [6, 7]. Developing high-yielding wheat cultivars under drought conditions in arid and semiarid regions is an important objective of breeding programs [5]. Although great efforts have been made in wheat drought resistance breeding, the decrease in agricultural productivity induced by drought stress still remains unsolved [8]. One reason is that the numerous germplasm resources were not effectively utilized in wheat breeding programs. However, with continuous collection of germplasm resources, the size of populations has been becoming bigger and bigger, which hindered the evaluation and utilization of the wheat germplasm resources.

Core collections provide an efficient way to evaluate and utilize germplasm resources. A core collection is a representative sample of the whole collection which has minimum repetitiveness and maximum genetic diversity of a plant species [9]. The core collection serves as a working

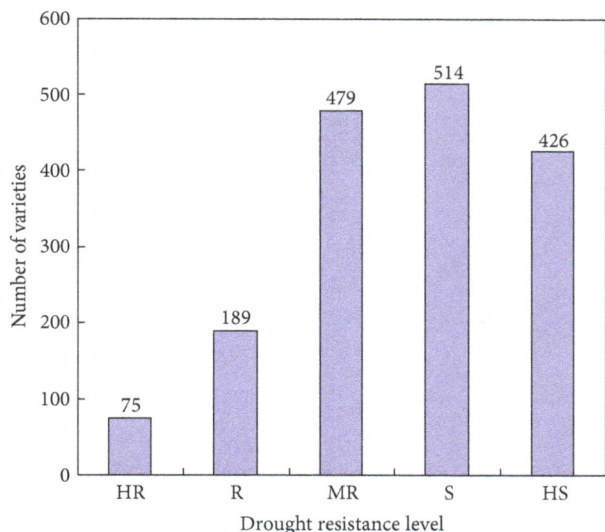

FIGURE 1: The distribution of 1,683 wheat varieties: HR: high resistant; R: resistant; MR: moderate resistant; S: susceptible; HS: high susceptible.

collection to be evaluated and utilized preferentially [10–13]. In this way, it is possible to preserve most of the genes in large germplasm populations using a small sample. Thus, the objectives of this research were (1) to investigate the ideal constructing strategy on wheat core collection based on data of agronomic traits combining drought resistance information and (2) to construct such a wheat core collection for the drought resistance breeding programs.

2. Materials and Methods

2.1. Materials. Wheat varieties were introduced from abroad. The drought resistance level combining four yield traits (plant height, spike length, grain numbers per spike, and 1000-grain weight) and four quality traits (crude protein content, lysine content, sedimentation, and hardness) of 1,683 varieties have been investigated. All data were downloaded from "Chinese Crop Germplasm Resources Information System" (http://icgr.caas.net.cn/).

2.2. Core Collection Construction. According to the drought resistance level, all 1,683 wheat varieties were divided into five germplasm groups (high resistant, HR; resistant, R; moderate resistant, MR; susceptible, S; and high susceptible, HS). The distribution of varieties was shown in Figure 1. The procedure for core collection construction was conducted by two steps. First, subcore collections were selected from each germplasm groups. Second, all the sub-core collections were combined together to construct a core collection.

The least distance stepwise sampling (LDSS) method [14] was adopted to construct sub-core collections from germplasm groups. The procedure was as follows. (1) The genetic distances among accessions were calculated, and accessions were classified by hierarchical cluster analysis

based on their genetic distance. (2) One accession from a subgroup with the least distance was randomly removed, and another accession of the subgroup was sampled. (3) The genetic distances among the remaining accessions were calculated, and the sampling was repeated in the same way. The stepwise samplings were performed until the percentage of the remaining accessions reached the desired one. This method performs sampling based on the subgroup with the least genetic distance, which can efficiently eliminate redundant accessions and ignore the effect of the cluster methods [15].

2.3. Genetic Distances and Evaluating Parameters. Six commonly used genetic distances (Euclidean distance, Euclid; Standardized Euclidean distance, Seuclid; Mahalanobis distance, Mahal; Manhattan distance, Manhat; Cosine distance, Cosine; and Correlation distance, Correlation) were used to assess genetic distances among accessions. Unweighted pair-group average (UPGMA) method was used for performing hierarchical cluster analysis [15].

Coincidence rate of range (CR) and variable rate of coefficient of variation (VR) [16, 17] were adopted to evaluate the representativeness of core collection. Those four parameters were formulated as follows: $\text{CR} = (1/n)\sum_{i=1}^{n}(R_{C(i)}/R_{I(i)}) \times 100$, where $R_{C(i)}$ is the range of the ith trait in the core collection; $R_{I(i)}$ is the range of the corresponding trait in the initial collection; n and is total number of traits, $\text{VR} = (1/n)\sum_{i=1}^{n}(\text{CV}_{C(i)}/\text{CV}_{I(i)}) \times 100$, where $\text{CV}_{C(i)}$ is the coefficient of variation of the ith trait in the core collection; $\text{CV}_{I(i)}$ is the coefficient of variation of the corresponding trait in the initial collection; n is total number of traits.

2.4. Data Analysis. The genetic distances calculation, the LDSS procedures, and the evaluating parameters calculation were performed using computer code programmed by the authors based on MATLAB software (version 6.5) [18].

3. Results

3.1. The Assessment of Subcore Collections Constructed by Different Genetic Distances. Subcore collections were constructed by different genetic distances at the sampling percentage of 10%, 20%, and 30% (Table 1). In any germplasm group, CR and VR of sub-core collections constructed by the genetic distance of Cosine and Correl were much lower than of those constructed by the other four genetic distances at the three sampling percentages (Table 1). In HR group, sub-core collections constructed by Manhat had larger CR and VR than those constructed by Euclid, Seuclid, and Mahal at the three sampling percentages (Table 1). In R group, sub-core collections constructed by Euclid had the largest CR at the sampling percentage of 10% and 30%, and those constructed by Manhat had the largest VR at the sampling percentage of 10% and 20% (Table 1). In MR group, sub-core collections constructed by Mahal had the largest CR at the sampling percentage of 20% and 30%, and those constructed by Seuclid had the largest VR at the sampling percentage

TABLE 1: The values of CR and VR of different subcore collection constructed by six genetic distances at the sampling percentage of 10%, 20%, and 30%.

Parameter	DRL[a]	SP[b]	Genetic distance					
			Euclid	Seuclid	Mahal	Manhat	Cosine	Correl
CR (%)	HR	10%	76.17	87.16	80.48	82.36	51.34	52.54
		20%	89.55	91.96	88.77	91.98	71.08	70.14
		30%	95.78	94.17	93.76	97.59	77.57	80.69
	R	10%	91.51	81.60	84.16	89.24	60.62	59.34
		20%	95.98	86.40	92.55	96.09	73.60	65.00
		30%	98.93	96.11	93.64	96.88	81.25	72.20
	MR	10%	94.20	92.61	94.58	95.37	58.36	70.18
		20%	97.41	96.27	98.00	97.10	66.23	78.06
		30%	98.37	97.35	99.31	97.11	75.63	86.26
	S	10%	96.46	96.91	96.05	95.24	69.75	65.32
		20%	97.81	99.58	98.92	99.46	75.42	72.74
		30%	98.04	99.58	99.35	99.49	82.47	77.19
	HS	10%	95.33	93.92	95.88	93.88	61.97	63.68
		20%	98.65	96.43	97.30	96.81	80.81	73.24
		30%	99.20	97.82	97.82	99.20	82.20	82.80
VR (%)	HR	10%	122.67	127.61	124.51	131.10	85.27	79.12
		20%	122.60	120.78	119.30	125.19	92.14	90.12
		30%	117.73	119.07	113.57	122.80	91.74	91.82
	R	10%	129.66	118.85	118.24	132.61	90.21	82.94
		20%	119.06	115.52	117.51	120.73	88.99	82.92
		30%	112.89	114.32	113.78	113.18	90.82	88.82
	MR	10%	128.26	128.60	126.81	128.51	80.95	90.70
		20%	119.56	122.15	120.58	120.39	85.01	92.93
		30%	116.79	116.78	113.45	115.29	89.06	92.51
	S	10%	130.14	129.53	126.03	127.10	90.40	85.16
		20%	120.64	120.67	119.72	121.22	88.94	88.29
		30%	115.44	115.40	115.53	114.36	92.36	90.24
	HS	10%	125.25	128.97	133.16	124.54	89.94	90.06
		20%	122.00	122.77	122.13	120.35	92.00	91.91
		30%	117.09	116.11	117.42	115.94	87.68	93.25

[a] DRL: drought resistance level (HR: high resistant; R: resistant; MR: moderate resistant; S: susceptible; and HS: high susceptible).
[b] SP: sampling percentage.

of 10% and 20%, but similar VR than that constructed by Euclid at the sampling percentage of 30% (Table 1). In S group, sub-core collections constructed by Seuclid had the largest CR at the three sampling percentage, while there was no significant pattern in VR (Table 1). In HS group, sub-core collections constructed by Euclid had the largest CR at the sampling percentage of 20% and 30%, and those constructed by Mahal had the largest VR at the sampling percentage of 10% and 30% (Table 1). Synthesizing the results above, five ideal combinations for sub-core collection were selected: HR-Manhat, R-Euclid, MR-Mahal, S-Seuclid, and HS-Euclid.

3.2. Selection of the Optimal Sampling Percentage.

In each germplasm group, sub-core collections were constructed based on the selected genetic distance with the sampling percentage increasing from 5% to 30%. The value of CR of each sub-core collection was calculated. Thus, 26 CRs were

calculated in each group. The constructing results of the five groups were summarized in Figure 2. In each group, the CR showed logarithmic changing. The CR increased drastically when the sampling percentage was small. With the sampling percentage increasing, CR increased steady (Figure 2). The rangeability in the group of HR and R was larger than that in the groups of MR, S, and HS (Figure 2).

Each curve of in Figure 2 was treated by curve fitting analysis, and the results were summarized in Table 2. The equations showed logarithmic form, and the coefficient of determination of fitted equations (R^2) of each equation was larger than 0.9 (Table 2). Based on the equations, the optimal sampling percentage was calculated by setting the value of CR (%) to 95.00 (Table 2).

3.3. Validation of the Ideal Constructing Strategy.

The principal component analysis was adopted to validate sub-core

TABLE 2: The logarithmic equations on the CR's value responded to the sampling percentage in five combinations of subcore collection construction. The optimal sampling percentage was calculated by the equation when the value of CR (%) was set to 95.00.

Combination[a]	Equation[b]	$R^{2[c]}$	Optimal sampling percentage (%)
HR-Manhat	$y = 12.19\ln(x) + 58.41$	0.9732	20.12
R-Euclid	$y = 6.52\ln(x) + 78.17$	0.9616	13.21
MR-Mahal	$y = 2.67\ln(x) + 90.16$	0.9451	6.13
S-Seuclid	$y = 2.80\ln(x) + 91.16$	0.9392	3.94
HS-Euclid	$y = 2.47\ln(x) + 91.14$	0.9498	4.77

[a] HR-Manhat: high resistant group combining Manhattan distance; R-Euclid: resistant group combining Euclidean distance; MR-Mahal: moderate resistant group combining Mahalanobis distance; S-Seuclid: susceptible group combining Standardized Euclidean distance; and HS-Euclid: high susceptible group combining Euclidean distance.

[b] x: the sampling percentage (%); y: the value of CR (%).

[c] R^2: coefficient of determination of fitted equations.

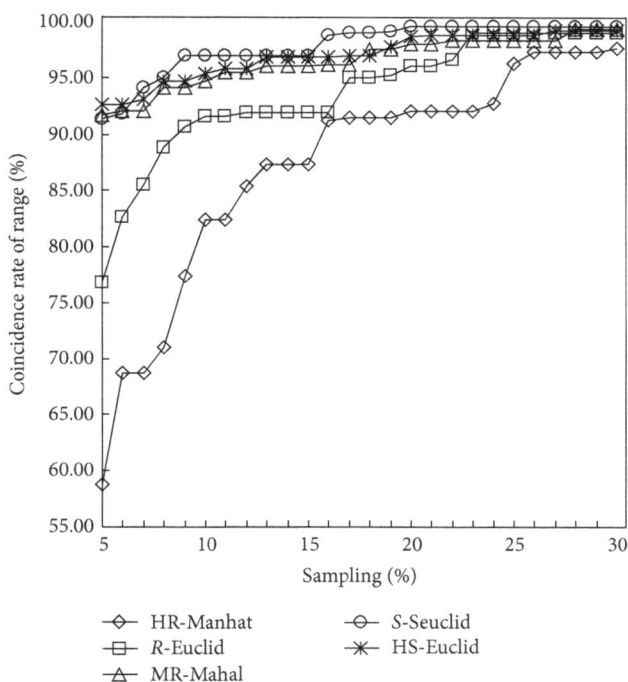

FIGURE 2: The coincidence rate of the range (CR) of subcore collections constructed by five combinations with the sampling percentage increasing from 5% to 30%. HR-Manhat, high resistant group combining Manhattan distance; R-Euclid, resistant group combining Euclidean distance; MR-Mahal, moderate resistant group combining Mahalanobis distance; S-Seuclid, susceptible group combining Standardized Euclidean distance; the HS-Euclid, high susceptible group combining Euclidean distance.

collections constructed by the ideal strategy selected by the present research. Principal component plots of core accessions and reserve accessions in each germplasm group were drown in Figure 3. The total genetic variation percentage of the first two principal components was 71.51% in HR group, 67.67% in R group, 66.90% in MR group, 68.45% in S group, and 71.83% in HS group. At the same sampling percentage, compared to the sub-core collections constructed by complete random selection, the core accessions selected by

the present strategy showed more symmetrical distribution in the whole germplasm group, and most extreme accessions were selected (Figure 3).

4. Discussion

Core collection has been studied for about twenty years [19, 20]. A valid core collection provides a high-efficient way to assess genetic diversity or to find beneficial genes [21–24]. Most core collection researches focused on finding efficient ways in sub-core collection selection [25–27]. However, there is not a widely accepted strategy for constructing sub-core collection up to now. One common approach for constructing a core collection is splitting the whole germplasm population into several groups, then, selecting representative core accessions from each group to form sub-core collections, and combining all sub-core collections to form the final core collection [16, 28]. The present research divided the whole wheat germplasm population into five groups based on drought resistance level. The results showed that the distribution pattern of accessions was various in different germplasm group, which might lead to different suitable strategy for sub-core collection construction. Therefore, different germplasm group required different constructing strategy, and it is needlessly to try to find a widely accepted constructing strategy.

The representativeness is the most important character for a core collection. The VR represents the difference of variance between core collection and the initial collection. The value of VR is affected greatly by the number of accessions in the core collection. In core collection construction based on a valid strategy, with the sampling percentage increasing, the variance decreased and the mean almost keeps unchanging, which led to the decrease of VR. However, at the same sampling percentage, bigger VR means more variation preserved in core collection. The CR shows the extent of preservation of the trait scope in a core collection. The value of CR is not affected greatly by the number of accessions. In the present research, the CR showed sensitivity to the representativeness of a sub-core collection. The CR has been reported to be an important parameter for the evaluation of the representativeness of the core collections [9, 29, 30]. Based on the above analysis, the ideal genetic distance for different

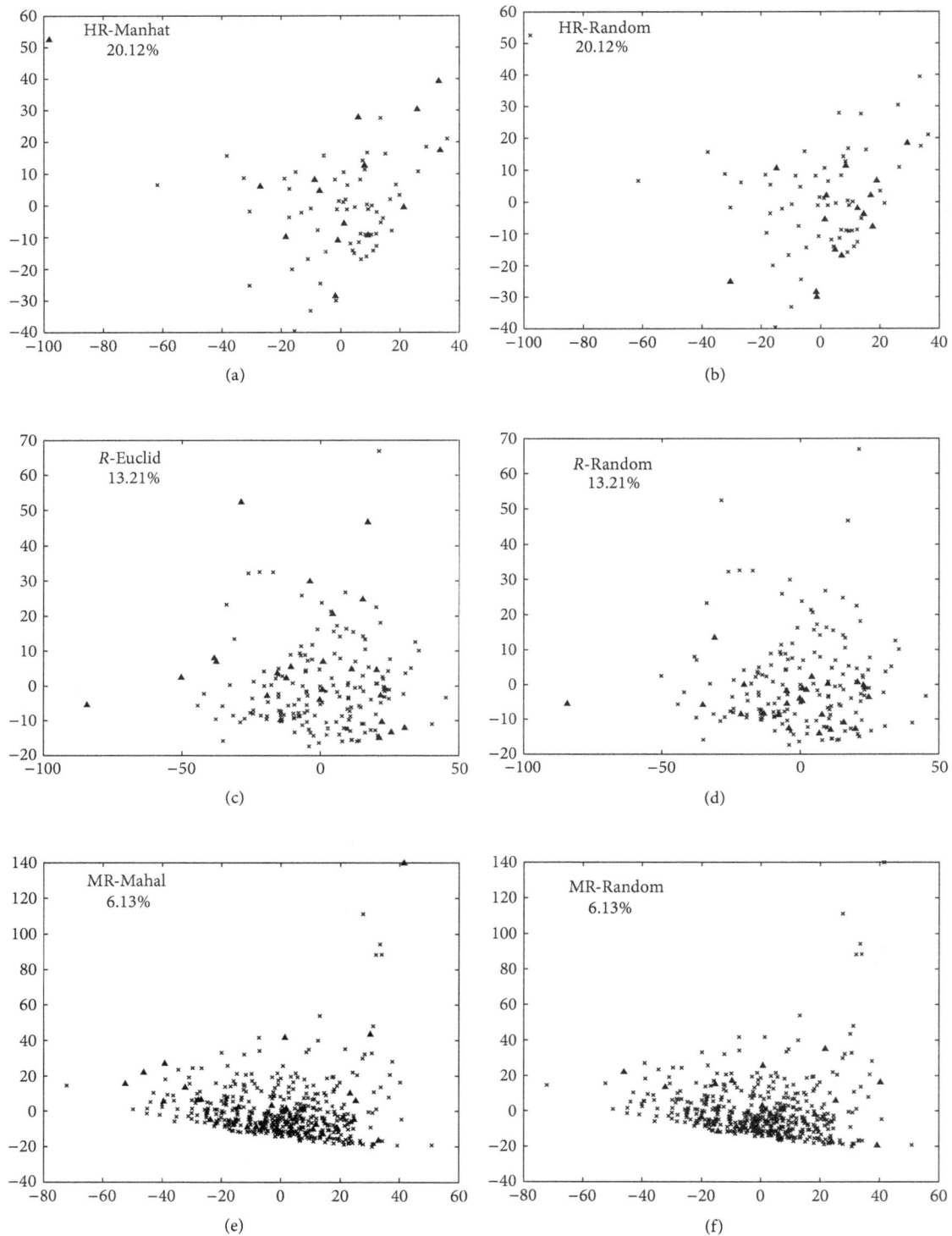

FIGURE 3: Continued.

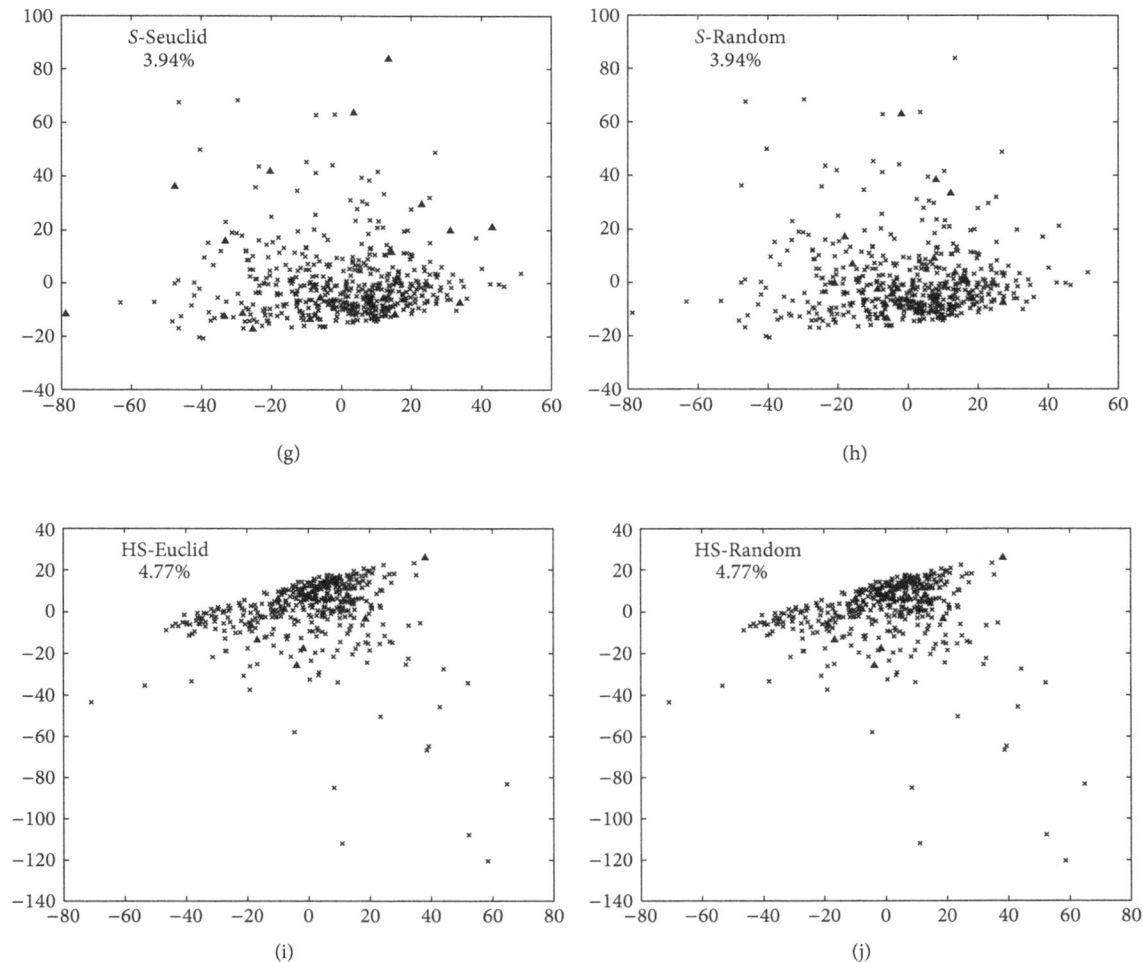

FIGURE 3: Principal component plots of core accessions and reserve accessions in the sampling percentage. The upward pointing triangles represented the core accessions; the crosses represented the reserved accessions. The left column showed plots for subcore collection constructed by LDSS method based on the selected genetic distance and sampling percentage; the right column showed plots for sub-core collection constructed by complete random selection based on the same sampling percentage. HR-Manhat, high resistant group combining Manhattan distance; R-Euclid, resistant group combining Euclidean distance; MR-Mahal, moderate resistant group combining Mahalanobis distance; S-Seuclid, susceptible group combining Standardized Euclidean distance; and HS-Euclid, high susceptible group combining Euclidean distance.

group was determined first by CR, then by VR. Moreover, a genetic distance that could make higher CR at low sampling percentage might be more valid than others.

In the present research, data of eight agronomic traits in 1,683 wheat varieties were downloaded from public database of "Chinese Crop Germplasm Resources Information System." Such a big number of wheat germplasm might not be planted within one area or one year. Therefore, the upper data might not be collected based on the same cultivating standards, which might affect the precision of the final core collection. However, there were more than one agronomic trait used in the present research. Data of eight agronomic traits were used to calculate CR and VR. The two evaluating parameters reflected the mean representativeness of the eight agronomic traits in the core collection, which reduced the error mentioned above. A wheat core collection for the

drought resistance breeding programs was constructed by the strategy selected in the present research based on the upper dataset. Table 2 showed the optimal genetic distance and the relative optimal sampling percentage for sub-core collection in each germplasm group. Therefore, the whole core collection was constructed by combining all sub-core collections. The principal component analysis showed that the genetic diversity was well preserved in that core collection. The method for the ideal constructing strategy selection suggested in the present research is also valuable in other crop's core collection construction.

Authors' Contribution

Jiancheng Wang and Yajing Guan contributed equally to this work.

Acknowledgments

The research was supported by the Special Fund for Agro-Scientific Research in the Public Interest of China (no. 201203052), the China Postdoctoral Science Foundation (no. 2012M521184), the Shandong Provincial Natural Science Foundation of China (no. ZR2010CQ016), and the Key Project of Natural Science Foundation of Zhejiang Province of China (no. Z3100150).

References

[1] C. Saint Pierre, J. L. Crossa, D. Bonnett, K. Yamaguchi-Shinozaki, and M. P. Reynolds, "Phenotyping transgenic wheat for drought resistance," *Journal of Experimental Botany*, vol. 63, no. 5, pp. 1799–1808, 2012.

[2] L. Simova-Stoilova, I. Vaseva, B. Grigorova, K. Demirevska, and U. Feller, "Proteolytic activity and cysteine protease expression in wheat leaves under severe soil drought and recovery," *Plant Physiology and Biochemistry*, vol. 48, no. 2-3, pp. 200–206, 2010.

[3] K. Bürling, Z. G. Cerovic, G. Cornic, J. M. Ducruet, G. Noga, and M. Hunsche, "Fluorescence-based sensing of drought-induced stress in the vegetative phase of four contrasting wheat genotypes," *Environmental and Experimental Botany*, vol. 89, no. 3, pp. 51–59, 2013.

[4] I. C. Dodd, W. R. Whalley, E. S. Ober, and M. A. J. Parry, "Genetic and management approaches to boost UK wheat yields by ameliorating water deficits," *Journal of Experimental Botany*, vol. 62, no. 15, pp. 5241–5248, 2011.

[5] A. Nouri, A. Etminan, J. A. T. da Silva, and R. Mohammadi, "Assessment of yield, yield-related traits and drought tolerance of durum wheat genotypes (*Triticum turjidum* var. durum Desf.)," *Australian Journal of Crop Science*, vol. 5, no. 1, pp. 8–6, 2011.

[6] M. Farooq, M. Irfan, T. Aziz, I. Ahmad, and S. A. Cheema, "Seed priming with ascorbic acid improves drought resistance of wheat," *Journal of Agronomy and Crop Science*, vol. 199, no. 1, pp. 12–22, 2013.

[7] C. Wendelboe-Nelson and P. C. Morris, "Proteins linked to drought tolerance revealed by DIGE analysis of drought resistant and susceptible barley varieties," *Proteomics*, vol. 12, no. 22, pp. 3374–3385, 2012.

[8] C. He, W. Zhang, Q. Gao, A. Yang, X. Hu, and J. Zhang, "Enhancement of drought resistance and biomass by increasing the amount of glycine betaine in wheat seedlings," *Euphytica*, vol. 177, no. 2, pp. 151–167, 2011.

[9] O. H. Frankel and A. H. D. Brown, "Plant genetics resources today: a critical appraisal," in *Crop Gentic Resources: Conservation and Evaluation*, J. H. W. Holden and J. T. Williams, Eds., pp. 249–257, George Allen and Unwin, London, UK, 1984.

[10] C. Silvar, A. M. Casas, D. Kopahnke et al., "Screening the Spanish Barley Core Collection for disease resistance," *Plant Breeding*, vol. 129, no. 1, pp. 45–52, 2010.

[11] A. Biabani, L. Carpenter-Boggs, C. J. Coyne, L. Taylor, J. L. Smith, and S. Higgins, "Nitrogen fixation potential in global chickpea mini-core collection," *Biology and Fertility of Soils*, vol. 47, no. 6, pp. 679–685, 2011.

[12] D. Pino Del Carpio, R. K. Basnet, R. C. H. de Vos, C. Maliepaard, R. Visser, and G. Bonnema, "The patterns of population differentiation in a *Brassica rapa* core collection," *Theoretical and Applied Genetics*, vol. 122, no. 6, pp. 1105–1118, 2011.

[13] C. Wang, S. Chen, and S. Yu, "Functional markers developed from multiple loci in GS3 for fine marker-assisted selection of grain length in rice," *Theoretical and Applied Genetics*, vol. 122, no. 5, pp. 905–913, 2011.

[14] J. C. Wang, J. Hu, H. M. Xu, and S. Zhang, "A strategy on constructing core collections by least distance stepwise sampling," *Theoretical and Applied Genetics*, vol. 115, no. 1, pp. 1–8, 2007.

[15] J. C. Wang, J. Hu, Y. J. Guan, and Y. F. Zhu, "Effect of the scale of quantitative trait data on the representativeness of a cotton germplasm sub-core collection," *Journal of Zhejiang University-SCIENCE B*, vol. 14, no. 2, pp. 162–170, 2012.

[16] J. C. Wang, J. Hu, X. X. Huang, and S. C. Xu, "Assessment of different genetic distances in constructing cotton core subset by genotypic values." *Journal of Zhejiang University-SCIENCE B*, vol. 9, no. 5, pp. 356–362, 2008.

[17] Y. J. Mei, J. P. Zhou, H. M. Xu, and S. J. Zhu, "Development of Sea Island cotton (*Gossypium barbadense* L.) core collection using genotypic values," *Australian Journal of Crop Science*, vol. 6, no. 4, pp. 673–680, 2012.

[18] The MathWorks Inc., *MATLAB Software*, The MathWorks Inc., Natick, Mass, USA, 2002.

[19] J. Bordes, C. Ravel, J. P. Jaubertie et al., "Genomic regions associated with the nitrogen limitation response revealed in a global wheat core collection," *Theoretical and Applied Genetics*, vol. 126, no. 3, pp. 805–822, 2013.

[20] A. Diederichsen, P. M. Kusters, D. Kessler, Z. Bainas, and R. Gugel, "Assembling a core collection from the flax world collection maintained by Plant Gene Resources of Canada," *Genetic Resources and Crop Evolution*, no. 1, pp. 1–7, 2012.

[21] Y. X. Zhang, X. R. Zhang, Z. Che, L. H. Wang, W. L. Wei, and D. H. Li, "Genetic diversity assessment of sesame core collection in China by phenotype and molecular markers and extraction of a mini-core collection," *BMC Genetics*, vol. 13, no. 1, p. 102, 2012.

[22] M. Sharma, A. Rathore, U. N. Mangala et al., "New sources of resistance to Fusarium wilt and sterility mosaic disease in a mini-core collection of pigeonpea germplasm," *European Journal of Plant Pathology*, vol. 133, no. 3, pp. 707–714, 2012.

[23] G. Saeidi, M. Rickauer, and L. Gentzbittel, "Tolerance for cadmium pollution in a core-collection of the model legume, Medicago truncatula L. at seedling stage," *Australian Journal of Crop Science*, vol. 6, no. 4, pp. 641–648, 2012.

[24] P. E. McClean, J. Terpstra, M. McConnell, C. White, R. Lee, and S. Mamidi, "Population structure and genetic differentiation among the USDA common bean (*Phaseolus vulgaris* L.) core collection," *Genetic Resources and Crop Evolution*, pp. 1–17, 2011.

[25] R. R. Coimbra, G. V. Miranda, C. D. Cruz, D. J. H. Silva, and R. A. Vilela, "Development of a Brazilian maize core collection," *Genetics and Molecular Biology*, vol. 32, no. 3, pp. 538–545, 2009.

[26] E. S. Rao, P. Kadirvel, R. C. Symonds, S. Geethanjali, and A. W. Ebert, "Using SSR markers to map genetic diversity and population structure of Solanum pimpinellifolium for development of a core collection," *Plant Genetic Resources*, vol. 1, no. 1, pp. 1–11, 2011.

[27] P. A. Reeves, L. W. Panella, and C. M. Richards, "Retention of agronomically important variation in germplasm core collections: implications for allele mining," *Theoretical and Applied Genetics*, vol. 124, no. 6, pp. 1155–1171, 2012.

[28] C. M. Díez, A. Imperato, L. Rallo, D. Barranco, and I. Trujillo, "Worldwide core collection of olive cultivars based on simple sequence repeat and morphological markers," *Crop Science*, vol. 52, no. 1, pp. 211–221, 2012.

[29] J. Hu, J. Zhu, and H. M. Xu, "Methods of constructing core collections by stepwise clustering with three sampling strategies based on the genotypic values of crops," *Theoretical and Applied Genetics*, vol. 101, no. 1-2, pp. 264–268, 2000.

[30] M. F. Oliveira, R. L. Nelson, I. O. Geraldi, C. D. Cruz, and J. F. F. de Toledo, "Establishing a soybean germplasm core collection," *Field Crops Research*, vol. 119, no. 2-3, pp. 277–289, 2010.

19

UV-B Radiation Impacts Shoot Tissue Pigment Composition in *Allium fistulosum* L. Cultigens

Kristin R. Abney,[1] **Dean A. Kopsell,**[1] **Carl E. Sams,**[1] **Svetlana Zivanovic,**[2] **and David E. Kopsell**[3]

[1] *Plant Sciences Department, The University of Tennessee, 2431 Joe Johnson Drive, Knoxville, TN 37996, USA*
[2] *Department of Food Science and Technology, The University of Tennessee, 2605 River Drive, Knoxville, TN 37996, USA*
[3] *Department of Agriculture, Illinois State University, Normal, IL 61790, USA*

Correspondence should be addressed to Dean A. Kopsell; dkopsell@utk.edu

Academic Editors: L. Fodorpataki, M. C. Martínez-Ballesta, and B. Muries

Plants from the *Allium* genus are valued worldwide for culinary flavor and medicinal attributes. In this study, 16 cultigens of bunching onion (*Allium fistulosum* L.) were grown in a glasshouse under filtered UV radiation (control) or supplemental UV-B radiation [$7.0\,\mu\text{mol·m}^{-2}\text{·s}^{-2}$ ($2.68\,\text{W·m}^{-2}$)] to determine impacts on growth, physiological parameters, and nutritional quality. Supplemental UV-B radiation influenced shoot tissue carotenoid concentrations in some, but not all, of the bunching onions. Xanthophyll carotenoid pigments lutein and β-carotene and chlorophylls *a* and *b* in shoot tissues differed between UV-B radiation treatments and among cultigens. Cultigen "Pesoenyj" responded to supplemental UV-B radiation with increases in the ratio of zeaxanthin + antheraxanthin to zeaxanthin + antheraxanthin + violaxanthin, which may indicate a flux in the xanthophyll carotenoids towards deepoxydation, commonly found under high irradiance stress. Increases in carotenoid concentrations would be expected to increase crop nutritional values.

1. Introduction

Fruits and vegetables have varying levels of phytonutrients, in addition to vitamins and minerals. Two important classes of phytonutrients are carotenoid and chlorophyll pigments. The primary carotenoids found in leaf tissue of most plant species include zeaxanthin, antheraxanthin, violaxanthin, lutein, β-carotene, and neoxanthin [1]. Chlorophylls are the dominant pigments in plants and serve primary roles in photosynthesis. These compounds are very effective antioxidants and help prevent certain types of cancers and aging eye diseases like macular eye degeneration [2, 3]. However, the chemical structures of these pigments also give them the ability to donate electrons and effectively become prooxidants under certain conditions [4]. *Allium* species contain chlorophyll and carotenoid pigments in shoot tissues [5] and also contain different levels of sulfur-containing compounds which prevent certain cancers [6]. While all higher plants contain chlorophylls and carotenoids, genetic variations for pigment accumulations exist both within and among plant species. Within any given crop species, there can be multiple landraces, accessions and cultivars, or, collectively, cultigens. These variations are important to advancements in plant development programs for increased nutrition, disease prevention, or other factors. However, cultigens will react differently under almost any given stress.

The absorption of light by chlorophyll and antenna pigments and the transfer of excitation energy to the reaction centers of PSII and PSI are the initial steps in photosynthesis. The photosynthetic apparatus has the ability to react to many different environmental stimuli, especially changes in light intensity. Under conditions of high light stress, photosynthetic systems are saturated, and excess energy needs to be diverted to avoid potential damage [7]. Carotenoids are unsaturated long chain polycarbons which protect the photosynthetic apparatus from high light excitation by quenching free radicals, functioning in nonphotochemical quenching, and dissipating excess thermal energy [7, 8]. The xanthophyll cycle, or violaxanthin cycle, is the mechanism by which plants regulate light energy available for photosynthesis. In intense light situations, violaxanthin is rapidly and reversibly converted to zeaxanthin, via antheraxanthin. Zeaxanthin is a direct quencher of chlorophyll excited states and can prevent photooxidative stress and lipid peroxidation [7].

Higher amounts of UV-A (380–320 nm) and UV-B radiation (280–320 nm) may influence the accumulation of plant compounds used to combat light stress. Carotenoid metabolites not only protect plants from excess UV radiation but also can protect humans from UV radiation when translocated to subdermal skin tissues [9]. What remains uncertain is the impact of increased UV radiation on growth and development and nutritional values of cultivated crops [10]. Previous studies have demonstrated impacts of UV radiation on plant performance, cellular structures, and pigment accumulations. In a study by Yuan et al. [11], 20 cultivars of wheat (*Triticum aestivum* L.) were grown under UV-B radiation stress to determine possible detrimental influences. Most wheat cultivars responded negatively to UV-B radiation; however, several cultivars showed increases in plant height and biomass. Structural changes like ruptured chloroplast envelopes have been noted in UV-sensitive rice cultivars (*Oryza sativa* L.) when exposed to UV stress [12]. Increases in UV radiation can delay flowering and harvest times among different cultigens of bush beans (*Phaseolus vulgaris* L.) [13]. Bush beans grown under UV radiation showed decreases in fruit size and yield when compared to cultivars not grown under UV radiation stress. Tomatoes (*Solanum lycopersicum* cv. DRW 5981) grown using UV-B blocking filters showed increases in lycopene and β-carotene, while fruits of the same variety showed decreases in lycopene, phytoene, and phytofluene when grown without UV-B blocking filters [14]. The tomato cultivar "HP1" accumulated more than twice the amount of lycopene in fruit tissues when grown under no UV-B radiation. Results from such studies demonstrate impacts on nutritional quality from excess UV radiation.

Allium species are valued worldwide for culinary flavor and medicinal attributes. Plants in this genus have been important to multiple cultures for centuries. *Alliums* have high levels of nutritionally important secondary plant metabolites which convey numerous health benefits. For example, bulb onions (*Allium cepa* L.) contain high levels of flavonoids [15], S-alk(en)yl-L-cysteine sulfoxides [16], and a variety of volatile antioxidant compounds [17]. However, no studies to date have measured the impact of UV radiation on the production of nutritionally important pigments in *Alliums*. *Allium fistulosum* is consumed, in part, for its shoot tissues as well as pseudostems. Carotenoid and chlorophyll compounds are present in the shoot tissues of *A. fistulosum* [5, 18]. Therefore, the objectives of this project were to examine both environmental and genetic responses to elevated UV-B radiation among a large subset of *A. fistulosum* cultigens. Responses were noted for plant height, shoot tissue biomass, photochemical efficiency (F_v/F_m), and concentrations of carotenoid and chlorophyll pigments in the shoot and pseudostem tissues.

2. Methods and Materials

2.1. Plant Culture. On December 16, 2008, seeds of 16 different *A. fistulosum* cultigens were sown in 15 cm pots holding soilless media in a glasshouse in Knoxville, TN USA (35°96′ N Lat.), which blocked UV-wavelengths (280–380 nm). The photosynthetically active radiation (PAR) in the glasshouse averaged 540 μmol·m^{-2}·s^{-2} (Apogee Nanologger model ANL, Apogee Instruments, Inc., Roseville, CA USA). The cultigens included eight accessions [PI 274254-05GI, PI 462345-05GI (Jionji Negi), PI 546343-90U01 (GA-C 76), PI 546228-06GI (Improved Beltsville Bunching), PI 280562-04GI (Pesoenyj), PI 436539-06GI (Zhang Qui Da Cong), PI 462357-06GI (Shounan), and G 30393-06GI] from the USDA-ARS National Plant Germplasm Repository (Geneva, NY USA); four cultivars ("Long White Bunching," "Feast, Performer," and "Parade") from Seedway, LLC (Hall, NY, USA); and four cultivars ("White Spear," "Evergreen Hardy White," "Deep Purple," and "Ishikura Improved F1") from Johnny's Selected Seeds (Winslow, ME, USA). The seedlings were watered daily for the duration of the experiment. On January 10, 2009, the seedlings were thinned to two plants per pot and fertilized with a nutrient solution containing (mg·L^{-1}): N (105), P (91.5), K (117.3), Ca (80.2), Mg (24.6), S (32.0), Fe (0.5), B (0.25), Mo (0.005), Cu (0.01), Mn (0.25), and Zn (0.025) [19]. Each pot was fertilized once a week for the duration of the experiment with 100 mL of nutrient solution. The experimental design was a split plot arranged as a randomized complete block. UV-B treatments were the main plots, and *A. fistulosum* cultigens were the subplots. Six individual plants per cultigen composed a replication, with four replications randomly assigned to each UV-B treatment.

Supplemental UV-B radiation (313 nm) was provided by banks of commercially available UV-B 313 lamps (Q-Panel Lab Products, Cleveland, OH, USA), and treatment began on January 27, 2009 delivering 7.0 μmol·m^{-2}·s^{-1} (2.68 W·m^{-2}) of UV-B (Spectroradiometer Model SPEC-UV/PAR, Apogee Instruments, Inc., Roseville, CA, USA) to the treated plants. To control pests in the greenhouse, the beneficial insect species of *Hypoaspis miles* and *Neoseiulus cucumeris* were used to control thrips, while *Orius insidiosus* was used to help control aphids. These insects were first released on January 23, 2009, and were released every two weeks thereafter.

On March 3, 2009, all of the bunching onion cultigens were harvested. Six plants were harvested from each replication. Fresh weights and plant heights were taken and averaged for each replication. One measure of F_v/F_m was taken from each of the harvest plants at the midpoint of plant height using a modulated fluorometer (OS1-F1 Modulated Fluorometer, Opti-Sciences, Hudson, NH, USA). The F_v/F_m value is an indication of photoinhibition and overall plant health. All plants were harvested, and pseudostem and leaf tissue were separated. The samples were immediately placed in a −20°C freezer before being moved to a −80°C freezer within 8 h.

2.2. Pigment Extraction and Determination. Tissue pigments were extracted according to Kopsell et al. [20] and analyzed according to Kopsell et al. [5]. The samples were freeze-dried and ground with a mortar and pestle with liquid nitrogen. A 0.10 g subsample was rehydrated with 0.8 mL of ultra-pure H_2O. The samples were incubated for 20 min, before 0.8 mL of ethyl-β-8′-apo-carotenotate (Sigma Chemical Co., St. Louis, MO, USA) was added as an internal standard to establish extraction efficiency. For pigment extraction, 2.5 mL of tetrahydrofluran was added to the sample. Using a Potter-Elvehjem tissue grinding tube (Kontes, Vineland, NJ, USA), the samples was homogenized in an ice bath to dissipate heat generated from maceration. The tubes were then centrifuged in a clinical centrifuge (Centrific Model 225, Fisher Scientific, Pittsburg, PA, USA) for 3 min at $500 g_n$. The supernatant was removed, and the pellet was rehydrated with 2.0 mL tetrahydrofluran. This procedure was repeated twice more until the supernatant was colorless. The combined supernatants were reduced to 0.5 mL under a stream of nitrogen gas and brought to a final volume of 5 mL with methanol. The samples were then filtered through a 0.2 μm Econofilter PTFE 25/20 polytetrafluoroethylene filter (Agilent Technologies, Wilmington, DE, USA) using a 5 mL syringe. A 1.5 mL aliquot was put into an amber vial and capped prior to high performance liquid chromatography (HPLC) analysis.

An Agilent 1200 series HPLC unit with a photodiode array detector (Agilent Technologies, Palo Alto, CA, USA) was used for pigment separation (Figure 1). The column used was a 250 × 4.6 mm i.d., 5 μm analytical scale polymeric RP-C_{30}, with a 10 × 4.0 mm i.d. guard cartridge and holder (ProntoSIL, MAC-MOD Analytical Inc., Chadds Ford, PA, USA), which allowed for effective separation of chemically similar compounds. The column was maintained at 30°C using a thermostated column compartment. All separations were achieved isocratically using a binary mobile phase of 11% methyl tert-butyl ether (MTBE), 88.99% MeOH, and 0.01% triethylamine (TEA) (v/v/v). The flow rate was 1.0 mL/min, with a run time of 53 min. There was a 2 min equilibration prior to the next injection. Eluted compounds from a 10 μL injection loop were detected at 453 nm (carotenoids, internal standard, chlorophyll b) and 652 nm (chlorophyll a). Data were collected, recorded, and integrated using ChemStation Software (Agilent Technologies). Peak assignments for each pigment were performed by comparing retention times and line spectra obtained from the photodiode array detection using external standards. Standards included antheraxanthin, neoxanthin, lutein, violaxanthin, zeaxanthin, and β-carotene, chlorophyll a and chlorophyll b (ChromaDex Inc., Irvine, CA, USA). The concentrations of the external standards were determined spectrophotometrically using a procedure by Davies and Köst [21]. Pigment data is presented on a fresh mass (FM) basis.

2.3. Statistical Analyses. Statistical analyses were completed using the GLM procedure of SAS (v. 9.1, SAS Institute, Cary, NC, USA). Cultigen means within each treatment were separated by least significant difference (LSD) at α = 0.05. Differences between cultigens means between treatments

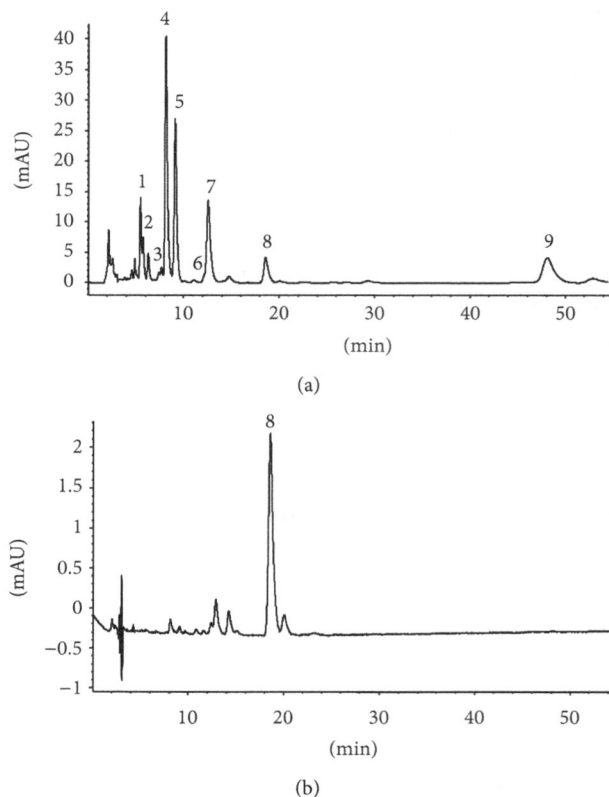

FIGURE 1: HPLC chromatogram of *Allium fistulosum* L. leaf (a) and pseudostem (b) tissues at 453 nm. Retention times (min) for the pigments were (1) violaxanthin, 5.52 min; (2) neoxanthin, 5.81 min; (3) antheraxanthin, 7.59 min; (4) chlorophyll b, 8.51 min; (5) lutein, 9.33 min; (6) zeaxanthin, 11.31 min; (7) chlorophyll a, 13.90 min; (8) ethyl-β-8′-apo-carotenoate (internal standard), 19.32 min; and (9) β-carotene, 48.46 min. HPLC conditions are described in the text.

were detected by using Student's t-test ($P = 0.05$) using JMP (v 7.0.1, SAS Institute).

3. Results and Discussion

3.1. Shoot Tissue Biomass. Significant differences were found among cultigens (F = 6.67, P < 0.001) for shoot tissue height, but no differences were found between the UV-B treatments or the interaction between the cultigen and UV-B radiation treatment (Table 1). Only one cultigen (GA-C 76) differed significantly between UV-B radiation treatments for shoot tissue height. "Long White Bunching" demonstrated the greatest growth in shoot tissue height under both UV-B radiation treatments, while "G 30393-06GI" had the shortest final shoot tissue height. There were differences in shoot tissue FM between UV-B radiation treatments (F = 238.10, P < 0.001) and among cultigens (F = 11.09, P < 0.001), but no difference in the treatment by cultigen interaction (Table 1). "Deep Purple," "Feast," "GA-C 76," "Ishikura Improved F1," "Improved Beltsville Bunching," "Jionji," "Long White Bunching," "Parade," "Performer," "Pesoenyj," "Shounan," "White Spear," "274254-05GI," and

"G 30393-06GI" all showed decreases in shoot tissue FM with exposure to the UV-B radiation treatment. Significant decreases in shoot tissue biomass from the UV-B treatment would indicate a radiational stress had occurred in the bunching onion cultigens in the current study. The cultigens with the greatest shoot tissue FM accumulations were "Long White Bunching" and "Improved Beltsville Bunching" (Table 1).

3.2. Shoot Tissue Carotenoid Pigment Concentrations. No carotenoid pigments were measured in the pseudostem tissues of any of the bunching onion cultigens (Figure 1; data not shown). Kopsell et al. [5] also reported no carotenoid pigmentation present in bunching onion pseudostem tissues. Shoot tissue zeaxanthin differed significantly among the bunching onion cultigens ($F = 4.07$; $P < 0.001$) (Table 2). However, there were no significant changes in shoot tissue zeaxanthin in response to UV-B treatment, or the interaction of the UV-B treatments and cultigens. Only the cultigens of "G 30393-06GI" and "Feast" showed an increase in shoot tissue zeaxanthin under the supplemental UV-B radiation, as compared to control. The ranges of zeaxanthin concentrations in the bunching onions under supplemental UV-B were from 0.08 mg/100 g FM for "Deep Purple" and "White Spear" to 0.16 mg/100 g FM for "Improved Beltsville Bunching". Cultigen "Pesoenyj" had the highest concentration of zeaxanthin among plants grown without supplemental UV radiation at 0.19 mg/100 g FM, while "Feast" and "Evergreen Hardy White" had the lowest zeaxanthin concentrations at 0.07 mg/100 g FM. Increases in zeaxanthin could be an indication that the plants experienced radiational stress from the UV-B treatment. Plant responses through increased zeaxanthin concentrations would be expected to help dissipate excess energy from the photosystems [7].

Shoot tissue violaxanthin responded significantly to both UV-B radiation treatment ($F = 6.76$; $P = 0.0109$) and cultigen ($F = 4.42$, $P < 0.001$), but not to the interaction between treatment and cultigen (Table 2). Many of the bunching onion cultigens showed higher concentrations of violaxanthin in response to UV-B radiational supplementation. However, only one cultigen had significant increases in violaxanthin concentrations (GA-C 76) in response to UV-B radiation treatment. Increases in violaxanthin in bunching onions grown under UV-B radiation may suggest that these cultigens may not be as susceptible to UV-B radiational damage as the other cultigens. Violaxanthin concentrations under supplemental UV-B radiation ranged from 2.04 mg/100 g FM for "GA-C 76" to 0.59 mg/100 g FM for "Performer." Cultigen "Pesoenyj" had the highest concentrations of violaxanthin (2.35 mg/100 g FM) for bunching onions grown without supplemental UV-B radiation, while "G 30393-06GI" had the lowest violaxanthin concentrations (0.53 mg/100 g FM).

Antheraxanthin, the intermediate compound in xanthophyll cycle, responded significantly to changes in UV-B radiation treatments ($F = 16.61$; $P < 0.0001$) and cultigens ($F = 4.68$; $P < 0.001$). The majority of cultigens had higher antheraxanthin concentrations in response to the UV-B radiation treatment; however, no cultigens had significantly higher levels as compared to the control treatment (Table 2).

The ranges for antheraxanthin concentrations in bunching onions grown under UV-B radiation treatment were from 1.38 mg/100 g FM for "Pesoenyj" to 0.79 mg/100 g FM for "274254-05GI." In the plants grown without UV-B radiation, "Pesoenyj" had the highest antheraxanthin concentrations (1.35 mg/100 g FM), while "Ishikura Improved F1" had the lowest concentrations (0.59 mg/100 g FM). While changes in this compound cannot directly tell which way the xanthophyll cycle is fluxing, increases or decreases may help predict potential energy flow.

Neoxanthin concentrations responded significantly to UV-B radiation treatment ($F = 12.13$; $P = 0.0008$), cultigen ($F = 3.20$; $P = 0.0003$), and the interaction of UV radiation treatment and cultigen ($F = 2.27$; $P = 0.0092$). There were significant increases in neoxanthin from the UV-B treatment for the cultigens "Feast," "GA-C 76," and "G 30393-06GI" when compared to the control treatment (Table 3). "Feast" showed the highest concentrations of neoxanthin under UV-B radiation treatment (1.86 mg/100 g FM), while "Deep Purple" had the lowest concentration of neoxanthin (0.73 mg/100 g FM). "Pesoenyj" showed the highest neoxanthin concentration (1.96 mg/100 g FM) compared to the other cultigens grown under the control treatment. "Hardy Evergreen White" had the lowest of all of the cultigens not grown under supplemental UV-B radiation at 0.40 mg/100 g FM.

The bunching onions showed significant changes in lutein in response to UV-B treatment ($F = 17.89$; $P < 0.0001$) and cultigen ($F = 2.34$; $P = 0.0070$). The majority of cultigens had higher lutein concentrations in response to the UV-B radiation treatment; however, only "Feast" and "GA-C 76" had significantly higher lutein (Table 3). "Pesoenyj" had the highest concentrations of lutein both with and without supplemental UV-B radiation at 8.01 and 9.23 mg/100 g FM, respectively. "Deep Purple" had the lowest concentration of lutein among bunching onions grown with supplemental UV-B radiation at 5.04 mg/100 g FM, and "Feast" had the lowest amount of lutein for bunching onions grown without supplemental UV-B radiation at 4.11 mg/100 g FM. Lutein acts as an accessory pigment and is the predominant carotenoid in photosystem (PS) II [7]. Research shows UV radiation will impact PSII functioning to a greater extent than PSI [22]. Therefore, increases in lutein concentrations for the cultigens in the current study may indicate increased radiational stress within PSII from the supplemental UV-B treatment.

Concentrations of β-carotene showed no changes in response to UV treatment or cultigen (Table 3). "Pesoenyj" had the highest concentrations of β-carotene in bunching onions grown without UV-B radiation, and "Ishikura Improved F1" had the lowest concentrations. The range of shoot tissue β-carotene levels for cultigens grown under supplemental UV-B radiation were 2.80 mg/100 g FM for "Shounan" to 0.88 mg/100 g FM for "Evergreen Hardy White." For the cultigens that were not grown under supplemental UV-B radiation, the ranges for β-carotene concentration were 3.45 mg/100 g FM (Pesoenyj) and 0.64 mg/100 g FM (Evergreen Hardy White). Reported mean value for β-carotene in shoot tissues of A. fistulosum is 0.60 mg/100 g FM, while the mean values for lutein and zeaxanthin are 1.14 mg/100 g

TABLE 1: Mean values[a] for shoot tissue height (cm) and fresh biomass (g) for *Allium fistulosum* L. cultigens grown under supplemental UV-B (313 nm) light [7.0 μmol·m^{-2}·s^{-2} (2.68 W·m^{-2}); UV-B] or UV-filtered (control) light in a glasshouse in Knoxville, TN, USA (35°96′N Lat.).

Cultigen	Shoot tissue height (cm)			Shoot tissue fresh biomass (g)		
	UV-B	Control	Pr > $|t|$[b]	UV-B	Control	Pr > $|t|$
Deep Purple	44.71 ± 5.49	45.19 ± 4.52	ns	64.54 ± 17.02	97.80 ± 17.44	P = 0.034
Evergreen Hardy White	39.11 ± 1.96	41.05 ± 5.38	ns	38.65 ± 7.80	70.78 ± 8.08	ns
Feast	38.58 ± 3.24	39.96 ± 2.40	ns	48.79 ± 15.99	90.46 ± 13.85	P = 0.008
GA-C 76	38.82 ± 1.91	44.13 ± 3.70	P = 0.043	41.90 ± 2.66	73.00 ± 15.53	P = 0.008
Ishikura improved F1	39.63 ± 1.12	41.17 ± 3.13	ns	54.93 ± 9.22	99.80 ± 9.47	P = 0.001
Improved Beltsville Bunching	44.26 ± 1.69	48.06 ± 2.19	ns	76.86 ± 9.82	110.42 ± 19.39	P = 0.021
Jionji Negi	36.20 ± 1.82	38.63 ± 1.54	ns	38.05 ± 4.73	62.04 ± 8.23	P = 0.002
Long White Bunching	49.68 ± 2.09	50.38 ± 4.62	ns	74.07 ± 8.73	116.01 ± 20.24	P = 0.009
Parade	42.13 ± 2.60	36.03 ± 9.40	ns	53.89 ± 10.73	90.81 ± 16.87	P = 0.010
Performer	39.37 ± 2.36	39.42 ± 3.10	ns	54.74 ± 11.08	86.25 ± 16.66	P = 0.020
Pesoenyj	39.44 ± 4.23	44.54 ± 4.03	ns	29.22 ± 9.30	58.76 ± 6.37	P = 0.002
Shounan	36.72 ± 3.02	37.14 ± 2.54	ns	39.17 ± 3.56	63.32 ± 4.02	P = 0.001
White Spear	40.06 ± 2.79	42.42 ± 1.72	ns	53.16 ± 14.41	87.42 ± 10.58	P = 0.009
Zhang Qui Da Cong	36.62 ± 2.88	37.87 ± 3.80	ns	53.57 ± 19.51	84.72 ± 16.67	ns
274254-05GI	42.48 ± 1.86	43.79 ± 4.40	ns	47.63 ± 4.34	84.65 ± 6.98	P = 0.001
G 30393-06GI	36.62 ± 3.40	35.67 ± 1.31	ns	50.96 ± 8.63	81.69 ± 12.01	P = 0.006
LSD$_{0.05}$[c]	5.77	5.80		10.72	19.33	

[a] Composition of n = 6 plant samples from 4 replications ± standard deviation. [b] Significance based on paired Student's t-test among treatments; ns: not significant. [c] LSD for differences between cultivar means α = 0.05.

TABLE 2: Mean values[a] for shoot tissue zeaxanthin, violaxanthin, and antheraxanthin (mg/100 g fresh mass) for *Allium fistulosum* L. cultigens grown under supplemental UV-B (313 nm) light [7.0 μmol·m^{-2}·s^{-2} (2.68 W·m^{-2}); UV-B] or UV-filtered (control) light in a glasshouse in Knoxville, TN, USA (35°96′N Lat.).

Cultigen	Zeaxanthin			Violaxanthin			Antheraxanthin		
	UV-B	Control	Pr > $|t|$[b]	UV-B	Control	Pr > $|t|$	UV-B	Control	Pr > $|t|$
				mg/100 g fresh mass					
Deep Purple	0.08 ± 0.02	0.10 ± 0.03	ns	1.25 ± 0.36	0.88 ± 0.42	ns	1.07 ± 0.31	0.78 ± 0.23	ns
Evergreen Hardy White	0.10 ± 0.03	0.07 ± 0.02	ns	1.73 ± 0.61	1.20 ± 0.25	ns	1.03 ± 0.18	0.78 ± 0.14	ns
Feast	0.11 ± 0.01	0.07 ± 0.01	P = 0.010	1.35 ± 0.82	0.66 ± 0.58	ns	1.15 ± 0.27	0.92 ± 0.19	ns
GA-C 76	0.12 ± 0.02	0.08 ± 0.02	ns	2.04 ± 0.19	1.34 ± 0.20	P = 0.002	1.92 ± 0.65	1.27 ± 0.47	ns
Ishikura Improved F1	0.10 ± 0.05	0.09 ± 0.02	ns	1.75 ± 0.94	1.10 ± 0.38	ns	0.81 ± 0.50	0.59 ± 0.13	ns
Improved Beltsville Bunching	0.16 ± 0.02	0.15 ± 0.04	ns	1.62 ± 0.54	1.25 ± 0.14	ns	0.99 ± 0.38	0.72 ± 0.21	ns
Jionji Negi	0.12 ± 0.06	0.13 ± 0.03	ns	1.54 ± 0.25	1.51 ± 0.35	ns	1.20 ± 0.25	0.78 ± 0.50	ns
Long White Bunching	0.12 ± 0.03	0.11 ± 0.03	ns	0.89 ± 0.18	0.81 ± 0.42	ns	0.82 ± 0.13	0.89 ± 0.14	ns
Parade	0.10 ± 0.02	0.11 ± 0.02	ns	1.23 ± 0.50	1.02 ± 0.61	ns	0.86 ± 0.16	0.79 ± 0.24	ns
Performer	0.09 ± 0.02	0.10 ± 0.02	ns	0.59 ± 0.52	1.45 ± 0.60	ns	0.87 ± 0.06	0.85 ± 0.32	ns
Pesoenyj	0.12 ± 0.03	0.19 ± 0.06	ns	1.93 ± 0.33	2.35 ± 0.82	ns	1.38 ± 0.30	1.35 ± 0.52	ns
Shounan	0.13 ± 0.05	0.09 = 0.01	ns	1.87 ± 0.81	1.04 ± 0.36	ns	1.18 ± 0.40	0.63 ± 0.30	ns
White Spear	0.08 ± 0.02	0.08 ± 0.03	ns	1.00 ± 0.60	0.77 ± 0.47	ns	0.74 ± 0.13	0.60 ± 0.09	ns
Zhang Qui Da Cong	0.11 ± 0.04	0.12 ± 0.02	ns	1.29 ± 0.48	1.30 ± 0.41	ns	0.81 ± 0.30	0.74 ± 0.06	ns
274254-05GI	0.13 ± 0.04	0.15 ± 0.03	ns	1.64 ± 0.42	1.27 ± 0.52	ns	0.79 ± 0.14	0.71 ± 0.33	ns
G 30393-06GI	0.13 ± 0.03	0.08 ± 0.02	P = 0.016	0.60 ± 0.51	0.50 ± 0.43	ns	1.07 ± 0.35	0.67 ± 0.14	ns
LSD$_{0.05}$[c]	ns	0.04		0.74	0.70		0.44	0.44	

[a] Composition of n = 6 plant samples from 4 replications ± standard deviation. [b] Significance based on paired Student's t-test among treatments; ns: not significant. [c] LSD for differences between cultivar means α = 0.05.

TABLE 3: Mean values[a] for shoot tissue neoxanthin, lutein, and β-carotene (mg/100 g fresh mass) for *Allium fistulosum* L. cultigens grown under supplemental UV-B (313 nm) light [7.0 μmol·m^{-2}·s^{-2} (2.68 W·m^{-2}); UV-B] or UV-filtered (control) light in a glasshouse in Knoxville, TN, USA (35°96′N Lat.).

Cultigen	Neoxanthin			Lutein			β-carotene		
	UV-B	Control	Pr > $|t|$[b]	UV-B	Control	Pr > $|t|$	UV-B	Control	Pr > $|t|$
	mg/100 g fresh mass								
Deep Purple	0.73 ± 0.51	1.04 ± 0.59	ns	5.04 ± 1.48	5.38 ± 0.73	ns	1.07 ± 0.81	1.09 ± 0.30	ns
Evergreen Hardy White	0.74 ± 0.47	0.40 ± 0.18	ns	7.10 ± 2.86	5.10 ± 1.49	ns	0.88 ± 0.71	0.64 ± 0.23	ns
Feast	2.09 ± 0.48	0.79 ± 0.57	P = 0.013	7.66 ± 0.90	4.11 ± 0.54	P = 0.001	1.85 ± 0.85	1.04 ± 0.74	ns
GA-C 76	1.53 ± 0.32	0.66 ± 0.18	P = 0.003	7.66 ± 0.38	5.57 ± 0.67	P = 0.002	1.48 ± 0.27	1.17 ± 0.09	ns
Ishikura Improved F1	0.63 ± 0.43	0.63 ± 0.30	ns	6.35 ± 3.18	4.80 ± 1.20	ns	2.20 ± 2.59	0.78 ± 0.41	ns
Improved Beltsville Bunching	0.82 ± 0.89	0.60 ± 0.16	ns	6.95 ± 1.34	5.62 ± 0.65	ns	1.64 ± 0.95	1.39 ± 0.30	ns
Jionji Negi	0.92 ± 0.25	0.91 ± 0.36	ns	7.35 ± 1.65	6.21 ± 1.49	ns	1.74 ± 1.20	2.29 ± 1.32	ns
Long White Bunching	1.76 ± 0.26	1.47 ± 0.35	ns	6.00 ± 0.58	5.05 ± 1.17	ns	1.69 ± 0.29	1.94 ± 0.56	ns
Parade	1.46 ± 0.81	0.87 ± 0.36	ns	6.36 ± 1.17	6.04 ± 1.47	ns	1.26 ± 0.14	2.54 ± 1.23	ns
Performer	1.14 ± 0.90	0.75 ± 0.47	ns	6.33 ± 0.96	5.60 ± 2.36	ns	1.28 ± 0.48	2.38 ± 1.65	ns
Pesoenyj	1.03 ± 0.19	1.96 ± 0.78	ns	8.01 ± 1.21	9.23 ± 2.59	ns	1.87 ± 0.20	3.45 ± 2.39	ns
Shounan	1.32 ± 0.71	0.54 ± 0.32	ns	7.66 ± 2.83	5.03 ± 1.26	ns	2.80 ± 1.46	1.10 ± 0.54	ns
White Spear	1.21 ± 0.62	0.93 ± 0.63	ns	6.08 ± 0.94	4.70 ± 1.37	ns	1.49 ± 0.51	1.08 ± 0.60	ns
Zhang Qui Da Cong	0.75 ± 0.42	0.65 ± 0.29	ns	5.65 ± 1.44	5.31 ± 1.47	ns	1.05 ± 0.24	1.20 ± 0.57	ns
274254-05GI	0.84 ± 0.37	0.72 ± 0.45	ns	6.18 ± 0.79	5.33 ± 1.82	ns	1.81 ± 0.86	1.86 ± 1.57	ns
G 30393-06GI	1.86 ± 0.44	0.85 ± 0.62	P = 0.040	6.02 ± 1.18	4.42 ± 0.69	ns	1.86 ± 0.77	1.28 ± 0.67	ns
LSD$_{0.05}$[c]	0.78	0.67		ns	2.14		ns	1.54	

[a]Composition of n = 6 plant samples from 4 replications ± standard deviation. [b]Significance based on paired Student's t-test among treatments; ns: not significant. [c]LSD for differences between cultivar means α = 0.05.

FM [23]. Umehara et al. [24] reported β-carotene values in the leaves of *A. fistulosum* L. cultigen "Kujyoasagikei" to be 4.63 mg/100 g FM. β-carotene is an accessory pigment and is the predominant carotenoid in PSI. β-carotene is present in PSII, but mostly in regions around the reaction center [7]. Since there were no impacts on β-carotene concentrations in the current study, it is possible that PSI is not under as much stress from the UV-B treatments imposed in this study [22].

The xanthophyll cycle pigments (zeaxanthin, antheraxanthin, and violaxanthin) are important for the dissipation of excess absorbed light, performed almost exclusively by ZEA. Photosynthetic rates are reduced under many environmental stressors, which increase the need for dissipation of excess absorbed light energy [7]. The ratio of zeaxanthin + antheraxanthin to zeaxanthin + antheraxanthin + violaxanthin (ZA/ZAV) responded significantly to cultigen (F = 3.01; P = 0.0006), but not to UV-B radiation treatment or the interaction between treatment and cultigen. Significant increases in response to supplemental UV-B were found for "Pesoenyj." "G 30393-06GI" had the highest ZA/ZAV ratio of cultigens grown under supplemental UV-B radiation, and "Ishikura Improved F1" had the lowest ZA/ZAV ratio at 0.34. For the cultigens not grown under UV-B radiation, "Feast" had the highest ZA/ZAV ratio at 0.65, while "Jionji Negi" had the lowest ZA/ZAV ratio at 0.35 (Table 4). Changes in the ZA/ZAV ratio can identify fluxes within the xanthophyll energy dissipation cycle. An increase in ZA/ZAV ratio shows a decrease in violaxanthin, which could mean these compounds are undergoing deepoxydation because of

high light energy [7]. A study by Niyogi et al. [25] helped demonstrate the importance of this photoprotective mechanistic cycle. In this study, mutant *Arabidopsis thaliana* was unable to undergo deepoxydation and converts violaxanthin to zeaxanthin, which resulted in an increased sensitivity to higher light levels. While the Niyogi et al. [25] study did not specifically look at how UV-B radiation affected xanthophyll cycle functioning, energy from UV wavelengths is higher than energy from PAR wavelengths and could be expected to change the flux between the xanthophyll pigments.

Kopsell et al. [5] grew many of the same bunching onion cultigens under field conditions in Knoxville, TN, USA, and Geneva, NY, USA, and reported similar levels of shoot tissue β-carotene and neoxanthin as found in the current study; however, values for violaxanthin, antheraxanthin, lutein, chlorophyll a, and chlorophyll b were much higher in the current study than previously reported. Differences in shoot tissue pigments for cultigens among the two studies may be attributed to differences in growing conditions (field versus glasshouse) and the time of year the cultigens were evaluated (summer versus winter).

Epidemiological data supports the positive association between increased dietary intake of plant foods high in carotenoids and greater carotenoid tissue concentrations with lower risks of certain chronic diseases. Many of these disease suppressing abilities can be attributed to the antioxidant properties of carotenoids. One of the most important physiological functions of carotenoids in human nutrition is as vitamin A precursors. Provitamin A carotenoid compounds

TABLE 4: Mean values[a] for the ratio of zeaxanthin + antheraxanthin to zeaxanthin + antheraxanthin + violaxanthin (Z + A/A + Z + V) and the ratio of chlorophyll a to chlorophyll b (chlorophyll a/chlorophyll b) in shoot tissues for *Allium fistulosum* L. cultigens grown under supplemental UV-B (313 nm) light [7.0 μmol·m^{-2}·s^{-2} (2.68 W·m^{-2}); UV-B] or UV-filtered (control) light in a glasshouse in Knoxville, TN, USA (35°96'N Lat.).

Cultigen	Z + A/A + Z + V			Chlorophyll a/chlorophyll b		
	UV-B	Control	Pr > $|t|$[b]	UV-B	Control	Pr > $|t|$
Deep Purple	0.48 ± 0.01	0.51 ± 0.19	ns	1.25 ± 0.90	1.50 ± 0.23	ns
Evergreen Hardy White	0.41 ± 0.06	0.41 ± 0.01	ns	1.15 ± 0.56	0.91 ± 0.16	ns
Feast	0.51 ± 0.16	0.65 ± 0.21	ns	2.14 ± 0.43	1.07 ± 0.63	P = 0.031
GA-C 76	0.49 ± 0.07	0.49 ± 0.10	ns	1.58 ± 0.18	0.93 ± 0.51	ns
Ishikura Improved F1	0.34 ± 0.03	0.40 ± 0.08	ns	1.69 ± 1.06	1.45 ± 0.70	ns
Improved Beltsville Bunching	0.41 ± 0.03	0.41 ± 0.07	ns	1.04 ± 0.92	1.29 ± 0.50	ns
Jionji Negi	0.46 ± 0.07	0.35 ± 0.15	ns	1.85 ± 0.70	2.21 ± 0.50	ns
Long White Bunching	0.52 ± 0.04	0.58 ± 0.17	ns	1.86 ± 0.06	2.25 ± 0.68	ns
Parade	0.45 ± 0.10	0.50 ± 0.24	ns	1.77 ± 0.33	2.16 ± 0.48	ns
Performer	0.66 ± 0.17	0.40 ± 0.02	ns	1.60 ± 0.11	2.07 ± 1.19	ns
Pesoenyj	0.44 ± 0.03	0.39 ± 0.01	P = 0.022	1.40 ± 0.42	1.95 ± 0.80	ns
Shounan	0.42 ± 0.03	0.40 ± 0.03	ns	1.78 ± 0.52	1.08 ± 0.53	ns
White Spear	0.50 ± 0.21	0.51 ± 0.22	ns	1.46 ± 0.42	1.08 ± 0.62	ns
Zhang Qui Da Cong	0.42 ± 0.05	0.41 ± 0.08	ns	1.11 ± 0.50	1.32 ± 0.46	ns
274254-05GI	0.36 ± 0.06	0.40 ± 0.00	ns	1.80 ± 0.71	2.14 ± 0.81	ns
G 30393-06GI	0.69 ± 0.20	0.55 ± 0.12	ns	1.84 ± 0.31	1.32 ± 0.64	ns
LSD$_{0.05}$[c]	0.15	0.20		ns	0.96	

[a]Composition of n = 6 plant samples from 4 replications ± standard deviation. [b]Significance based on paired Student's t-test among treatments; ns: not significant. [c]LSD for differences between cultivar means α = 0.05.

(β-carotene, α-carotene, and cryptoxanthins) support the maintenance of healthy epithelial cell differentiation, normal reproductive performance, and visual functions [26]. Both provitamin A carotenoids and nonprovitamin A carotenoids (lutein, zeaxanthin, and lycopene) function as free radical scavengers, enhance the immune response, suppress cancer development, and protect eye tissues [27]. Humans cannot synthesize carotenoids and therefore must rely on dietary sources to provide sufficient levels. Studies indicate that high intakes of a variety of vegetables, providing a mixture of carotenoids, were more strongly associated with reduced cancer and eye disease risk than intake of individual carotenoid supplements [28]. There is clear evidence that cultural practices that maintain or enhance tissue carotenoid levels would be beneficial to humans when regularly consumed in the diet.

3.3. Shoot Tissue Chlorophyll Pigment Concentrations. No chlorophyll pigments were measured in the pseudostem tissues of any of the bunching onion cultigens (Figure 1; data not shown). Chlorophyll a responded significantly to UV radiation treatments (F = 4.35; P = 0.0398), but not to cultigens or the interaction between treatment and cultigen. "Feast" had the highest concentration of chlorophyll a at 59.56 mg/100 g FM for cultigens grown under supplemental UV-B radiation, while "Deep Purple" had the lowest at 27.75 mg/100 g FM. For the cultigens grown without supplemental UV-B radiation, "Pesoenyj" had the highest concentration of chlorophyll a at 63.27 mg/100 g FM, while "Evergreen Hardy White" had the lowest at 16.52 mg/100 g FM

(Table 5). Values for chlorophyll a for cultigens are in close agreement with Dissanayake et al. [29] who reported values of ~75.00 mg chlorophyll a/100 g FM for the A. fistulosum cultigen "Kujyo-hoso."

The bunching onions showed significant differences in chlorophyll b caused by UV-B treatment (F = 19.04; P < 0.0001) and cultigen (F = 2.08; P = 0.0179), but there were no influences from their interaction. Values for chlorophyll b for cultigens are in close agreement with Dissanayake et al. [29] who reported values of ~17.00 mg chlorophyll b/100 g FM for the A. fistulosum cultigen "Kujyo-hoso." Significant increases in chlorophyll b in response to UV-B radiation were found for cultigens "Feast," "GA-C 76," and "Shounan" (Table 5). The concentrations of chlorophyll b for cultigens grown under supplemental UV-B radiation ranged from 29.24 mg/100 g FM for "GA-C 76" to 18.49 mg/100 g FM for "Improved Beltsville Bunching." For cultigens grown without supplemental UV-B radiation, chlorophyll b concentrations ranged from 29.74 mg/100 g FM for "Pesoenyj" to 15.78 mg/100 g FM for "Improved Beltsville Bunching."

Concentrations of total chlorophyll (chlorophyll a + b) in bunching onions were found to differ between UV-B treatments (F = 6.82; P = 0.0105), but not among cultigens. "Feast" and "GA-C 76" were the only bunching onion cultigens to show differences between UV-B treatments (Table 5). Total chlorophyll concentrations ranged from 88.82 mg/100 g FM for "Feast" to 45.62 mg/100 g FM for "Zhang Qui Da Cong" for bunching onions grown under supplemental UV-B radiation. For the plants grown without UV-B radiation, ranges from total chlorophyll varied from 93.01 mg/100 g FM

TABLE 5: Mean values[a] for shoot tissue chlorophyll *a*, chlorophyll *b*, and total chlorophyll (chlorophyll *a* + chlorophyll *b*) (mg/100 g fresh mass) for *Allium fistulosum* L. cultigens grown under supplemental UV-B (313 nm) light [7.0 μmol·m^{-2}·s^{-2} (2.68 W·m^{-2}); UV-B] or UV-filtered (control) light in a glasshouse in Knoxville, TN, USA (35°96′N Lat.).

Cultigen	Chlorophyll *a*			Chlorophyll *b*			Total chlorophyll		
	UV-B	Control	Pr > |t|[b]	UV-B	Control	Pr > |t|	UV-B	Control	Pr > |t|
					mg/100 g fresh mass				
Deep Purple	27.8 ± 22.1	25.3 ± 5.6	ns	20.6 ± 4.4	16.9 ± 2.4	ns	48.3 ± 25.6	42.1 ± 7.5	ns
Evergreen Hardy White	31.4 ± 21.1	16.5 ± 3.2	ns	25.5 ± 5.1	18.2 ± 0.7	ns	56.9 ± 26.1	34.7 ± 3.5	ns
Feast	59.6 ± 20.3	19.0 ± 10.5	P = 0.012	27.3 ± 4.0	17.7 ± 1.9	P = 0.005	86.8 ± 24.3	36.7 ± 11.0	P = 0.009
GA-C 76	47.0 ± 16.4	17.2 ± 10.6	ns	29.2 ± 7.1	18.1 ± 1.3	P = 0.021	76.2 ± 23.3	35.4 ± 11.5	P = 0.020
Ishikura Improved F1	49.9 ± 53.6	23.1 ± 5.9	ns	25.6 ± 11.6	17.6 ± 4.7	ns	74.5 ± 65.1	40.7 ± 2.6	ns
Improved Beltsville Bunching	23.9 ± 28.4	20.6 ± 8.9	ns	18.5 ± 7.6	15.8 ± 1.0	ns	42.4 ± 36.0	36.1 ± 9.8	ns
Jionji Negi	48.3 ± 23.4	47.2 ± 18.9	ns	25.5 ± 3.1	20.9 ± 4.8	ns	73.8 ± 26.0	68.0 ± 23.4	ns
Long White Bunching	36.9 ± 6.6	37.2 ± 7.4	ns	19.7 ± 2.8	17.1 ± 2.7	ns	56.5 ± 9.2	54.3 ± 6.3	ns
Parade	38.6 ± 13.2	44.4 ± 18.8	ns	21.2 ± 4.1	19.9 ± 4.4	ns	59.8 ± 17.2	64.3 ± 23.0	ns
Performer	36.8 ± 3.1	49.6 ± 31.9	ns	23.0 ± 2.1	20.9 ± 7.1	ns	59.8 ± 4.8	70.5 ± 38.9	ns
Pesoenyj	36.8 ± 10.2	63.3 ± 36.9	ns	26.3 ± 2.4	29.7 ± 8.7	ns	26.3 ± 2.4	93.0 ± 45.6	ns
Shounan	44.3 ± 24.9	20.6 ± 19.1	ns	24.3 ± 8.3	16.6 ± 7.6	ns	68.6 ± 32.6	37.2 ± 26.7	ns
White Spear	30.4 ± 9.9	18.2 ± 9.6	ns	20.8 ± 1.9	17.8 ± 1.0	P = 0.033	51.2 ± 10.7	35.4 ± 8.9	ns
Zhang Qui Da Cong	25.3 ± 18.3	27.3 ± 13.5	ns	20.4 ± 7.4	19.3 ± 4.2	ns	45.6 ± 25.6	47.1 ± 17.6	ns
274254-05GI	36.6 ± 23.1	40.0 ± 18.1	ns	18.9 ± 5.3	18.7 ± 4.9	ns	55.5 ± 28.3	58.6 ± 21.5	ns
G 30393-06GI	39.8 ± 11.3	22.6 ± 13.8	ns	21.3 ± 4.1	17.0 ± 3.7	ns	61.1 ± 15.1	39.6 ± 16.0	ns
LSD$_{0.05}$[c]	ns	26.0		ns	6.8		ns	31.6	

[a] Composition of n = 6 plant samples from 4 replications ± standard deviation. [b] Significance based on paired Student's *t*-test among treatments; ns: not significant. [c] LSD for differences between cultivar means α = 0.05.

for "Pesoenyj" to 34.74 mg/100 g FM for "Evergreen Hardy White."

The ratio of chlorophyll *a* to chlorophyll *b* in the bunching onions showed significant changes based on cultigen (F = 2.26; P = 0.0094), but not for UV-B radiation treatments. In general, cultigens were evenly divided in their responses to UV-B radiation, with half the cultigens displaying higher chlorophyll *a*/chlorophyll *b* under the supplemental UV-B radiation treatment (Table 4). However, only the cultigen "Feast" had a significantly higher chlorophyll *a*/chlorophyll *b* ratio under UV-B radiation. "Long White Bunching" had the highest chlorophyll *a*/chlorophyll *b* ratio in the bunching onions grown without supplemental UV-B at 2.25, and "GA-C 76" had the lowest ratio at 0.91. Under UV-B radiation treatment, "Feast" has the highest chlorophyll *a*/chlorophyll *b* ratio at 2.14, while "Improved Beltsville Bunching" has the lowest ratio at 1.04.

3.4. Shoot Tissue Photochemical Efficiency (F_v/F_m). Photochemical efficiency (F_v/F_m) showed significant differences between UV treatments (F = 13.89, P = 0.0003) and cultigen (F = 2.11, P = 0.0152), but no difference due to treatment and cultigen interaction (data not shown). Values for F_v/F_m for all of the cultigens evaluated in the study averaged 0.82. One previous study by Tsormpatsidis et al. [30] showed that while "Lollo Rosso" lettuce (*Lactuca sativa* L.) had decreased vegetative growth under UV light treatments, there was no

difference in photochemical efficiency. By contrast, when the agronomic crop wheat was exposed to UV radiation, decreases in F_v/F_m occurred under the UV light treatment [31]. None of the cultigens in this study showed differences in F_v/F_m; however, most of the cultigens differed in shoot tissue fresh biomass when exposed to UV-B radiation.

4. Conclusion

Data from multiple studies demonstrates cultigens within a given plant species can react differently under variable stress conditions. Most often, harsh stress conditions negatively impact plant biomass. In the current study, decreases in bunching onion shoot tissue biomass confirmed that a radiational stress from the UV-B treatment had occurred. The bunching onion cultigens demonstrated genetic variability in response to UV-B radiation (Tables 1–5). Changes in plant pigments associated with light harvesting and photoprotection can be expected when bunching onion cultigens experience greater levels of UV-B radiation in the growing environment. In the current study, the cultigens with the greatest stimulation in carotenoid pigments from UV-B exposure were "Feast" and the accession G 30393-06GI. Data presented here may be valuable to improve abiotic stress tolerance to increasing UV-B radiation for specialty crop breeding programs.

Conflict of Interests

The authors of the paper do not have a direct financial relation with the commercial identities mentioned in the paper that might lead to conflict of interests.

Acknowledgments

Mention of trade names or commercial products in this publication is solely for the purpose of providing specific information and does not imply recommendation or endorsement by the University of Tennessee Institute of Agriculture. Authors would like to acknowledge funding and support for this work by the University of Tennessee Agricultural Experiment Station.

References

[1] G. Sandmann, "Genetic manipulation of carotenoid biosynthesis: strategies, problems and achievements," *Trends in Plant Science*, vol. 6, no. 1, pp. 14–17, 2001.

[2] J. T. Landrum and R. A. Bone, "Lutein, zeaxanthin, and the macular pigment," *Archives of Biochemistry and Biophysics*, vol. 385, no. 1, pp. 28–40, 2001.

[3] M. G. Ferruzzi and J. Blakeslee, "Digestion, absorption, and cancer preventative activity of dietary chlorophyll derivatives," *Nutrition Research*, vol. 27, no. 1, pp. 1–12, 2007.

[4] Y. Endo, R. Usuki, and T. Kaneda, "Antioxidant effects of chlorophyll and pheophytin on the autoxidation of oils in the dark. II. The mechanism of antioxidative action of chlorophyll," *Journal of the American Oil Chemists' Society*, vol. 62, no. 9, pp. 1387–1390, 1985.

[5] D. A. Kopsell, C. E. Sams, D. E. Deyton, K. R. Abney, D. E. Kopsell, and L. Robertson, "Characterization of nutritionally important carotenoids in bunching onion," *HortScience*, vol. 45, no. 3, pp. 463–465, 2010.

[6] K. A. Steinmetz and J. D. Potter, "Vegetables, fruit, and cancer. II. Mechanisms," *Cancer Causes and Control*, vol. 2, no. 6, pp. 427–442, 1991.

[7] B. Demmig-Adams, A. M. Gilmore, and W. W. Adams III, "In vivo functions of carotenoids in higher plants," *The FASEB Journal*, vol. 10, no. 4, pp. 403–412, 1996.

[8] R. Croce, S. Weiss, and R. Bassi, "Carotenoid-binding sites of the major light-harvesting complex II of higher plants," *The Journal of Biological Chemistry*, vol. 274, no. 42, pp. 29613–29623, 1999.

[9] J. A. Mares-Perlman, A. E. Millen, T. L. Ficek, and S. E. Hankinson, "The body of evidence to support a protective role for lutein and zeaxanthin in delaying chronic disease. Overview," *Journal of Nutrition*, vol. 132, no. 3, pp. 517S–524S, 2002.

[10] G. J. F. MacDonald, Ed., *Biological Impacts of Increased Intensities of Solar Ultraviolet Radiationedition*, National Academy of Sciences, Washington, DC, USA, 1st edition, 1973.

[11] L. Yuan, Z. Yanqun, C. Haiyan, C. Jianjun, Y. Jilong, and H. Zhide, "Intraspecific responses in crop growth and yield of 20 wheat cultivars to enhanced ultraviolet-B radiation under field conditions," *Field Crops Research*, vol. 67, no. 1, pp. 25–33, 2000.

[12] M. Caasi-Lit, M. I. Whitecross, M. Nayudu, and G. J. Tanner, "UV-B irradiation induces differential leaf damage, ultrastructural changes and accumulation of specific phenolic compounds in rice cultivars," *Australian Journal of Plant Physiology*, vol. 24, no. 3, pp. 261–274, 1997.

[13] M. Saile-Mark and M. Tevini, "Effects of solar UV-B radiation on growth, flowering and yield of Central and Southern European bush bean cultivars (Phaseolus vulgaris L.)," *Plant Ecology*, vol. 128, no. 1-2, pp. 114–125, 1997.

[14] D. Giuntini, G. Graziani, B. Lercari, V. Fogliano, G. F. Soldatini, and A. Ranieri, "Changes in carotenoid and ascorbic acid contents in fruits of different tomato genotypes related to the depletion of UV-B radiation," *Journal of Agricultural and Food Chemistry*, vol. 53, no. 8, pp. 3174–3181, 2005.

[15] M. Marotti and R. Piccaglia, "Characterization of flavonoids in different cultivars of onion (*Allium cepa* L.)," *Journal of Food Science*, vol. 67, no. 3, pp. 1229–1232, 2002.

[16] D. A. Kopsell and W. M. Randle, "Selenium affects the S-alk(en)yl cysteine sulfoxides among short-day onion cultivars," *Journal of the American Society for Horticultural Science*, vol. 124, no. 3, pp. 307–311, 1999.

[17] M. Takahashi and T. Shibamoto, "Chemical compositions and antioxidant/anti-inflammatory activities of steam distillate from freeze-dried onion (*Allium cepa* L.) sprout," *Journal of Agricultural and Food Chemistry*, vol. 56, no. 22, pp. 10462–10467, 2008.

[18] A. Denny and J. Buttriss, *Synthesis Report no. 4, Plant Foods and Health: Focus on Plant Bioactive*, European Food Information Resource (EuroFIR) Consortium. EuroFIR Project/British Nutrition Foundation, Institute of Food Research,, Norwich, UK, 2007.

[19] D. R. Hoagland and D. I. Arnon, "The water culture method for growing plants without soil," *California Agricultural Experiment Station Circular*, vol. 347, pp. 4–32, 1950.

[20] D. A. Kopsell, D. E. Kopsell, M. G. Lefsrud, J. Curran-Celentano, and L. E. Dukach, "Variation in lutein, β-carotene, and chlorophyll concentrations among *Brassica oleracea* cultigens and seasons," *HortScience*, vol. 39, no. 2, pp. 361–364, 2004.

[21] B. H. Davies and H. P. Köst, "Chromatographic methods for the separation on carotenoids," in *Plant Pigments. Fat Soluble Pigments*, H. P. Köst, G. Zweig, and J. Sherma, Eds., vol. 1 of *Handbook of Chromatography*, pp. 1–85, CRC Press, Boca Raton, Fla, USA, 1988.

[22] V. G. Kakani, K. R. Reddy, D. Zhao, and K. Sailaja, "Field crop responses to ultraviolet-B radiation: a review," *Agricultural and Forest Meteorology*, vol. 120, no. 1-4, pp. 191–218, 2003.

[23] U.S. Department of Agriculture, *USDA National Nutrient Database for Standard Release*, SR25, http://ndb.nal.usda.gov.

[24] M. Umehara, T. Sueyoshi, K. Shimomura et al., "Interspecific hybrids between *Allium fistulosum* and Allium schoenoprasum reveal carotene-rich phenotype," *Euphytica*, vol. 148, no. 3, pp. 295–301, 2006.

[25] K. K. Niyogi, A. R. Grossman, and O. Björkman, "Arabidopsis mutants define a central role for the xanthophyll cycle in the regulation of photosynthetic energy conversion," *Plant Cell*, vol. 10, no. 7, pp. 1121–1134, 1998.

[26] G. F. Combs, "Vitamin A," in *The Vitamins: Fundamental Aspects in Nutrition and Health*, pp. 107–153, Academic Press, San Diego, Calif, USA, 2nd edition, 1998.

[27] K. J. Yeum and R. M. Russell, "Carotenoid bioavailability and bioconversion," *Annual Review of Nutrition*, vol. 22, pp. 483–504, 2002.

[28] E. J. Johnson, B. R. Hammond, K. J. Yeum et al., "Relation among serum and tissue concentrations of lutein and zeaxanthin and macular pigment density," *American Journal of Clinical Nutrition*, vol. 71, no. 6, pp. 1555–1562, 2000.

[29] P. K. Dissanayake, N. Yamauchi, and M. Shigyo, "Chlorophyll degradation and resulting catabolite formation in stored Japanese bunching onion (*Allium fistulosum* L.)," *Journal of the Science of Food and Agriculture*, vol. 88, no. 11, pp. 1981–1986, 2008.

[30] E. Tsormpatsidis, R. G. C. Henbest, F. J. Davis, N. H. Battey, P. Hadley, and A. Wagstaffe, "UV irradiance as a major influence on growth, development and secondary products of commercial importance in Lollo Rosso lettuce "Revolution" grown under polyethylene films," *Environmental and Experimental Botany*, vol. 63, no. 1–3, pp. 232–239, 2008.

[31] X. C. Lizana, S. Hess, and D. F. Calderini, "Crop phenology modifies wheat responses to increased UV-B radiation," *Agricultural and Forest Meteorology*, vol. 149, no. 11, pp. 1964–1974, 2009.

20

The Waterlogging Tolerance of Wheat Varieties in Western of Turkey

Ilkay Yavas, Aydin Unay, and Mehmet Aydin

Kocarli Vocational High School, Adnan Menderes University, 09100 Aydın, Turkey

Correspondence should be addressed to Ilkay Yavas, iyavas@adu.edu.tr

Academic Editors: M. Cresti and M. N. V. Prasad

This research was conducted to determine the wheat varieties against waterlogging which was clearly increased in recent years. For this purpose, this study was performed at Field Crops and Soil Science Department of Agricultural Faculty of Adnan Menderes University during wheat growth stages of 2007-2008. The experimental design was randomized complete block design with split split plot arrangements. The main plots were temperature applications (heat and normal), the growth periods (Zadoks scale; GS14, GS32, GS14 + GS32, and control) were split plots and varieties were split-split plots. The eight different wheat varieties were evaluated in the pots. The waterlogging was performed during GS14, GS32 and GS14 + GS32. In a pot experiment, plants were subjected to waterlogging to the soil surface for 10 days. All applications and varieties decreased the single plant yield. The waterlogging caused a yield loss compared with wheat grown on well-drained soil. In this study, the crop loss due to waterlogging is highly temperature dependent. The severity of the effects of the waterlogging depends on the growth stage of the plot. When all applications were compared with control by means of yield performance, Sagittario and Basribey varieties were less affected than the others.

1. Introduction

Cereals are the basic products used in human nutrition. Wheat is great importance in cereals. The world annual wheat production is around 689 million tons and plant area is 223 million hectares. The wheat production is 17.8 million tons and plant area is 7.6 million hectares in Turkey [1].

Worldwide, about 10% of all irrigated land suffers from waterlogging. Soaking of agricultural land is caused by a rising water table or excessive irrigation. Waterlogging compacts soil, deprives roots of oxygen, and contributes to salinization. Waterlogging is common in wheat following rice, with plants turning yellow due to oxygen stress. Waterlogging has been shown to limit wheat yields in many regions of the world; an area about 10 million ha is waterlogged each year in developing countries [2].

Wheat has high adaptability for all kinds of climates and regions. Waterlogging which causes oxygen deficieny, will prevent to root and shoot growth, reduce the accumulation of dry matter and as a result of these yield will reduce. Especially, February and March rainfalls in western Turkey lead to the breakdown of cultivated fields and reduction of the yields. Wheat (*Triticum aestivum* L.) is one of the most intolerant crops to soil waterlogging [3]. Prolonged periods of rainfall combined with poor soil drainage often cause low oxygen in the soil. Decrease in soil O_2 content, under flooded conditions, is often accompanied by increase in soil CO_2 and ethylene content. Such changes have detrimental effects on root and shoot growth of wheat [4]. Productivity from soils susceptible to waterlogging may be increased by drainage and the introduction of waterlogging-tolerant genotypes [3]. High temperatures tend to exacerbate the negative effects of waterlogging. Yield is affected differently, depending on the crop's stage of development at the time stress is applied [5]. On waterlogged sites, plants show chlorosis and necrotic spots on older leaves. Both Mn toxicity and N deficiency may be induced by the low redox potential in waterlogged soils that produces plant-available Mn^{2+} and promotes denitrification of NO_3^-. Under these anaerobic conditions, root metabolism and root growth are inhibited, since the lack of O_2 affects the energy status of the plant [6].

The colder soil temperatures associated with waterlogging in winter-wheat growing areas reduce the amount of oxygen required for root respiration. Thus yield reductions

TABLE 1: The variance analysis results of soil Fe, Mn, and P contents at three growing stages in different waterlogging applications.

source	df	Mean squares		
		Fe	Mn	P
Conditions	1	144.95*	0.484	32.711
Growing stage	3	89.78*	58.41**	267.18
Interaction	3	35.89	4.54	30.58
Error	64	32.07	11.28	103.83
Total	71			

$^*P < 0.05$, $^{**}P < 0.01$.

TABLE 2: The average values of soil Fe, Mn, and P contents at three growing stages in different waterlogging applications.

		Fe	Mn	P
Conditions	Heat	19.85 b*	11.70	27.46
	Normal	22.69 a	11.15	26.12
	LSD$_{(0.05)}$	2.67		
Growing	Tillering + Jointing (GS14 + GS32)	23.52 a	13.45 a	28.83
	Jointing (GS32)	22.10 ab	11.21 ab	24.49
Stages	Tillering (GS14)	20.77 ab	11.91 ab	31.10
	Control	18.68 b	9.11 c	22.73
	LSD$_{(0.05)}$	3.78	2.33	

* Within each column, means followed by a different letter are significantly different at 5% level.

associated with waterlogging in colder areas are not as great as those in the more temperate and tropical areas of the world [7]. Similarly, some studies show that soil oxygen decline under waterlogging is rapid at most temperature ranges [8]. Chlorosis of lower leaves [9], decreased plant height and lower number of spike-bearing tillers [10], reduced root and shoot growth [11] in the wheat waterlogging areas. The uptake of trace metals by two plant species (French bean and maize) has been measured on two soils subjected to various waterlogging regimes. Uptake of both manganese and iron was increased due to soil waterlogging, although reoxidation of the soil affected iron more than manganese. The abilities of these species to take up trace metals from soil followed the pattern predicted by selective extraction of soil for manganese, iron, and cobalt, but not for zinc and copper [12]. In waterlogged acidic soil, shoot concentrations of aluminum (Al), manganese (Mn), and iron (Fe) increased by two- to 10-fold, and in some varieties they were above critical concentrations compared with plants in drained soil. These elements decreased or remained the same in shoots of plants grown in waterlogged neutral soil [13]. The greater P uptake per unit of root biomass was a consequence of (1) an increase in soil P availability induced by waterlogging, (2) a change in root morphology, and/or (3) an increase in the intrinsic uptake capacity of each unit of root biomass. Soil P content was higher during waterlogging periods and the roots of waterlogged plants showed a higher physiological capacity to absorb P. Soil P availability was higher during waterlogging periods, roots of waterlogged plants showed a morphology more favorable to nutrient uptake (finer roots) and these roots showed a higher physiological capacity to absorb P [14].

Waterlogging is a serious environmental stress on winter wheat grown in western Turkey. In this study, the aim was the determination of tolerant varieties to waterlogging conditions, effects of temperature, and growing stages.

2. Materials and Methods

This study was conducted in 2007-2008 at the Field Crops Department of the Faculty of Agriculture at Adnan Menderes University in Aydin, Turkey. Aydin province is situated at 37° 39′ E and 27° 52′ N in the west Aegeon Region of Turkey, and typical Mediterranean climatic conditions are dominant in Aydin. To evaluate losses from waterlogging under pot

experiments, 8 wheat genotypes were grown under no waterlogging (control) and 10 days of continuous flooding, GS14, GS32, and GS14 + GS32 [15]. After flooding the pots were allowed to come normal dry conditions by evaporation and plant use but without draining. Sufficient water was applied to waterlog from the top down, by applying water in excess of the rate at which it could infiltrate the soil, indicated by surface pooling [16]. Soil iron and soil manganese content (DTPA analysis, [17]), and soil phosphorus content [18] were determined.

The experimental design was randomized complete block design with split-split plots. The main plots were temperature applications, the growth plots were Split Plots and varieties were split-split plots. Data on plant height, tiller number, root biomass, shoot biomass, single plant yield, soil iron content, soil manganese content, and soil phosphorus content were taken on five plants per pot, and means were used for data analysis. The experiments were designed in split-split plot design with 3 replications. To increase the temperature, low tunnels were used. The growing conditions (low tunnel and normal) were the main plots, growing periods (GS14, GS32, and GS14 + GS32) were the subplots and cultivars (Golia, Gonen, Basribey, Adana 99, Cumhuriyet 75, Sagittario, Pamukova, Negev) were sub-subplots. The wheat varieties were sown on 27.11.2007 and harvested at 23.05.2008. At maturity, plants were harvested and threshed manually. The data regarding plant height, tiller number, root biomass, shoot biomass, single plant yield, soil iron content, soil manganese content and soil phosphorus content were recorded. The significance of main effects and interactions was determined at the 0.05 and 0.01 probability levels by the F-test. Means of the significant ($P \leq 0.05$) main effects and interactions were separated using Fischer's protected LSD test at $P = 0.05$. The data were statistically analyzed by using a standard analysis of variance technique for a split-split plot design using Tarist software [19].

3. Results

The soil iron, manganese, and phosphorus contents were given at Table 1. The difference between heat and normal

TABLE 3: Results of variance analysis for some characters in normal conditions.

	df	PH	TN	SB	RB	SPY
			Means of square			
Stage	3	4645.7**	15.100**	1.834**	0.001	5.567**
Cultivar	7	423.802**	2.645**	1.789**	0.146**	1.949**
Stage × Cultivar	21	51.102**	1.063**	0.075**	0.127**	0.262**

$*P < 0.05$, $**P < 0.01$. Plant height, TN: tiller number, SB: shoot biomass, RB: root biomass, SPY: single plant yield.

TABLE 4: Results of variance analysis for heat conditions.

	df	PH	SB	RB	SPY
			Means of square		
Stage	3	4207.480**	2.082**	0.005	6.556**
Cultivar	7	199.154**	0.767**	0.093*	0.561**
Stage × Cultivar	21	102.052**	0.112**	0.033	0.293**

$*P < 0.05$, $**P < 0.01$. Plant height, SB: Shoot Biomass, RB: Root Biomass, SPY: Single plant yield.

TABLE 5: The tiller number at GS14 and control.

	Growing stage 14		
	Heat temperature	Normal	Control
Golia	0	0	3
Gonen	3	0	2
Basribey	0	0	3
Adana 99	0	0	2
Cumhuriyet 75	2	0	4
Sagittario	2	0	6
Pamukova	0	0	4
Negev	3	1	6

conditions for soil Fe and the accumulation of Fe and Mn for different growing stages were significant (Table 1).

The soil Fe content in normal condition was found significantly higher compared with heat condition (Table 2). During waterlogging, O_2 in the soil is rapidly depleted, and the soil may become hypoxic or anoxic within a few hours. Moreover, some waterlogged soils become rich in Mn^{2+} and Fe^{2+}. Waterlogging stress leads to changes in soil conditions which may affect plant growth through oxygen deficiency, reduced nitrogen availability, or manganese toxicity and so on [20].

Table 3 showed the results of variance analysis of the plant height, tiller number, shoot biomass, root biomass, and single plant yield. The results showed that the stage and cultivar factors and their interactions were found significant except for root biomass at growing stage factor.

Table 4 was that the the results of variance analysis for heat conditions. The results showed that stage and cultivar interactions were significant except for root biomass. And the root biomass for different stage was nonsignificant. Also, the differences among wheat varieties were significant for observed characters in high temperature.

The waterlogging conditions reduced all cultivar's height in GS14 + GS32 than control conditions. The tiller number at different stage was given at Table 5. The waterlogging reduced the number of tiller in wheat. For all treatments, grain losses were much less than expected from the extent of tiller loss in winter, losses after single waterlogging events ranged from 2% (after 47 days with the water table at 5 cm) to 16% (after 80 days with the water table at the soil surface). Yield losses after three waterlogging at the seedling, tillering, and stem elongation stages of growth were additive, and totaled 19% [21]. Transient waterlogging during winter and spring reduces wheat yield. Yield reductions from waterlogging are associated with reduced production and survival of tillers, fewer and smaller fertile tillers, and smaller grain size [22]. Waterlogging induced a transient N deficiency. The N concentration of the youngest expanded leaf on the mainstem and tillers declined markedly during waterlogging, but its recovery 14 days after the waterlogging was ended was independent of treatment, reaching the critical minimum concentration of 3.5%. The growth of primary tillers 1 and 2 was severely inhibited by waterlogging while the exertion of new tillers was delayed by 9 days [23].

Figure 1 pointed out plant height at different stages of wheat. Waterlogging significantly reduced plant height and tillering, delayed ear emergence, and resulted in 8, 17, 27, and 39% reduction in grain yield, respectively [24].

Figure 2 showed shoot biomass at different stage of wheat. The waterlogging conditions reduced the shoot biomass values.

The other important characteristic for waterlogging is root biomass examined at Figure 3. The lowest root biomass value was found in Adana 99 at GS14. On the other hand the best cultivar was Cumhuriyet 75. Doubled haploid lines grown in waterlogged soil for 49 days ranged from 82% reductions in biomass at the end of waterlogging in acidic soil [25].

The single plant yield values for different stages were given at Figure 4. Results indicated that significant linear responses were found for yield. Reduction of the number of spike after the waterlogging caused a significant decrease of single plant yield.

4. Discussion

Sparrow and Uren [26] and Wagatsuma et al. [27] stated that soil Mn concentration that could be toxic to plant growth increased in waterlogging. Similarly, mineral Fe coating of epidermal surface of roots increased under waterlogging [28]. In addition, Belford et al. [29] revealed that decreased soil oxygen was generally greater at warmer temperature. Also, in our study the soil Fe and Mn contents of nonwaterlogging pots (control) had significantly lower values, whereas waterlogging applications in both jointing and tillering significantly increased soil Fe and Mn contents. The differences in soil Fe and Mn between conditions and

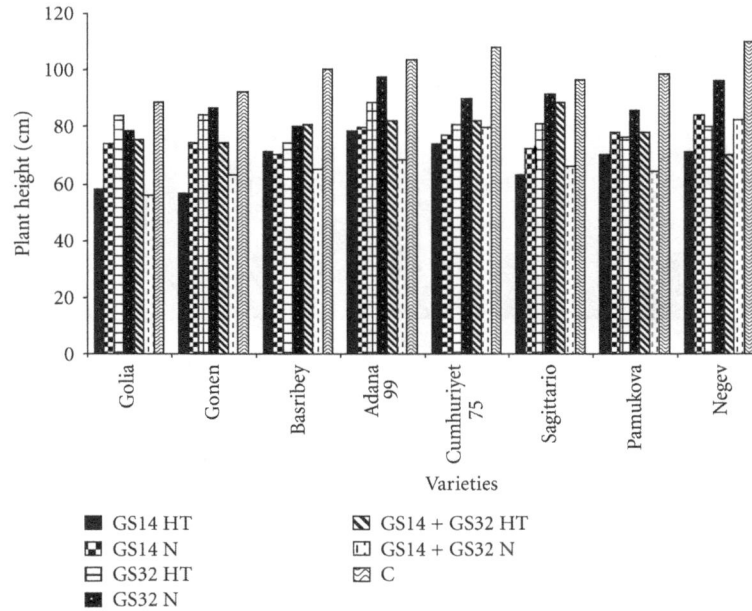

FIGURE 1: The plant height at GS14, GS32, GS14 + GS32, and control. HT: high temperature, N: normal, C: control.

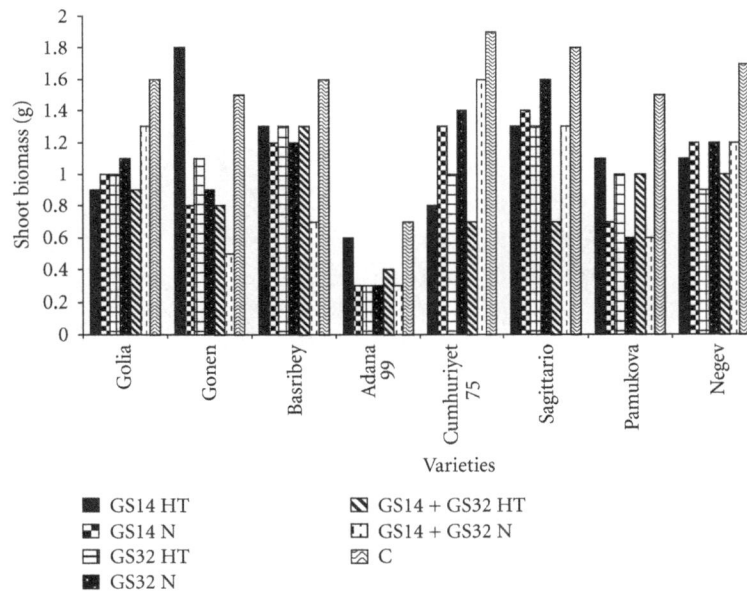

FIGURE 2: Shoot biomass at GS14, GS32, GS14 + GS32, and control. HT: high temperature, N: normal, C: control.

growing stages prepared a suitable condition to evaluate wheat varieties response to waterlogging.

There are significant differences in waterlogging tolerance for different growing stages. Diversity occurs in the timing, duration, and severity at waterlogging stress [30].

The highest plant height was found in Adana 99, Golia, Gonen, Sagittario, Basribey and Pamukova were negatively affected in the GS14 (high temperature) and GS14 + GS32 (normal conditions) for plant height. The effects of waterlogging on the Basribey and Adana 99 occurred in the GS14 + GS32 (normal conditions) negatively affected in the early growing stages and increased temperatures (GS14; high temperature). Negev was influenced by waterlogging which occurred during GS14 and GS32 periods, together with heat application. Flooding and increased temperature application were most significant effects for all cultivars. Losses caused by waterlogging need to be measured to determine the importance of traits used as waterlogging tolerance indicators. Kernels per spike and plant height had a less severe reduction from waterlogging, about 30 and 19%, respectively [31].

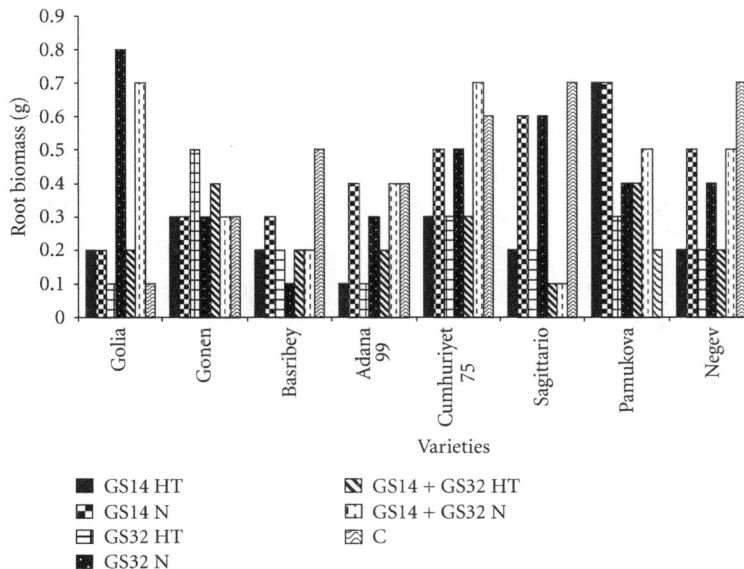

FIGURE 3: Root biomass at GS14, GS32, GS14 + GS32, and control. HT: high temperature, N: normal, C: control.

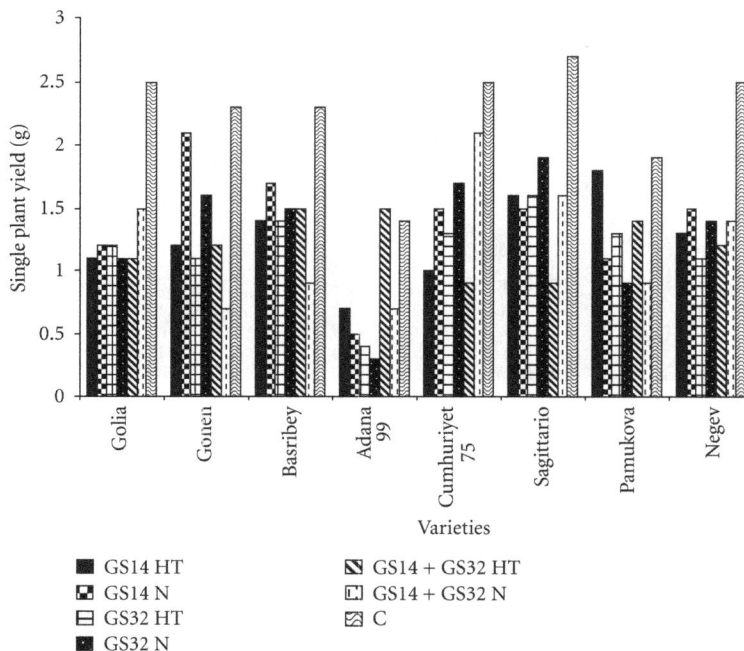

FIGURE 4: Single plant yield at GS14, GS32, GS14 + GS32, and control. HT: high temperature, N: normal, C: control.

Taeb et al. [32] revealed that tiller number were decreased during waterlogging, tiller production, shoot dry matter, and root penetration were used for screening Tritieae species for tolerance. When these criteria were used, many wild species expressed a level of tolerance to waterlogging that was better than that of wheat.

By the end of the experiment, shoot mass remained lower in plants from waterlogged treatments compared with continuously drained control, due to lower tiller numbers. Comparisons between alternately waterlogged and contin-uously waterlogged plants showed that in the alternaely waterlogged plants, shoot weights were heavier [3].

The lowest yield was observed in Adana 99 which was exposed to waterlogging at GS32. Correspondingly, the highest value was obtained from Cumhuriyet 75 at GS14 + GS32. In a field study, wheat (*Triticum aestivum* L.) crop to waterlogging for 1, 2, 4, and 6 days at the time of first irrigation (25-day-old plants) significantly reduced tillering and plant height, delayed ear emergence, and resulted in 8, 17, 27, and 39% reduction in grain yield, respectively [24].

The wheat varieties used as material showed different response to waterlogging. The significant negative effects of waterlogging on cultivars were both tillering and jointing stage (GS14 + GS32) with high temperature. Soil Fe and Mn contents significantly increased in tillering and jointing stage waterlogging applications. Basribey and Sagittario wheat varieties should be suggested as waterlogging tolerance varieties. This study will play an important role for wheat growers in the future as they will be able to make better crop management decisions when they encounter incidence of waterlogging.

Huang et al. [11] observed that waterlogging reduced shoot nitrogen content, shoot, and root growth.

Abbreviations

GS14:   growing stage 14
GS32:   growing stage 32
GS14 + GS32: growing stage 14 + 32.

Acknowledgments

The authors express to the Adnan Menderes University Research Fund for providing financial support, project number is BAP: ZRF 07022.

References

[1] Anonymous, "FAO statistical databases, 2008," 2010, http://faostat.fao.org/site/567/default.aspx#ancor.

[2] K. D. Sayre, M. Van Ginkel, S. Rajaram, and I. Ortiz-Monasterio, "Tolerance to waterlogging losses in spring bread wheat. Effect of time on onset of expression," in *Annual Wheat Newsletter*, vol. 40, pp. 165–171, 1994.

[3] C. J. Thompson, T. D. Colmer, E. L. J. Watkin, and H. Greenway, "Tolerance of wheat (*Triticum aestivum* cvs. Gamenya and Kite) and triticale (Triticosecale cv. Muir) to waterlogging," *New Phytologist*, vol. 120, pp. 335–344, 1992.

[4] G. B. Gunther, P. Manske, and L. G. Vlek, "Root architechture. Wheat as a model plant," in *Plant Roots the Hidden Half*, W. Yoav, A. Eshel, and U. Kafkaf, Eds., p. 120, 2002.

[5] N. Brisson, B. Rebière, D. Zimmer, and P. Renault, "Response of the root system of a winter wheat crop to waterlogging," *Plant and Soil*, vol. 243, no. 1, pp. 43–55, 2002.

[6] D. Steffens, B. W. Hütsch, T. Eschholz, T. Lošák, and S. Schubert, "Water logging may inhibit plant growth primarily by nutrient deficiency rather than nutrient toxicity," *Plant, Soil and Environment*, vol. 51, no. 12, pp. 545–552, 2005.

[7] A. Samad, C. A. Meisner, M. Saifuzzaman, and M. van Ginkel, "Waterlogging tolerance," in *Application of Physiology in Wheat Breeding*, M. P. Reynolds, J. I. Ortiz-Monasterio, and A. McNab, Eds., pp. 136–144, CIMMYT, Mexico, DF, Mexico, 2001.

[8] M. C. T. Trought and M. C. Drew, "The development of waterlogging damage in young wheat plants in anaerobic solution cultures," *Journal of Experimental Botany*, vol. 31, no. 6, pp. 1573–1585, 1980.

[9] M. van Ginkel, S. Rajaram, and M. Thijssen, "Waterlogging in wheat: germplasm evaluation and methodology development," in *Proceedings of the 7th Regional Wheat Workshop for Easthern, Central and Southern Africa*, D. G. Tanner and W. Mwangi, Eds., pp. 115–124, CIMMYT, Nakuru, Kenya, 1992.

[10] J. G. Wu, S. F. Liu, F. R. Li, and J. R. Zhou, "Study on the effect of wet injury on growth and physiology winter wheat," *Acta Agriculture Universitatis Henanensis*, vol. 26, pp. 31–37, 1992.

[11] B. Huang, J. W. Johnson, S. Nesmith, and D. C. Bridges, "Growth, physiological and anatomical responses of two wheat genotypes to waterlogging and nutrient supply," *Journal of Experimental Botany*, vol. 45, no. 271, pp. 193–202, 1994.

[12] K. L. Iu, I. D. Pulford, and H. J. Duncan, "Influence of soil waterlogging on subsequent plant growth and trace metal content," *Plant and Soil*, vol. 66, no. 3, pp. 423–427, 1982.

[13] H. Khabaz-Saberi, T. L. Setter, and I. Waters, "Waterlogging induces high to toxic concentrations of iron, aluminum, and manganese in wheat varieties on acidic soil," *Journal of Plant Nutrition*, vol. 29, no. 5, pp. 899–911, 2006.

[14] G. Rubio, M. Oesterheld, C. R. Alvarez, and R. S. Lavado, "Mechanisms for the increase in phosphorus uptake of waterlogged plants: soil phosphorus availability, root morphology and uptake kinetics," *Oecologia*, vol. 112, no. 2, pp. 150–155, 1997.

[15] J. C. Zadoks, T. T. Chang, and C. F. Konzak, "A decimal code for the growth stages of cereals," *Weed Research*, vol. 14, pp. 415–421, 1974.

[16] E. Dickin and D. Wright, "The effects of winter waterlogging and summer drought on the growth and yield of winter wheat (Triticum aestivum L.)," *European Journal of Agronomy*, vol. 28, no. 3, pp. 234–244, 2008.

[17] W. A. Norvell and W. L. Lindsay, "Reactions of DTPA complexes of iron, zinc, copper, and manganese with soil," *Soil Science Society of America Proceedings*, vol. 36, pp. 773–788, 1972.

[18] B. Kacar, *Bitki ve Toprağın Kimyasal Analizleri. III. Toprak Analizleri*, Egitim Araştirma Gelistirme Vakfi Yayinlari, 1994.

[19] N. Acikgoz, K. Ozcan, M. E. Akkas, and A. F. Moghaddam, "PC'ler için veri tabani esasli Türkçe istatistik paket: Tarist," in *Proceedings of the Tarla Bitkileri Kongresi*, pp. 25–29, London, UK, April 1994.

[20] C. M. Zaicou, L. C. Campbell, and J. Angus, "Nitrogen application and waterlogging in wheat," in *Proceedings of the 5th Australian Agronomy Conference (AAC '89)*, 1989.

[21] R. K. Belford, "Response of winter to prolonged waterlogging under outdoor conditions," *The Journal of Agricultural Science*, vol. 97, pp. 557–568, 1981.

[22] A. G. Condon and F. Giunta, "Yield response of restricted-tillering wheat to transient waterlogging on duplex soils," *Australian Journal of Agricultural Research*, vol. 54, no. 10, pp. 957–967, 2003.

[23] D. Robertson, H. Zhang, J. A. Palta, T. Colmer, and N. C. Turner, "Waterlogging affects the growth, development of tillers, and yield of wheat through a severe, but transient, N deficiency," *Crop and Pasture Science*, vol. 60, no. 6, pp. 578–586, 2009.

[24] D. P. Sharma and A. Swarup, "Effect of short-term waterlogging on growth, yield and nutrient composition of wheat in alkaline soil," *The Journal of Agricultural Science*, vol. 112, pp. 191–197, 1989.

[25] T. L. Setter, I. Waters, S. K. Sharma et al., "Review of wheat improvement for waterlogging tolerance in Australia and India: the importance of anaerobiosis and element toxicities associated with different soils," *Annals of Botany*, vol. 103, no. 2, pp. 221–235, 2009.

[26] L. A. Sparrow and N. C. Uren, "The role of manganese toxicity in crop yellowing on seasonally waterlogged and strongly

acidic soils in northeastern Victoria," *Australian Journal of Experimental Agriculture*, vol. 27, pp. 303–307, 1987.

[27] T. Wagatsuma, T. Nakashima, K. Tawaraya, S. Watanbe, A. Kamio, and A. Ueki, "Relationship between wet tolerance, anatomical structure of aerenchyma and gas exchange ability among several plant species," *Agricultural Sciences*, vol. 11, no. 1, pp. 121–132, 1990 (Japanese).

[28] N. Ding and M. E. Musgrave, "Relationship between mineral coating on roots and yield performance of wheat under waterlogging stress," *Journal of Experimental Botany*, vol. 46, no. 289, pp. 939–945, 1995.

[29] R. K. Belford, R. Q. Cannell, and R. J. Thomson, "Effect of single and multiple waterlogging on the growth and yield of winter wheat on a clay soil," *Journal of the Science of Food and Agriculture*, vol. 36, pp. 142–156, 1985.

[30] T. L. Setter and I. Waters, "Review of prospects for germplasm improvement for waterlogging tolerance in wheat, barley and oats," *Plant and Soil*, vol. 253, no. 1, pp. 1–34, 2003.

[31] A. Collaku and S. A. Harrison, "Losses in wheat due to waterlogging," *Crop Science*, vol. 42, no. 2, pp. 444–450, 2002.

[32] M. Taeb, R. M. D. Koebner, and B. P. Forster, "Genetic variation for waterlogging tolerance in the Triticeae and the chromosomal location of genes conferring waterlogging tolerance in *Thinopyrum elongatum*," *Genome*, vol. 36, no. 5, pp. 825–830, 1993.

21

Quality and Trace Element Profile of Tunisian Olive Oils Obtained from Plants Irrigated with Treated Wastewater

Cinzia Benincasa,[1] Mariem Gharsallaoui,[2] Enzo Perri,[1] Caterina Briccoli Bati,[1] Mohamed Ayadi,[2] Moncen Khlif,[2] and Slimane Gabsi[3]

[1] *Centro di Ricerca per l'Olivicoltura e l'Industria Olearia, CRA, via Li Rocchi 111, 87036 Rende, Italy*
[2] *Olive Tree Institute, University of Sfax, Route de l'aeroport Km 1.5, BP 1087, 3000 Sfax, Tunisia*
[3] *National School of Engineering, University of Gabes, Rue Omar Ibn El Khattab 6029, Tunisia*

Correspondence should be addressed to Cinzia Benincasa, cinzia.benincasa@entecra.it

Academic Editor: María Luisa de la Torre

In the present work the use of treated wastewater (TWW) to irrigate olive plants was monitored. This type of water is characterized by high salinity and retains a substantial amount of trace elements, organic and metallic compounds that can be transferred into the soil and into the plants and fruits. In order to evaluate the impact of TWW on the overall quality of the oils, the time of contact of the olives with the soil has been taken into account. Multi-element data were obtained using ICP-MS. Nineteen elements (Li, B, Na, Mg, Al, K, Ca, Sc, Cr, Mn, Fe, Co, Ni, Cu, Zn, Sr, Mo, Ba and La) were submitted for statistical analysis. Using analysis of variance, linear discriminant analysis and principal component analysis it was possible to differentiate between oils produced from different batches of olives whose plants received different types of water. Also, the results showed that there was correlation between the elemental and mineral composition of the water used to irrigate the olive plots and the elemental and mineral composition of the oils.

1. Introduction

Tunisia is a very important country in the olive oil producing world, the largest, African exporter and fourth worldwide after Spain, Italy and Greece with an annual average export over 10,000 metric tonnes [1]. The olive tree (*Olea europaea* L.) is present practically in every region of the country up to the border of the southern desert. Tunisia belongs to the Middle East and North Africa (MENA), which is considered one of the driest regions in the world [2]; in fact, only 30% of the cultivated area in the region is irrigated but produces about 75% of the total agricultural production. Alternative water resources are therefore needed to satisfy further increases in demand. Tunisia launched a national water reuse program in the early 1980s to increase usable water resources. Most municipal wastewater is from domestic sources and receives secondary biological treatment. In 2003, 187 million m³ (78%) of the 240 million m³ of wastewater collected in Tunisia received treatment. About 30–43% of the treated wastewater (TWW) was used for agricultural and landscape irrigation [3]. Reusing wastewater for irrigation is viewed as a way to increase water resources, provide supplemental nutrients, and protect coastal areas, water resources, and sensitive receiving bodies. Reclaimed water is used on 8000 ha to irrigate industrial (cotton, tobacco, sugarbeet, etc.) and fodder crops (alfalfa, sorghum, etc.), cereals, vineyards, citrus, and other fruit trees (olives, peaches, pears, apples, pomegranates, etc.). Regulations allow the use of secondary-treated effluent on all crops except vegetables, whether eaten raw or cooked [4]. As it is demonstrated that the olive trees respond favourably and efficiently to the irrigation management [5–7] and as the water resources in Tunisia are limited, the use of nonconventional water can be a good alternative. However, due to the particular characteristics of the water and potential health risks associated with their use in agriculture, it is important that the effects of irrigation with TWW are objectively evaluated. The primary and secondary wastewater treatments improve distinctly the water quality as the treated wastewater still retains a substantial amount of organic, and metallic compounds [8, 9] such as carbon, nitrogen, phosphorous, and potassium which have a favourable effect on the growth of certain crops [10].

Moreover, the reuse of the TWW can also have important consequences for the irrigated soils as the TWW can change their characteristics and quality due to the process of salinization and pollution by some mineral, organic, and bacteriological materials [10]. Several authors have studied the impact of TWW on the quality of the soils [11–19]. The possibility of transfer metals and trace elements by using TWW into the soil and from the soil to the plant is not well studied. The most commonly used techniques for the determination of metals in oil samples are inductively coupled plasma atomic emission spectrometry (ICP-AES) and atomic absorption spectrometry (AAS) [20]. Fats and oils are particularly difficult to analyze for their trace metal contents since some of them are present at very low concentration levels. For this reason inductively coupled plasma mass spectrometry (ICP-MS) is considered an interesting tool because of its well-known high sensitivity and because it allows a simultaneous quantitative determination of multielements in such a complex matrix [21]. However, since ICP-MS analysis faces many drawbacks, sample needs to be treated in order to eliminate the organic matrix that causes the extinction of the plasma. The aim of this study was, therefore, to determine the impact of TWW used to irrigate olive trees on the quality of the olive oils produced. Since farmers in Tunisia collect fallen olives, especially if the quantity that has dropped is consistent and mix them with olives harvested by hand directly from the tree, it is essential to control also the quality of oils extracted from olives fallen on field irrigated with TWW [22]. Hence, this paper proposes a quantitative analysis of elements in olive oils from Chemlali cultivar produced in Tunisia by means of ICP-MS.

2. Experimental

2.1. Olive Sampling and Oil Extraction.
The olive samples utilized in this study were collected from experimental plots located in the region of Sfax in central eastern Tunisia in the experimental station of El Hajeb. These experimental plots are characterized by sandy soil, and the use of the TWW is for the irrigation of annual crops in insertion with the olive trees. If olive trees are in insertion with alfalfa they receive water at an annual rate of $10,000\,m^3$ per ha by means of a continuous drip irrigation system. During the experimental years, irrigation was performed every month (continuous irrigation) with the exception of the harvest period that occurred on December and January. If olive trees are in insertion with oats, they receive water at an annual rate of $5,000\,m^3$ per ha by means of a continuous drip irrigation system that was performed from April to May and from October to December (alternate irrigation). Twenty-two-year-old trees spaced $24 * 24\,m$ were used in block design with three different irrigation systems: (a) olive trees intercropped with alfalfa receiving continuous irrigation with TWW since 2001; (b) olive trees intercropped with oats receiving an alternating irrigation with TWW since 2001; (c) olive trees irrigated with conventional water (CW), (d) olive trees not irrigated (control plants). For each irrigation system, fallen olives were picked from under the trees. The residence time of the olives on the ground is from a few days to several months,

and, normally, their moisture percentage content reflects their residence time under the olive trees and then in contact with the soil. Therefore, once collected, olives were sorted according to their moisture percentage content and called as follows: first fall 50.73% (±5.31%), second fall 33.45% (±5.15%), third fall 14.89% (±2.62%) and forth fall 4.22% (±0.72%). The moisture percentage of the olives collected directly from the plant resulted to be 44.22% (±7.45%). For each irrigation system, 3 kg of olives were milled by means of a laboratory scale hammer mill. After 20 minute of malaxation, the oil was separated by centrifugation.

2.2. Chemical Characteristics of Treated Wastewater and Conventional Water [23].
In order to remove the biodegradable matters of TWW, biological processes are performed by the action of aerobic microorganisms that, in the presence of oxygen, are able to metabolize the organic matter producing more microorganisms and inorganic end products such as CO_2, NH_3, and H_2O. Conventional water (CW) characteristics are quite different from those of TWW. Calculated values of pH resulted in 7.60 and 7.95 for CW and TWW, respectively, and, as they fall within the range of 6 to 9, their use can be employed in agriculture [24]. The electrical conductivity for TWW was $6.30\,dS/m$ while $4.60\,dS/m$ for CW indicating, respectively, a high and moderate level of salinity [25]. The concentration of NH_4^+, NO_3^-, and P_{total} was high (37.9, 15.9, and 10.3 mg/mL compared to 2.24, 1.11 and 0.8 mg/mL of CW, resp.). The concentrations of Na^+ and Cl^- in TWW were 470 mg/mL and 1999 mg/mL while, in CW, was 355 mg/mL and 1580 mg/mL. In both TWW and CW, chloride concentration was higher than the threshold reported by Chartzoulakis in the guidelines for olive irrigation [26, 27]. The concentrations of Ca^{2+} (96 mg/mL) and Mg^{2+} (84 mg/mL) were almost half compared with those present in CW. The concentrations of Zn^{2+} (0.42 mg/mL) and Mn^{2+} (0.5 mg/mL) were four and five times higher than in CW. Although the values of COD and BOD were very high in TWW (73 mg/mL and 22 mg/mL) compared to the values in CW, both biological and chemical oxygen demands were below the Tunisian thresholds for water reuse (30 and 90 mg/mL, resp.).

2.3. Physicochemical Analysis of Olive Oils.
The olive oils were analysed for their most important physicochemical parameters, that is, free acidity, peroxide index, UV absorption characteristics at 232 and 270 nm, and fatty acid methyl esters according to the official methods of European Union [28].

2.4. Total Phenols Analysis.
Total phenols content were determined as described in our previous paper [29]. In brief, 2.5 g of olive oil were dissolved in hexane and extraction with a solution of methanol and water was performed. The phenolic fraction was determined by mean of Folin-Ciocalteu reagent and quantitation achieved by external calibration curve ($r^2 = 0.996$) made with caffeic acid purchased from Sigma Aldrich.

2.5. Fatty Acid Methyl Esters (FAMEs) Analysis.
Fatty acid methyl esters (FAMEs) analysis were carried out as described

in our previous paper [30]. FAMEs were analyzed by gas chromatography, and peaks were identified by comparing their retention times with those of authentic reference compounds. The fatty acid composition was expressed as relative percentages of each fatty acid calculated considering the internal normalization of the chromatographic peak area.

2.6. ICP-MS Analysis

2.6.1. Materials and Apparatus for ICP-MS Analysis. The ultrapure HNO_3 (normaton ultrapure, VWR prolabo) used in this work was acquired by analytical-reagent grade certified for the impurities. Single- and multielement standards (Certipur, Merk, Darmstadt, Germany) were also analytical-reagent grade. Aqueous solutions were prepared using ultrapure water, with a resistivity of 18.2 Mcm, obtained from a Milli-Q plus system (Millipore, Saint Quentin Yvelines, France). All glassware were decontaminated with nitric acid (2%, v/v) for at least two hours, rinsed with ultrapure water, and dried. The experimental work was carried out using a Milestone MLS-1200 MEGA oven system for the microwave digestion. The determination of the elements of interest was carried out by means of an Agilent 7500e ICP-MS instrument (Agilent Technologies, Santa Clara, USA) where oil samples were introduced by means of a quartz nebulizer. The ICP torch was a standard torch equipped with platinum injector. As the performance of the ICP-MS instrument strongly depends on the operating conditions [31], a solution containing Rh, Mg, Pb, Ba and Ce (10 g/L) was used to optimize the instrument in terms of sensitivity, resolution, and mass calibration. The $^{140}Ce^{16}O^+/^{140}Ce^+$ ratio was used to check the level of oxide ions in the plasma that could interfere in the determination of some elements; also, instrumental parameters such as RF power and carrier gas flow were optimized and the level of doubly charged ion monitored by means of the signal $^{137}Ba^{2+}/^{137}Ba^+$. ICP-MS analysis was performed following the operating program and parameters as follows: plasma power: 1150 W; nebuliser: glass concentric type; carrier gas flow rate: 1.05 l/min; oxide ratios: < 3% (CeO/Ce); double charged species: <2% (Ba^{2+}/Ba^+); sample uptake rate: 1.2 mL/min; coolant argon flow: 17 l/min; sweeps/reading: 45; readings/replicate: 3; replicates: 3; dwell time: 55 ms; scan mode: peak hopping; isotopes monitored: $^7Li^+$, $^9Be^+$, $^{10}B^+$, $^{23}Na^+$, $^{25}Mg^+$, $^{27}Al^+$, $^{29}Si^+$, $^{31}P^+$, $^{34}S^+$, $^{39}K^+$, $^{43}Ca^+$, $^{45}Sc^+$, $^{49}Ti^+$, $^{51}V^+$, $^{53}Cr^+$, $^{55}Mn^+$, $^{57}Fe^+$, $^{59}Co^+$, $^{62}Ni^+$, $^{65}Cu^+$, $^{66}Zn^+$, $^{74}Ge^+$, $^{75}As^+$, $^{82}Se^+$, $^{88}Sr^+$, $^{89}Y^+$, $^{90}Zr^+$, $^{93}Nb^+$, $^{95}Mo^+$, $^{111}Cd^+$, $^{121}Sb^+$, $^{139}La^+$, $^{140}Ce^+$, $^{141}Pr^+$, $^{145}Nd^+$, $^{147}Sm^+$, $^{151}Eu^+$, $^{158}Gd^+$, $^{159}Tb^+$, $^{163}Dy^+$, $^{165}Ho^+$, $^{169}Tm^+$, $^{173}Yb^+$, $^{175}Lu^+$, $^{181}Re^+$, $^{205}Tl^+$, $^{208}Pb^+$, $^{232}Th^+$.

2.6.2. Sample Preparation. Sample preparation was performed as reported in Benincasa et al. [21]. Briefly, each olive oil sample was homogenized by vigorous shaking, and 0.5 g was weighed directly into the digestion vessel where nitric acid was added.

Olive oil samples were digested and analysed three times in order to check the sensitivity and reproducibility of the digestion procedure. The operating program for the microwave digestion system consisted of 6 steps, and the total time of the digestion procedure lasted 31 minutes. After cooling at room temperature, all the digestion liquors were quantitatively transferred into volumetric flask and diluted to volume (30 mL) with ultrapure water. An analytical batch contained 3 procedural blanks and two procedural blanks spiked with a standard solution containing 48 elements. A mid-range calibration standard was measured at the end of each analytical run, for quality control purposes, that is, to assess instrumental drift throughout the run. Limits of quantitations (LOQs) were defined as 10 times the standard deviation of the signal from reagent blanks, after correction for sample weight and dilution. All the elements that were below this value were not accepted for statistical analysis.

2.6.3. Calibration Procedure. In order to quantify the elements in the oils, external calibration curves were build on five different concentrations. Standard solutions were prepared by diluting a multielement solution of Ce, Dy, Er, Eu, Gd, Ho, La, Lu, Nd, Pr, Sm, Sc, Tb, Th, Tm, Y, and Yb at 10 mg/mL; a multielement solution of Ag, Al, B, Ba, Bi, Ca, Cd, Co, Cr, Cu, Fe, K, Li, Mg, Mn, Mo, Na, Ni, Pb, Sr, Tl, and Zn at 100 mg/mL; a multielement solution of Au, Ge, Pt, Sn, Ti, and Zr at 10 mg/mL; a solution of Si, S, and P at 1000 mg/mL. The concentration range for the elements were between 0.01–100 mg/mL and 0.2–2000 mg/mL. Two spike solutions were used as a recovery test.

2.7. Statistical Analysis. All statistical treatment was performed by STATGRAPHICS Plus Version 5.1 (Statistical Graphics Corporation, Professional Edition—Copyright 1994–2001).

The statistical approach that has been chosen to analyse the set of data obtained by ICP-MS was principal component analysis (PCA) and linear discriminant analysis (LDA).

Also, in order to check possible differences between the oils, two-way analysis of variance (ANOVA) was performed considering, as main factors, the irrigation regimes and the type of irrigation water. Moreover, to evaluate significant differences between averages, Tukey test was performed on the oil quality parameters. Differences were considered statistically significant for $P \geq 0.01$ (capital letters) and $P \geq 0.05$ (small letters). The values obtained for free acidity and FAMEs were statistically analyzed after arcsine transformation in order to meet assumptions for ANOVA. However, the results are presented in their original scale of measurement and reported as the averages of three repetitions ($n = 3$).

3. Result and Discussions

3.1. Physicochemical Analysis of Olive Oils. The most important quality parameters of the oils analysed in this study are listed in Table 1. In general, oils obtained from olives whose plants were irrigated with TWW were found to be "lampanti" characterized by high values of free acidity and low values of polyphenols. This result could be explained by considering that the quality of edible oils and fats is effected by their concentrations of trace metals [32, 33]. Traces of

TABLE 1: Quality parameters of the olive oils under investigation.

Harvesting mode based on the olives percentage humidity	Palmitic acid (%)	Stearic acid (%)	Oleic acid (%)	Linoleic acid (%)	Free acidity (% oleic acid)	Polyphenols (mg kg^{-1})	K232	K270
From the tree	19.72 A	1.24 a A	50.53 d B	24.97 a A	0.95 E	70.01 E	2.35 A c	0.13 B D
First fall	19.93 A	1.04 c CD	51.66 c B	24.17 ab AB	1.52 D	42.80 D	2.42 B c	0.17 B C
Second fall	19.98 A	1.11 b B	51.49 c B	23.99 abc AB	3.27 C	25.92 C	2.70 C b	0.20 A B
Third fall	19.09 B	0.92 d D	53.16 b A	23.82 bc AB	5.70 B	18.72 B	2.92 D a	0.24 A A
Forth fall	18.70 B	1.12 b BC	54.27 a A	23.04 c B	6.42 A	12.32 A	2.89 E ab	0.23 A A
Water regimes								
TWW (alternate)	19.35 B	1.39 A	50.71 B	24.90 AB	6.19 B	21.29 D	2.83 D a	0.26 AB A
TWW (continuous)	17.40 C	0.94 C	57.32 A	21.48 C	6.80 A	26.36 C	2.32 C c	0.21 C B
Not irrigated	19.67 B	0.71 D	51.49 B	25.58 A	0.72 C	50.87 A	2.63 A b	0.15 B C
CW	21.52 A	1.30 B	49.37 C	24.03 B	0.58 C	37.28 B	2.84 B a	0.16 A C

Means, for each factor, in the same column followed by the same letter do not differ according to Tukey's test (capital letters $P < 0.01$; small letters $P < 0.05$). Values are the mean of three replications.

metals in edible oils are known to have an effect on the rate of oil oxidation, decreasing the shelf life of commercial products [34]. Apart from causing premature rancidity, these oxidation processes may generate peroxides, aldehydes, ketones, acids, epoxides, and other compounds that may cause effects in the digestive system and also react with other food components (proteins and pigments), sensitising the action of some carcinogens [35].

3.2. Free Fatty Acid. Analysis of variance (ANOVA) applied to the values of free acidity has clearly made a distinction between the oils under investigation. Considering as main factors the harvesting mode, oils obtained from olives that remained in contact with the soil for a short period gave a value of free acidity comprised in a range between 1.52 and 3.27%, while oils obtained from olives that stayed for longer periods gave values up to 6.42%. Oils obtained from olives harvested directly from the plants gave a value of 0.95%.

Considering the second factor being the water regime, oils obtained from olives whose plants were irrigated with TWW were found to be of poor quality and characterized by high values of free acidity. However, we must point out that oils produced from olives harvested directly from the plant (control plants) gave a value of free acidity of 6.19% in an alternate irrigation regime with TWW and 6.80% in a continuous one; 0.72% in the rain-fed regime and 0.58% in the conventional water one (Table 1).

3.3. Polyphenols. Considering the first factor analysed being the harvesting mode, it is highly evident that the most abundant content of polyphenols is found in the oils produced from olives harvested directly from the plants. This value decreases significantly increasing the contact of the fruits with the soil. In fact, the polyphenols content of oils produced by processing the fruits collected from the ground in the last period of the harvest was very low (12.32 mg/kg). Considering the second factor being the water regime; the content of polyphenols found in the oils obtained from olives belonging to the rainfed regime resulted to be statistically different from all the other values (50.87 mg/kg). This value decreases significantly within the water regime, in fact, oils coming from plant irrigated with TWW gave values of 26.36 and 21.29 mg/kg for the continuous and alternate system, respectively, while 37.28 mg/kg for the conventional one (Table 1).

3.4. Fatty Acid Methyl Esters (FAMEs). In general, the fatty acid content is typical of Tunisian oils [36, 37]. However, the oils under analysis were dominated by palmitic acid (C16: 0), stearic acid (C18: 0), oleic acid (C18: 1), and linoleic acid (C18: 2) (Table 1). In this study, the observed values do not show a particular pattern that can explain the permanence of the olives with the soil and the incidence of the different water irrigation regimes on the quality of the oils obtained. In fact, the fatty acids of an olive oil are not dependent on the processing and harvesting of the olives but rather depend on genetic factors [38, 39].

3.5. Quality Control and Quality Assurance Data of ICP-MS Analysis. Initially, 48 elements were investigated but only 19 were submitted for statistical analysis. The criteria utilized to select those elements were as follows: recovery data were accepted if results were in the range of 70–120%, with 80% within 80–110% and for CRM values, within 20%. The results must be not below the limit of detection (LOD). The replicate agreement was considered acceptable if the value of the RSD was minor of 10%. Table 2 details the LOD values and the percentage recovery of a known amount of analytes spiked for all the elements submitted for statistical analysis. Results from spikes and recovery experiments at levels of 30, 80, and 300 ng/mL were in the range 92–104%, for almost all the elements. Li, Ca, and Ge gave a value of 74, 80, and 87%, respectively, whereas Ba and Tl gave a value of 111 and 120%, respectively. Furthermore, it can be seen from Table 2, the values of LODs which were in the range of 0.000–0.051 mg/kg, for almost all the elements. Fe, Ca, Al and Mg gave a value of 0.294, 0.587, 0.953, and 5.118 mg/kg respectively. Values of LOQs were in the range of 0.000–0.400 for almost all the elements except for Fe, K, Ca, Al and Mg that gave a value of 0.980, 1.068, 1.956, 3.178, and 17.061 mg/kg respectively. The relative standard deviations of the elements were less than 2% for all the elements.

3.6. Statistical Analysis of Multielement Data. At first, the trace elements profile and the minerals content present in the oils obtained from olives harvested from the ground irrigated with TWW have been monitored (Table 3). In order to develop a model to discriminate among the 5 levels of type of olive (see *Olive Sampling and Oil Extraction*), 90 cases (30 olive oil samples three time replicated) were used and 19 predictor variables (Li, B, Na, Mg, Al, K, Ca, Sc, Cr, Mn, Fe, Co, Ni, Cu, Zn, Sr, Mo, Ba, and La) were entered. The LDA plot resulting is showed in Figure 1, while Table 4 lists the summing up of the analyses of the discriminating functions.

The scores of the first two functions produced from LDA showed a separation into 3 groups: a first group is represented only by oils produced from olives hand picked from the plant (control plants) independently of irrigation regime, a second group is formed by oils produced from olives that stayed in contact with the ground for a short period and having a moisture containing comprised between 50.73% and 33.45% (1st and 2nd fall), and a third group formed by oils produced from olives that stayed in contact with the ground for very long time and having a moisture containing comprised in a range of 14.89% and 4.22% (3rd and 4th fall).

In particular, the control plants produced oils with a lower concentration of trace elements and minerals. These oils are lying in the lower left portion of the plot function characterized by negative values.

The oils produced from olives collected from the ground after a short permanence are characterized by higher values of Na, K, Ni, La, Sr, B, and Mo, while those produced from olives collected from the ground after a longer period are characterized by higher values of Li, Cu, Fe, Mn, and Sc (Table 4).

In order to verify the reliability of the model, the method has been tested using known samples as unknown variables.

TABLE 2: Limit of quantitations (LOQs), limit of detections (LODs), percentage recovery of spike solutions analysed by ICP-MS.

	Spike 1 (rec.)	Spike 2 (rec.)	LOD	LOQ
	%	%	mg kg^{-1}	mg kg^{-1}
^{7}Li	84	74	0,005	0,016
^{10}B	96	89	0,051	0,170
^{23}Na	95	90	0,104	0,347
^{25}Mg	109	95	5,118	17,061
^{27}Al	96	98	0,953	3,178
^{39}K	100	97	0,319	1,063
^{43}Ca	80	74	0,587	1,956
^{45}Sc	103	96	0,000	0,000
^{49}Ti	104	103	0,025	0,084
^{53}Cr	104	103	0,000	0,000
^{55}Mn	104	102	0,012	0,039
^{57}Fe	92	90	0,294	0,980
^{59}Co	105	104	0,005	0,015
^{62}Ni	105	105	0,011	0,038
^{65}Cu	104	104	0,000	0,000
^{74}Ge	87	69	0,000	0,000
^{88}Sr	106	103	0,006	0,019
^{90}Zr	101	101	0,039	0,130
^{95}Mo	103	101	0,028	0,092
^{111}Cd	92	92	0,000	0,000
^{118}Sn	99	93	0,000	0,000
^{137}Ba	111	108	0,007	0,023
^{139}La	103	102	0,017	0,056
^{140}Ce	102	101	0,000	0,000
^{141}Pr	103	102	0,000	0,000
^{145}Nd	103	104	0,000	0,000
^{147}Sm	104	103	0,000	0,000
^{151}Eu	103	102	0,000	0,000
^{158}Gd	104	103	0,000	0,000
^{159}Tb	102	101	0,000	0,000
^{163}Dy	104	102	0,000	0,000
^{165}Ho	101	100	0,000	0,000
^{166}Er	104	103	0,000	0,000
^{169}Tm	102	101	0,000	0,000
^{173}Yb	104	102	0,000	0,000
^{175}Lu	102	101	0,000	0,000
^{205}Tl	120	115	0,000	0,000
^{208}Pb	109	105	0,000	0,000
^{209}Bi	108	103	0,000	0,000
^{232}Th	101	100	0,006	0,021

TABLE 3: Summing up of the analyses of the discriminating functions.

(a)

Discriminant function	Eigenvalue	Relative percentage	Canonical correlation
1	5.71366	73.38	0.92252
2	0.994461	12.77	0.70612
3	0.711944	9.14	0.64488
4	0.366289	4.7	0.51777

(b)

Functions derived	Wilks λ	Chi-Square	DF	P value
1	0.0319289	265.2069	76	0.0000
2	0.2143590	118.5878	54	0.0000
3	0.4275310	65.4290	34	0.0010
4	0.7319100	24.0316	16	0.0888

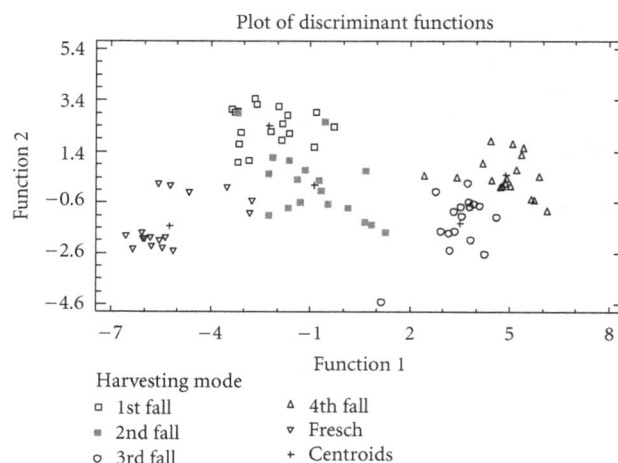

FIGURE 1: LDA plot for 90 olive oil samples (30 samples repeated three times) based on the concentration of 19 elements and using as input a priori five groups, that is, those corresponding to the olives harvested directly from the plants, called fresh, and those that have fallen naturally and stayed in contact with the soil irrigated with TWW. The moisture percentage of the olives were as follows: 50.73% ($\pm$5.31%) called 1st fall; 33.45% ($\pm$5.15%) called 2nd fall; 14.89% ($\pm$2.62%) called 3rd fall; 4.22% ($\pm$0.72%) called 4th fall.

In particular, a set of 5 samples, composed of one sample for each category of olives, was randomly removed for five times and the model was recalculated. Amongst the 90 observations used to fit the model, 78 or 86.67% were correctly classified.

Accordingly, by the PCA applied to the concentration of the 19 elements of each single sample, 5 principal components have been extracted, having Eigenvalues greater than or equal to 1.0, and together they account for 71% of the variability in the original data. The elements that mainly contributed to the separation of the groups are Al, Sr, Ba, and Mg on PC1 and Na, K, and Cr on PC 2.

In a second time, LDA was applied to the data set considering as input a priori the irrigation regimes. The bidimensional plot of the first two functions shows a clean separation into 2 groups: oils produced from olives whose plants were irrigated with TWW (in a continuous and alternate regime) and oils obtained from olives whose plants were irrigated with CW (Figure 2).

TABLE 4: Concentration of elements (mg/kg) in olive oil samples analysed by ICP-MS.

Sample code	Harvesting mode	Water regime	Li mg/kg	B mg/kg	Na mg/kg	Mg mg/kg	Al mg/kg	K mg/kg	Ca mg/kg	Sc mg/kg	Cr mg/kg	Mn mg/kg	Fe mg/kg	Co mg/kg	Ni mg/kg	Cu mg/kg	Zn mg/kg	Sr mg/kg	Mo mg/kg	Ba mg/kg	La mg/kg
K150	From the tree	TWW (alternate)	0,080	0,000	0,202	56,586	16,332	0,000	0,869	0,011	0,152	0,058	3,775	0,060	0,050	0,006	133,349	0,014	0,049	0,185	0,074
K173	From the tree	TWW (continuous)	0,077	0,000	0,303	57,543	16,462	0,000	1,257	0,013	0,126	0,059	3,684	0,059	0,041	0,016	134,715	0,014	0,068	0,179	0,075
K640	From the tree	CW	0,079	0,053	0,537	53,690	14,701	0,431	1,330	0,013	0,129	0,331	3,578	0,059	0,038	0,061	131,816	0,020	0,056	0,177	0,071
K661	From the tree	not irrigated	0,079	0,000	4,654	50,292	14,402	3,953	1,753	0,012	0,107	0,102	3,882	0,059	0,037	0,076	122,980	0,013	0,038	0,178	0,071
K151	First fall	TWW (alternate)	0,077	0,079	0,429	59,414	17,790	0,000	1,461	0,013	0,143	0,072	4,229	0,063	0,045	0,008	134,205	0,015	0,044	0,191	0,063
K174	First fall	TWW (continuous)	0,078	0,155	0,738	55,682	17,367	0,000	1,582	0,014	0,139	0,084	4,804	0,062	0,029	0,019	135,066	0,037	0,041	0,185	0,074
K644	First fall	CW	0,079	0,000	5,791	31,538	8,973	0,000	0,990	0,018	0,112	0,061	4,062	0,060	0,072	0,064	141,022	0,016	0,036	0,176	0,077
K664	First fall	not irrigated	0,084	0,088	0,143	55,673	16,160	0,000	0,214	0,013	0,127	0,061	4,071	0,058	0,045	0,079	68,014	0,010	0,040	0,175	0,073
K152	Second fall	TWW (alternate)	0,080	0,094	1,110	57,455	16,879	5,467	0,000	0,017	0,133	0,062	4,136	0,063	0,035	0,003	135,414	0,011	0,048	0,183	0,076
K175	Second fall	TWW (continuous)	0,077	0,000	3,568	59,915	16,947	0,000	1,287	0,014	0,141	0,061	3,876	0,060	0,048	0,022	132,173	0,016	0,041	0,187	0,072
K648	Second fall	CW	0,029	0,000	3,898	61,493	17,427	0,000	3,000	0,018	0,121	0,063	3,689	0,060	0,034	0,067	134,367	0,016	0,047	0,188	0,072
K668	Second fall	not irrigated	0,073	0,022	4,089	59,968	17,397	51,288	4,461	0,009	0,160	0,072	3,883	0,060	0,065	0,082	134,062	0,019	0,045	0,188	0,072
K153	Third fall	TWW (alternate)	0,022	0,083	3,943	60,040	17,003	24,273	3,674	0,016	0,128	0,068	4,228	0,061	0,062	0,010	126,257	0,019	0,042	0,184	0,076
K176	Third fall	TWW (continuous)	0,053	0,327	16,586	61,289	21,159	15,623	3,409	0,015	0,114	0,067	4,349	0,086	0,059	0,025	142,977	0,075	0,157	0,204	0,078

TABLE 4: Continued.

Sample code	Harvesting mode	Water regime	Li mg/kg	B mg/kg	Na mg/kg	Mg mg/kg	Al mg/kg	K mg/kg	Ca mg/kg	Sc mg/kg	Cr mg/kg	Mn mg/kg	Fe mg/kg	Co mg/kg	Ni mg/kg	Cu mg/kg	Zn mg/kg	Sr mg/kg	Mo mg/kg	Ba mg/kg	La mg/kg
K652	Third fall	CW	0,021	0,167	3,793	57,979	16,859	0,000	4,075	0,009	0,167	0,066	3,630	0,061	0,034	0,070	162,894	0,019	0,042	0,184	0,054
K672	Third fall	Not irrigated	0,022	0,129	1,555	63,346	25,625	0,000	3,344	0,010	0,160	0,076	4,411	0,076	0,061	0,085	158,817	0,042	0,051	0,201	0,076
K154	Forth fall	TWW (alternate)	0,080	0,143	4,629	57,887	17,235	53,418	2,551	0,038	0,137	0,072	4,144	0,059	0,036	0,031	124,821	0,023	0,033	0,190	0,072
K187	Forth fall	TWW (continuous)	0,022	0,098	7,846	61,749	17,337	98,075	1,536	0,021	0,146	0,073	4,104	0,062	0,044	0,028	127,282	0,045	0,048	0,194	0,080
K656	Forth fall	CW	0,016	0,090	2,642	58,431	18,083	0,000	1,890	0,021	0,159	0,066	4,146	0,060	0,051	0,073	95,394	0,015	0,042	0,185	0,070
K677	Forth fall	Not irrigated	0,075	0,088	0,523	61,106	18,244	0,000	4,268	0,050	0,160	0,066	17,596	0,068	0,046	0,088	129,939	0,018	0,043	0,187	0,073

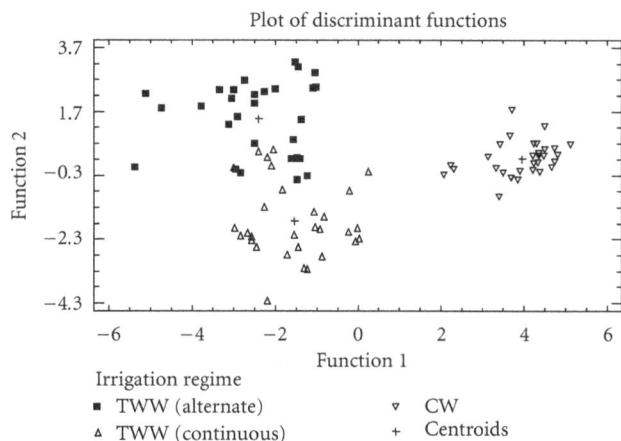

Plot of discriminant functions

Irrigation regime
■ TWW (alternate)   ▽ CW
△ TWW (continuous)   + Centroids

FIGURE 2: LDA plot for 90 olive oil samples (30 samples repeated three times) based on the concentration of 19 elements and using as input a priori three groups, that is, those corresponding to the irrigation regimes: alternate and continuous using TWW and CW.

Amongst the 90 observations used to fit the model, 79 or 88% were correctly classified. When PCA was applied, 7 components have been extracted having Eigenvalues greater than or equal to 1.0, and together they account for 74% of the variability in the original data. The elements that mainly contributed to the separation of the groups are mainly Sc, Mn, Li, Cu, Fe, La, Ni, K, and Na on PC1 and Mg, Ba, Al, Ca, Sr and B on PC 2.

4. Conclusions

The development of rapid and accurate analytical methods for the determination of metal concentrations in edible oils and fats is still a challenge in terms of quality control analysis, owing to the low concentration levels of some elements and the difficulties that arise due to the characteristics of the matrix [21, 40]. The metals analysed in this study, particularly Fe, Cu, Ca, Mg, Ni, and Mn, are known to increase the rate of oil oxidation [41, 42].

Statistical analysis clearly showed how the way to harvest the olives strongly influences the quality of oils produced [32]. Oils obtained from olives harvested from the ground were found in fact to be of poor quality. This data is not surprising considering that these olives are source of fermentation processes that lead to high values of free acidity and very low values of phenolic compounds in the oil. Moreover, if we consider the contact of the olives with the moist soil, surely the produced oil will be marked by the negative attribute of mould and ground. All the ICP-MS results and statistical evaluations performed have showed that oils produced from olives harvested from the ground have a richer trace elemental and mineral profile than oils produced from olives picked directly from the plant. In particular, olive trees irrigated with treated wastewater produced oils even richer in trace elements and minerals.

The European Community sets limit only for PB in edible oils, which is 0.1 mg/kg (European Commission 2006) [43];

for the other elements, no limits are set and their concentrations could be a concern for consumers.

Therefore, this study has pointed out that the irrigation with TWW adversely affects the quality of the oils. In fact, the olive oils were found to be "lampanti" characterized by high values of free acidity and very low values of polyphenols. Although the availability of wastewater is a very interesting alternative for urban agriculture, the health risks associated with this practice can constitute a real obstacle to the development of this activity.

Acknowledgments

This work was supported by "Riom II-Risorse aggiuntive" project sponsored by MIPAF (D.M. MiPAF no. 92691 del 18.12.03), the Tunisian Ministry of Higher Education, Scientific Research and Technology, and the Institution of Agricultural Research and Higher Education (IRESA). The authors thank M. J. Duff for English revision. Cinzia Benincasa and Mariem Gharsallaoui have participated equally to this work.

References

[1] IOOC, Olive oil exportations. International Olive Oil Council, 2004, http://www.internationaloliveoil.org.

[2] *World Bank. From Scarcity to Security: Averting a Water Crisis in the Middle East and North Africa*, World Bank, Washington, DC, USA, 1995.

[3] M. Qadir, D. Wichelns, L. Raschid-Sally et al., "The challenges of wastewater irrigation in developing countries," *Agricultural Water Management*, vol. 97, no. 4, pp. 561–568, 2010.

[4] A. N. Angelakis, M. H. F. Marecos Do Monte, L. Bontoux, and T. Asano, "The status of wastewater reuse practice in the Mediterranean basin: need for guidelines," *Water Research*, vol. 33, no. 10, pp. 2201–2217, 1999.

[5] C. Briccoli Bati, P. Basta, C. Tocci, and D. Turco, "Influence of irrigation with brackish water on young olive plants," *Olivae*, vol. 53, pp. 35–38, 1994.

[6] G. Celano, B. Dichio, G. Montanaro, V. Nuzzo, A. M. Palese, and C. Xiloyannis, "Distribution of dry matter and amount of mineral elements in irrigated and non-irrigated olive trees," *Acta Horticulturae*, vol. 474, pp. 381–384, 1999.

[7] H. Chehab, D. Boujnah, M. Braham, and S. Ben Elhadj, "Effets de trois régimes hydriques sur les comportements biologiques et agronomiques de deux variétés d'olivier de table dans le centre Tunisien," *Annales de l'INRGREF*, vol. 6, pp. 111–125, 2004.

[8] I. Nagel, F. Lang, M. Kaupenjohann, K. H. Pfeffer, F. Cabrera, and L. Clemente, "Guadiamar toxic flood: factors that govern heavy metal distribution in soils," *Water, Air, and Soil Pollution*, vol. 143, no. 1–4, pp. 211–224, 2003.

[9] T. W. Speir, A. P. Van Schaik, H. J. Percival, M. E. Close, and L. Pang, "Heavy metals in soil, plants and groundwater following high-rate sewage sludge application to land," *Water, Air, and Soil Pollution*, vol. 150, no. 1–4, pp. 319–358, 2003.

[10] A. Bahri, "Utilization of treated wastewaters and sewage sludge in agriculture in Tunisia," *Desalination*, vol. 67, pp. 233–244, 1987.

[11] USSL (United States Salinity Laboratory Staff), *Diagnosis and Improvement of Saline and Alkali Soils*, vol. 60 of *USDA Handbook*, USSL, Washington, DC, USA, 1954.

[12] A. M. El-Gazzar, E. M. El-Azab, and M. Sheha, "Effect of irrigation with fractions of sea water and drainage water on growth and mineral composition of young grapes, guavas, oranges and olives," *Alexandria Journal of Agricultural Research*, vol. 27, pp. 207–219, 1979.

[13] M. Benlloch, F. Arboleda, D. Barranco, and R. Fernandezescobar, "Response of young olive trees to sodium and boron excess in irrigation water," *Hortscience*, vol. 26, pp. 867–870, 1991.

[14] D. Charfi, A. Trigui, and K. Medhioub, "Effects of irrigation with treated wastewater on olive trees cv Chemlali of Sfax at the station of el hajeb," *Acta Horticulturae*, vol. 474, pp. 385–390, 1999.

[15] N. M. Al-Gazzaz, *Long-term irrigation effect of Khirbit Es-Samra Effluent Water on soil and olive* (Olea europaea L.) quality, M.S. thesis, Faculty of Science, University of Jordan, 1999.

[16] J. M. Murillo, R. López, J. E. Fernández, and F. Cabrera, "Olive tree response to irrigation with wastewater from the table olive industry," *Irrigation Science*, vol. 19, no. 4, pp. 175–180, 2000.

[17] M. H. Loupassaki, K. S. Chartzoulakis, N. B. Digalaki, and I. I. Androulakis, "Effects of salt stress on concentration of nitrogen, phosphorus, potassium, calcium, magnesium, and sodium in leaves, shoots, and roots of six olive cultivars," *Journal of Plant Nutrition*, vol. 25, no. 11, pp. 2457–2482, 2002.

[18] K. Al-Absi, M. Qrunfleh, and T. Abu-Sharar, "Mechanism of salt tolerance of two olive (*Olea europaea* L.) cultivars as related to electrolyte and toxicity," *Acta Horticulturae*, vol. 618, pp. 281–290, 2003.

[19] K. M. Al-Absi, F. M. Al-Nasir, and A. Y. Mahadeen, "Mineral content of three olive cultivars irrigated with treated industrial wastewater," *Agricultural Water Management*, vol. 96, no. 4, pp. 616–626, 2009.

[20] M. Zeiner, I. Steffan, and I. J. Cindric, "Determination of trace elements in olive oil by ICP-AES and ETA-AAS: a pilot study on the geographical characterization," *Microchemical Journal*, vol. 81, no. 2, pp. 171–176, 2005.

[21] C. Benincasa, J. Lewis, E. Perri, G. Sindona, and A. Tagarelli, "Determination of trace element in Italian virgin olive oils and their characterization according to geographical origin by statistical analysis," *Analytica Chimica Acta*, vol. 585, no. 2, pp. 366–370, 2007.

[22] A. Rhouma, M. Gharsallaoui, and M. Khlif, "Etude de l'effet de stockage des olives des deux variétés chemleli et chetoui sur les caractéristiques de leur huile," *Revue Ezzitouna*, vol. 10, pp. 52–59, 2005.

[23] ONAS, Office National d'Assanissement, Sfax, Tunisia, 2003.

[24] R. K. Rattan, S. P. Data, P. K. Chhonkar, K. Suribabu, and A. K. Singh, "Long term impact of irrigation with sewage effluents on heavy metal content in soil, crops and groundwater—a case study," *Agriculture, Ecosystems & Environment*, vol. 13, pp. 236–242, 2005.

[25] Z. Wiesman, D. Itzhak, and N. Ben Dom, "Optimization of saline water level for sustainable Barnea olive and oil production in desert conditions," *Scientia Horticulturae*, vol. 100, no. 1–4, pp. 257–266, 2004.

[26] K. S. Chartzoulakis, G. Psarras, S. Vemmos, and M. Loupassaki, "Effects of salinity and potassium supplement on photosynthesis, water relations and Na, Cl, K and carbohydrate concentration of two olive cultivars," *Agricultural Research*, vol. 27, pp. 75–84, 2004.

[27] K. S. Chartzoulakis, "Salinity and olive: growth, salt tolerance, photosynthesis and yield," *Agricultural Water Management*, vol. 78, no. 1-2, pp. 108–121, 2005.

[28] "ECC regulation No 1989/2003 of 6 November 2003 amending Regulation (EEC) No 2568/91 on the characteristics of olive oil and olive-pomace oil and on the relevant methods of analysis," *Official Journal of the European Union*, vol. 295, pp. 57–77, 2003.

[29] M. Gharsallaoui, C. Benincasa, M. Ayadi, E. Perri, M. Khlif, and S. Gabsi, "Study on the impact of wastewater irrigation on the quality of oils obtained from olives harvested by hand and from the ground and extracted at different times after the harvesting," *Scientia Horticulturae*, vol. 128, no. 1, pp. 23–29, 2011.

[30] W. W. Christie, *The Preparation of Derivatives of Fatty Acids. In Gas Chromatography and Lipids*, The Oily Press, Ayr, Scotland, 1998.

[31] S. J. Huang and S. J. Jiang, "Determination of Zn, Cd and Pb in vegetable oil by electrothermal vaporization inductively coupled plasma mass spectrometry," *Journal of Analytical Atomic Spectrometry*, vol. 16, no. 6, pp. 664–668, 2001.

[32] M. Khlif, N. Grati-Kammoun, and M. El Euch, "Effet de la durée de stockage des olives sur la qualité de l'huile," *Ezzaitouna*, vol. 2, no. 1-2, article 102, pp. 1–9, 1996.

[33] A. Rhouma, M. Gharsallaoui, and M. Khlif, "Etude de l'effet de stockage des olives des deux variétés chemleli et chetoui sur les caractéristiques de leur huile," *Revue Ezzitouna*, vol. 10, pp. 52–59, 2005.

[34] M. Murillo, Z. Benzo, E. Marcano, C. Gomez, A. Garaboto, and C. Marin, "Determination of copper, iron and nickel in edible oils using emulsified solutions by ICP-AES," *Journal of Analytical Atomic Spectrometry*, vol. 14, no. 5, pp. 815–820, 1999.

[35] J. R. Castillo, M. S. Jiménez, and L. Ebdon, "Semiquantitative simultaneous determination of metals in olive oil using direct emulsion nebulization," *Journal of Analytical Atomic Spectrometry*, vol. 14, no. 9, pp. 1515–1518, 1999.

[36] S. Dabbou, I. Rjiba, A. Nakbi, N. Gazzah, M. Issaoui, and M. Hammami, "Compositional quality of virgin olive oils from cultivars introduced in Tunisian arid zones in comparison to Chemlali cultivars," *Scientia Horticulturae*, vol. 124, no. 1, pp. 122–127, 2010.

[37] M. Issaoui, G. Flamini, F. Brahmi et al., "Effect of the growing area conditions on differentiation between Chemlali and Chétoui olive oils," *Food Chemistry*, vol. 119, no. 1, pp. 220–225, 2010.

[38] H. Manaï, F. M. Haddada, A. Trigui, D. Daoud, and M. Zarrouk, "Compositional quality of virgin olive oil from two new Tunisian cultivars obtained through controlled crossings," *Journal of the Science of Food and Agriculture*, vol. 87, no. 4, pp. 600–606, 2007.

[39] S. Dabbou, S. Dabbou, H. Chehab et al., "Chemical composition of virgin olive oils from Koroneiki cultivar grown in Tunisia with regard to fruit ripening and irrigation regimes," *International Journal of Food Science and Technology*, vol. 46, no. 3, pp. 577–585, 2011.

[40] R. Ansari, T. G. Kazi, M. K. Jamali et al., "Variation in accumulation of heavy metals in different verities of sunflower seed oil with the aid of multivariate technique," *Food Chemistry*, vol. 115, no. 1, pp. 318–323, 2009.

[41] A. N. Anthemidis, V. Arvanitidis, and J. A. Stratis, "On-line emulsion formation and multi-element analysis of edible oils by inductively coupled plasma atomic emission spectrometry," *Analytica Chimica Acta*, vol. 537, no. 1-2, pp. 271–278, 2005.

[42] D. Mendil, O. D. Uluozlu, M. Tuzen, and M. Soylak, "Investigation of the levels of some element in edible oil samples produced in Turkey by atomic absorption spectrometry," *Journal of Hazardous Materials*, vol. 165, no. 1–3, pp. 724–728, 2009.

[43] European Commission. Regulation (EC), Official Journal of the European Union, 1881/2006; Section 3—Metals, item 3.1.14, p. 21, 2006.

22

Combining Ability Analysis in Complete Diallel Cross of Watermelon (*Citrullus lanatus* (Thunb.) Matsum. & Nakai)

M. Bahari,[1] M. Y. Rafii,[2,3] G. B. Saleh,[2] and M. A. Latif[2,4]

[1] *Malaysian Agricultural Research and Development Institute, Bukit Tangga, 06050 Bukit Kayu Hitam, Kedah, Malaysia*
[2] *Department of Crop Science, Faculty of Agriculture, Universiti Putra Malaysia, 43400 UPM Serdang, Selangor, Malaysia*
[3] *Institute of Tropical Agriculture, Universiti Putra Malaysia, 43400 UPM Serdang, Selangor, Malaysia*
[4] *Plant Pathology Division, Bangladesh Rice Research Institute (BRRI), Gazipur-1701, Bangladesh*

Correspondence should be addressed to M. Y. Rafii, mrafii@putra.upm.edu.my

Academic Editor: Gerald E. Brust

The experiments were carried out in two research stations (MARDI Bukit Tangga, Kedah, and MARDI Seberang Perai, Penang) in Malaysia. The crossings were performed using the four inbred lines in complete diallel cross including selfs and reciprocals. We evaluated the yield components and fruit characters such as fruit yield per plant, vine length, days to fruit maturity, fruit weight, total soluble solid content, and rind thickness over a period of two planting seasons. General combining ability and its interaction with locations were statistically significant for all characteristics except number of fruits per plant across the environments. Results indicated that the additive genetic effects were important to the inheritance of these traits and the expression of additive genes was influenced greatly by environments. In addition, specific combining ability effect was statistically evident for fruit yield per plant, vine length, days to first female flower, and fruit weight. Most of the characters are simultaneously controlled by additive and non-additive gene effects. This study demonstrated that the highest potential and promising among the crosses was cross P2 (BL-14) × P3 (6372-4), which possessed prolific plants, with early maturity, medium fruit weight and high soluble solid contents. Therefore this hybrid might be utilized for developing high yielding watermelon cultivars and may be recommended for commercial cultivation.

1. Introduction

Watermelon (*Citrullus lanatus* (Thunb.) Matsum. & Nakai) is a very popular short-term nonseasonal fruit in Malaysia and has been classified under major fruits by the Ministry of Agriculture and Agro-Based Industry. Its refreshing and diuretic properties, associated with the pleasant taste and being a low-calorie fruit, make it a great alternative for the most varied diets. Malaysia occupied the 41st place in all watermelon-producing countries, with a production of 154,416 tones and a harvested area of approximately 9,241 hectares [1]. The main producing state is Johor, which contributes over 50% of national productions. At present, approximately 70% of productions are exported to Singapore, Taiwan, and Hong Kong. The world's largest producer of watermelon is China which usually accounts for over half of the world production [2].

Breeding for hybrid watermelon has so far achieved limited success in Malaysia. Several constraints in the breeding programme were faced, such as lack of genetic resources and crop failure attributed to high humidity, rainfall, and disease outbreaks [3]. Consequently there is limited choice of breeding material to work with and totally depends on commercial hybrids released by foreign seed companies. There is an ongoing need for watermelon improvement in country, especially for producing new hybrid cultivars with better fruit qualities comparable with the imported cultivars.

Estimating combining ability can be used to determine the usefulness of the inbred lines in hybrid combinations [4]. Selection of superior parents of hybridization is very important for watermelon improvement programmed, because the performance of a hybrid is related to the general (GCA) and specific (SCA) combining abilities of the inbred lines involved in the cross. According to Cruz and Vencovsky [5], the most promising hybrids are those that coming from the crossing of divergent parents, where at least one of them presented high GCA. Chaudhary [6] stated that diallel crosses are the most popular mating design and are used to

TABLE 1: Details of environmental conditions of two locations.

Descriptions	MARDI Seberang Prai (MSP)	MARDI Bukit Tangga (MBT)
Latitude	5°8′N	6°28′N
Longitude	100°32′E	100°32′E
Temperature	30°C	33°C
Annual rainfall	2670 mm	2480 mm
Humidity	90%	85%
Soil type	Sogo series	Kuah series

MARDI: Malaysian Agricultural Research Development Institute.

obtain information on value of parents and to assess the gene action in various characters.

Some methods have been proposed to estimate the combining ability of genotypes to be used in breeding programs. Diallel analysis repeated over environments proposed by Matizinger et al. [7] is useful to estimate the combining ability of the parents. Information on the general and specific combining abilities and their interactions with environments will be helpful in the analysis and interpretation of the genetic basis of important traits.

Several studies on GCA and SCA in watermelon had been done by many researchers such as Brar and Sukhija [8], Brar and Nandpuri [9], and Sidhu and Brar [10]. In another study, Souza et al. [11] worked on 3 × 3 diallel analysis in watermelon and reported that most of the characters studied are simultaneously controlled by additive and nonadditive gene effects. Their work demonstrated that the potential of the cross is between "Sugar Baby" and "Kodama" which had prolific plants, small fruit and high soluble solids content. While Ferreira et al. [12] reported that GCA effects were more important than SCA effects for number and weight of fruit per plant, as well as for flesh colour, thickness, and soluble solid content, for number of days to the first female flower and number of seeds, indicating the predominance of non-additive gene effects was found.

The main aim of the research was to identify breeding lines having good ability effects for yield components and fruit characters. Hence, the present study was to estimate the combining ability of four watermelon inbred lines and identify the important characters to support the breeding program of watermelon in the country.

2. Material and Methods

The materials for this study consisted of four watermelon inbred lines and their F_1 hybrids including the reciprocals. The four inbred lines used as parents were CS-19-S_7 (P1), BL-14-S_7 (P2), 6372-4-S_7 (P3), and CH-8-S_7 (P4). Four parents were crossed in complete diallel mating scheme in greenhouse at Bukit Tangga MARDI station in the second half of 2008. Two varieties, C1 and C2, were used as check.

The 18 genotypes (4 parents, 6 F_1 hybrids (P1 × P2, P1 × P3, P1 × P4, P2 × P3, P2 × P4, and P3 × P4), 6 F_1 reciprocal hybrids (P2 × P1, P3 × P1, P4 × P1, P3 × P2, P4 × P2, and P4 × P3), and two check varieties) were evaluated in a randomized block design with 4 replications at Bukit Tangga MARDI Research Station and Seberang Prai

MARDI Research Station, over a period of two planting seasons. Each plot consisted of 7 plants spaced 1.2 m apart in rows and 2.5 m between rows. Details of two environments are presented in Table 1.

The seeding was carried out in trays, and the seedlings were transplanted 14 days after sowing. Fertilization was performed by applying, in foundation, the rates of 500 kg/ha of NPK Green. Cultural practices were conducted according to technical recommendations for the culture in the country. The identification of ripe fruit was based on observation of drying of the tendril adjacent to the stalk and the sound emitted by woody fruit when struck by his fingertips. Plants were evaluated on the yield components (fruit yield, number of fruits per plant, vine length, number of days for the appearance of the first female flowers, and days to fruit maturity) and fruit characters (fruit weight, total soluble solids and rind thickness). To obtain the means of the variables in each plot, we sampled five plants and five fruits, taken at random during the harvest.

Estimates of the general and specific combining ability effects were obtained using the methodology proposed by Griffing [13] for analysis of diallel with parents, F_1 and F_1 reciprocal (Method I), considering fixed effect of treatments. The analysis of variance of diallel was performed according to the scheme presented by Zhang and Kang [14]. This program is considered as an effective method in analyzing and interpreting diallel cross data which are conducted in a number of environments [15].

3. Results

3.1. Combined ANOVA. A combined analysis of variance was performed on the data of yield components and fruit traits to estimate the amount of variability for these characteristics among parents, their F_1 and F_1 reciprocals. Analysis of variance for diallel cross of watermelon genotypes over the four environments is presented in Table 2. Results showed that significant differences were found among the environments (E) and among the genotypes (parents and off springs) for the yield components (fruit yield, number of fruits, vine length, days to the first female flowers, and days to fruit maturity) and fruit characters (fruit weight, total soluble solids, and rind thickness). Mean data for yield components and fruit characters are given in Table 3.

Analysis of variance of combining ability indicated that effects of general combining ability (GCA) were found to be significant for all yield components and fruit characters

TABLE 2: Mean squares of combined ANOVA for yield components and fruit traits in diallel cross of four watermelon inbred lines over four environments.

Source of variation	d.f.	Fruit yield	Number of fruits	Vine length	Days to first female flower	Days to fruit maturity	Fruit weight	Total soluble solids	Rind thickness
Environments (E)	3	441.08**	6.42**	38.19**	579.05**	2162.11**	53.20**	116.21**	4.47**
Reps (environment)	12	15.99	0.45	0.18	2.76	8.00	1.44	0.47	0.15**
Genotypes (G)	15	46.15**	0.82*	1.34**	31.51**	113.10**	11.31**	1.49**	0.14**
GCA	3	114.40**	1.02	5.86**	92.09**	484.82**	46.26**	3.84**	0.20**
SCA	6	26.22*	0.38	0.36*	32.34**	31.21**	4.37**	0.93	0.21**
Reciprocal (R)	6	31.97**	1.15**	0.07	2.66	9.13*	0.77	0.87	0.04
G × E	45	15.55	0.45	0.29**	8.91**	37.23**	1.78*	0.95*	0.14**
GCA × E	9	16.37	0.24	0.55**	12.10**	77.13**	3.41**	1.29*	0.53**
SCA × E	18	14.62	0.63	0.28*	12.81**	42.42**	1.39	0.62	0.06
R × E	18	16.07	0.39	0.17	3.20	12.10*	1.36	1.11*	0.02
Error	180	10.04	0.39	0.15	3.24	6.76	1.24	0.65	0.039

*, ** Significant at $P \leq 0.05$ and $P \leq 0.01$, respectively.

TABLE 3: Mean performance of parents and their hybrids for yield components and fruit traits in watermelon.

Genotype	Fruit yield (kg)	Number of fruits	Vine length (m)	Days to first female flower	Days to fruit maturity	Fruit weight (kg)	Total soluble solids (^{0}Brix)	Rind thickness (cm)
P1 (CS-19)	14.80	2.94	3.48	37.38	77.75	7.94	8.91	1.53
P2 (BL-14)	10.31	2.25	2.60	31.06	67.56	5.20	9.26	1.34
P3 (6372-4)	10.76	2.38	2.76	31.55	71.13	5.47	9.02	1.31
P4 (CH-8)	11.77	2.56	2.42	31.75	67.06	5.56	9.65	1.50
P1 × P2	14.98	2.13	3.06	33.00	71.94	6.21	8.71	1.26
P1 × P3	11.97	2.56	3.04	33.82	72.94	6.42	8.82	1.25
P1 × P4	12.71	2.63	2.70	32.50	71.25	6.97	9.13	1.42
P2 × P3	14.98	2.81	2.98	31.94	71.88	6.20	9.07	1.29
P2 × P4	12.21	2.31	2.96	32.67	67.56	5.67	9.49	1.39
P3 × P4	10.69	2.44	2.61	32.06	70.38	5.22	9.49	1.33
P2 × P1	13.68	2.63	3.23	33.56	72.19	5.91	8.69	1.35
P3 × P1	13.66	2.88	2.92	32.63	68.94	5.80	9.11	1.44
P4 × P1	11.14	2.69	2.53	32.44	69.06	5.43	9.04	1.32
P3 × P2	15.19	2.90	2.72	32.63	72.19	7.48	9.67	1.55
P4 × P2	10.12	2.38	2.54	32.69	68.69	5.03	9.25	1.34
P4 × P3	11.10	2.44	2.56	33.06	69.88	5.44	9.05	1.30
C1 (check)	9.28	2.50	2.35	32.06	70.44	4.21	9.50	1.31
C2 (check)	8.11	3.63	2.63	31.88	67.56	2.51	11.16	1.70
Mean	11.88	2.59	2.78	32.69	70.47	5.70	9.28	1.40
CV (%)	28.17	24.30	21.00	6.76	5.03	20.45	9.59	18.58
LSD (5%)	1.53	0.44	0.27	1.22	2.47	0.81	0.62	0.32

except number of fruits (Table 2). However the interaction of GCA effects and environment (GCA × E) was statistically evident for vine length, days to first female flower, days to fruit maturity, fruit weight, total soluble solids, and rind thickness.

For specific combining ability (SCA) analysis, significant effects were found for all traits measured except number of fruits and total soluble solids (Table 2). However the inter-

action of SCA effects and environment (SCA × E) was significantly differe for vine length, days to first female flower and days to fruit maturity. Analysis of variance for reciprocal effects (R) showed that significant effects were found for fruit yield, number of fruits, and days to maturity. Meanwhile, the interaction of reciprocal effects and environment (R × E) was nonsignificant for all traits measured except days to maturity and total soluble solid content (Table 2).

TABLE 4: Estimation of general combining ability (GCA) of parental lines for characters measured in the F_1 watermelon hybrids.

Character	CS-19 (P1)	BL-14 (P2)	6372-4 (P3)	CH-8 (P4)
Fruit yield	1.413**	−0.568*	−0.373	−0.472
Number of fruits	0.113*	−0.105*	−0.004	−0.004
Vine length	0.266**	−0.036	0.025	−0.255**
Days to first female flower	1.238**	−0.645**	−0.168	−0.426**
Days to fruit maturity	2.836**	−1.336**	−0.297	−1.203*
Fruit weight	0.886**	−0.431**	−0.294**	−0.161
Total soluble solids	−0.159**	0.058	−0.117	0.218
Rind thickness	−0.015	−0.073	−0.080	0.168**

*, ** Significant at $P \leq 0.05$ and $P \leq 0.01$, respectively.

TABLE 5: Estimation of specific combining ability (SCA) and reciprocal (R) effects for characters measured in the F_1 watermelon hybrids.

Hybrid	Fruit yield	Number of fruits	Vine length	Days to female flower	Days to maturity	Fruit weight	Total soluble solids	Rind thickness
P1 × P2	0.080	−0.074	−0.032	−0.910**	−0.242	−0.246	−0.157	−0.093
P1 × P3	−0.500	−0.051	0.022	−0.168	−0.625	−0.423*	−0.113	−0.062
P1 × P4	0.727	−0.051	−0.120*	−1.035**	−0.563	0.501*	0.192	0.052
P2 × P3	1.594**	0.168	0.128*	0.684**	−0.766	0.463*	0.210	0.110*
P2 × P4	−0.837	−0.020	0.050	0.660**	1.422	−0.280	−0.053	−0.044
P3 × P4	−0.321	0.035	−0.044	0.559*	0.320	−0.104	−0.205	−0.059
P2 × P1	−1.769**	−0.344**	0.040	0.531	0.031	0.006	−0.180	−0.018
P3 × P1	−0.856	−0.031	−0.094	0.125	0.375	0.253	0.064	−0.050
P4 × P1	−1.238	0.031	−0.008	−0.063	−0.469	−0.257	−0.273	−0.065
P3 × P2	−0.721	−0.281	0.020	0.031	−0.688	−0.068	0.191	−0.027
P4 × P2	0.285	0.031	0.034	−0.313	0.844	0.098	0.123	−0.003
P4 × P3	0.022	0.125	−0.013	−0.311	−0.406	−0.004	−0.006	0.012

*, ** Significant at $P \leq 0.05$ and $P \leq 0.01$, respectively.

3.2. Analysis of Combining Ability

3.2.1. Yield Components.
The parent with the highest ranking for fruit yield was P1 (14.8 kg). This parent also showed significant positive GCA effects across environment (1.41) for fruit yield. Among the F_1 hybrids, P3 × P2 (15.19 kg), followed by P2 × P3 (14.98 kg), displayed the best performance for this character (Table 3). The estimated specific combining ability (SCA) effects revealed that hybrid P2 × P3 gave significant positive SCA effects of 1.59 for fruit yield per plant across the environments. It had the higher positive reciprocal effects values for number of fruit, days to female flower, total soluble solids, and rind thickness (Table 5).

With regards to number of fruits per plant, the check variety C2 (3.6 fruits) was significantly higher than all genotypes (Table 3). However, the parent with the highest ranking for number of fruits was P1 (2.94), which had the highest positive and significant GCA value of 0.11 (Table 4). Among the F_1 hybrids, the highest number of fruits was recorded at P3 × P2 (2.9 fruits), followed by P3 × P1 (2.8 fruits) and P4 × P1 (2.7 fruits). The highest SCA positive value for number of fruits was found at P2 × P3 (0.17). Meanwhile the highest positive reciprocal effect (R) value was observed in reciprocal hybrid, P4 × P3 (0.13) (Tables 3 and 5).

For vine length, parental line P1 (3.48 m) was significantly longer than all genotypes and had the highest positive GCA value of 0.27 (Tables 3 and 4). Of the hybrids, P2 × P1 (3.23 m) ranked the highest among the crosses for vine length (Table 3). However the estimated SCA effects showed that hybrid P2 × P3 had the highest positive SCA effects (0.13) for vine length (Table 5). Meanwhile the highest positive reciprocal effect (R) value was 0.04 and was observed in hybrid P2 × P1.

For the appearance of the first female flowers, the genotype with the earliest flowering was P2 (BL-14) (31.1 days) and the GCA effect was −0.64 (Tables 3 and 4). Among the hybrids, P1 × P4 was the earliest to the first female flower appearance and had the highest negative SCA effects (−1.04) (Table 5). The highest negative R effects for days to first female flowers across the environments were observed in cross P4 × P2 that was nonsignificant among the reciprocal hybrids.

Parental line P4 showed the earliest fruit maturity (67.1 days), which had the highest negative GCA effects (−1.34) (Tables 3 and 4). However, the estimated SCA effects showed that hybrid P2 × P3 had the highest negative SCA effects (−0.77) (Table 5). Meanwhile the highest negative R effects value was observed in hybrid P3 × P2 (−0.69).

3.2.2. Fruit Characters. The parental line with the highest ranking for fruit weight was P1 (7.94 kg) which also gave significant positive GCA effects across the environments (0.886) (Tables 3 and 4). The highest SCA effect was measured at P1 × P4 (0.501), followed by P2 × P3 (0.463), which was significantly higher than that of the other crosses (Table 5). However, the estimates of R effects for fruit weight across the environments showed that no hybrid revealed positive and significant R effects.

The result showed that check variety C2 was the highest ranking genotype for total soluble solids (11.16 °Brix), and it was significantly different from other genotypes (Table 3). There was no parental inbred line which revealed significant and positive GCA effects for total soluble solids. However, inbred CH-8 (P4) gave the highest positive effect for total soluble solids (0.218) across the environments (Table 4). Estimates of SCA and R effects for total soluble solid content across the environments showed that no hybrid revealed positive and significant SCA effect (Table 5).

For rind thickness, inbred line CH-8 showed significant positive GCA effect for rind thickness (0.168) (Table 4). As a result, this parent contributed to produce hybrids with thick rind. The highest SCA effect was measured for P2 × P3 (0.11), which was significantly higher than that of other crosses (Table 5).

4. Discussion

The inbred and their hybrids were tested over different environments, and performance of all characters was investigated. These results indicated those environmental factors which varied from location to location and planting cycle to planting cycle and influenced the expression of the characteristics of watermelon parents and their progenies. Interaction of genotypes with environment (G × E) was found to be significant for the vine length, days to first female flower, days to fruit maturity, fruit weight, total soluble solids, and rind thickness, indicating that parents and their progenies were susceptible to environmental conditions.

Parental line P1 was the highest fruit yielder, and this parent also showed significant positive GCA effects across the environment. This inbred line is a good combiner for fruit yield. In combining ability analysis, the effects of general combining ability (GCA) were found to be significant for all yield components and fruit characters except number of fruit yield. Results indicate that there is presence of additive and nonadditive gene actions. Mean squares were higher in general combining ability than that of mean squares of specific combining ability for majority traits. It is suggested that the additive gene effects are more important than the nonadditive gene effects. This finding is similar to that of the report of Souza et al. [11] as described in their study on three intercrossings of watermelon genotypes. The significant effect of GCA × E was found among the traits, vine length, days to first female flower, days to fruit maturity, fruit weight, total soluble solids, and rind thickness which revealed that the action of additive genes was influenced by the environmental variation. In addition, variation of these traits was influenced greatly by environments.

Inbred P1 gave significant positive GCA effects across the environments for fruit yield, number of fruits, and vine length. Results suggested that this inbred contributed to achieving plants more compact, prolific, productive, and producing large fruit. Inbred P2 gave significant and that are highest negative values across environments for early in days to first female flower and early in days to maturity period, while inbred lines P3 and P4 showed significant positive effect across the environments for total soluble solids and rind thickness, respectively. It indicated that these parents contributed to producing hybrids with thick rind and sweeter fruits. The inbred lines that revealed strong GCA effects could be utilized further as sources for population improvement towards the accumulation of favorable additive genes in population for watermelon variety improvement in Malaysia.

Significant SCA effects were found for most traits except number of fruits, total soluble solids, and rind thickness. Results from this study were not in agreement with Ferreira et al. [12] who reported that the number of fruits, total soluble solids, and rind thickness were significant in their study. So, our study revealed that hybrid P2 × P3 had the highest SCA effects for most of traits except days to flower. It indicated that the significant role of nonadditive effects was involved in the inheritance of this trait. There was no hybrid which revealed significant and positive R effects for rind thickness, indicating that reciprocal effect is not important.

Reciprocal effects (R) showed that significant effect was found for yield components, that is, fruit yield, number of fruit, and days to maturity. Similar results were reported by Feyzian et al. [16] and pointed out that reciprocal effects were significant for fruit yield and fruit maturity on melon. The interaction of reciprocal effects and environment (R × E) was nonsignificant for all traits measured except days to maturity and total soluble solids content. Therefore, extra chromosomal inheritance or maternal effects may be involved in gene control of these characters. The study of this type of gene interaction is important in breeding programs which allow the determination of parents to be used as donors or recipients of pollen.

This study strongly suggests that those with strong SCA effects such as P2 × P3 could be advanced for hybrid variety release after other yield stability factors have been considered.

Acknowledgment

The work was supported by the Science Fund Grant 05-03-08-SF159 to MARDI from the Ministry of Science, Technology and Environment, Malaysia.

References

[1] DOA, Crops acreage, 2009, http://www.doa.gov.my/web/guest/home.

[2] M. Arney, S. R. Fore, and R. Brancucci, *Watermelon Reference Book*, National Watermelon Promotion Board, Orlando, Fla, USA, 2006.

[3] R. S. Zainab and K. A. K. Hasnah, "Breeding for hybrid watermelon, local assessments of the parental lines," in *Proceeding*

of the 4rth National Congress on Genetics, Genting Highlands, Pahang, Malaysia, 2000.

[4] G. F. Sprague and L. A. Tatum, "General vs. specific combining ability in single crosses of corn," *Journal American Society Agronomy*, vol. 34, pp. 923–932, 1942.

[5] C. D. Cruz and R. Vencovsky, "Comparison of some methods of diallel analysis," *Revista Brasileira de Genética*, vol. 12, pp. 425–438, 1989.

[6] R. C. Chaudhary, *Introduction to Plant Breeding*, Oxford and IBH publishing Co. PVT. LD, New Delhi, India, 1982.

[7] D. F. Matizinger, G. F. Sprague, and C. C. Cockerman, "Diallel crosses of maize in experiments repeated over locations and years," *Agronomy Journal*, vol. 51, pp. 346–350, 1959.

[8] J. S. Brar and B. S. Sukhija, "Hybrid vigour in inter-varietal crosses in watermelon (*Citrullus lanatus* (Thunb. Mansf.)," *Indian Journal of Horticulture*, vol. 34, pp. 277–283, 1977.

[9] J. S. Brar and K. S. Nandpuri, "Inheritance of fruit weight in watermelon [*Citrulluslanatus* (Thunb.) Mansf.]," *Journal of Research Punjab Agriculture University*, vol. 11, pp. 140–144, 1974.

[10] A. S. Sidhu and J. S. Brar, "Genetic divergence and hybrid performance in watermelon," *Indian Journal of Agriculture Science*, vol. 55, pp. 459–461, 1985.

[11] F. F. Souza, F. C. Gama, and M. A. Queiroz, "Combining ability analysis in dialelic crosses among three watermelon genotypes," *Horticultural Brasileira, Brasilia*, vol. 22, no. 4, pp. 789–793, 2004.

[12] M. A. J. Ferreira, L. T. Braz, A. M. de Queiroz, M. G. C. Churata-Masca, and R. Vencovsky, "Combining ability of seven watermelon populations," *Pesquisa Agropecuaria Brasileira*, vol. 37, no. 7, pp. 963–970, 2002.

[13] B. Griffing, "Concept of general and specific combining ability in relation to diallel crossing systems," *Australian Journal Biological Science*, vol. 9, no. 4, pp. 463–493, 1956.

[14] Y. Zhang and M. S. Kang, "DIALLEL-SAS: a program for Griffing's diallel method," in *Handbook of Formulus and Software for Plant Geneticists and Breeder*, pp. 1–9, 2003.

[15] Y. Zhang, M. S. Kang, and K. R. Lamkey, "DIALLEL-SAS05: a comprehensive program for griffing's and Gardner-Eberhart analyses," *Agronomy Journal*, vol. 97, no. 4, pp. 1097–1106, 2005.

[16] E. Feyzian, H. Dehghani, A. M. Rezai, and M. J. Javaran, "Diallel cross analysis for maturity and yield-related traits in melon (*Cucumis melo* L.)," *Euphytica*, vol. 168, no. 2, pp. 215–223, 2009.

23

Nitrogen and Potassium Concentrations in the Nutrients Solution for Melon Plants Growing in Coconut Fiber without Drainage

Luiz Augusto Gratieri,[1] **Arthur Bernardes Cecílio Filho,**[2]
José Carlos Barbosa,[2] **and Luiz Carlos Pavani**[2]

[1] *Instituto Federal de Educação, Ciência e Tecnologia Sul de Minas Gerais (IFSULDEMINAS), Campus Muzambinho, Morro Preto s/n, 37890-000 Muzambinho, MG, Brazil*
[2] *Universidade Estadual Paulista, Via de Acesso Prof. Paulo D. Castellane s/n, 14884-900 Jaboticabal, SP, Brazil*

Correspondence should be addressed to Arthur Bernardes Cecílio Filho; rutra@fcav.unesp.br

Academic Editors: A. M. De Ron, A. Ferrante, and W. Ramakrishna

With the objective of evaluating the effects of N and K concentrations for melon plants, an experiment was carried out from July 1, 2011 to January 3, 2012 in Muzambinho city, Minas Gerais State, Brazil. The "Bonus no. 2" was cultivated at the spacing of 1.1 $\times$ 0.4. The experimental design was a randomized complete block with three replications in a 4×4 factorial scheme with four N concentrations (8, 12, 16, and 20 mmol L^{-1}) and four K concentrations (4, 6, 8, and 10 mmol L^{-1}). The experimental plot constituted of eight plants. It was observed that the leaf levels of N and K, of $N\text{-}NO_3$ and of K, and the electrical conductivity (CE) of the substrate increased with the increment of N and K in the nutrients' solution. Substratum pH, in general, was reduced with increments in N concentration and increased with increasing K concentrations in the nutrients' solution. Leaf area increased with increments in N concentration in the nutrients solution. Fertigation with solutions stronger in N (20 mmol L^{-1}) and K (10 mmol L^{-1}) resulted in higher masses for the first (968 g) and the second (951 g) fruits and crop yield (4,425 $g m^{-2}$).

1. Introduction

The fruit of the melon plant is highly appreciated all over the world. In 2010, among fruits in general, melon was the first item of the Brazilian exportations. According to Agrianual [1] data, 177,829 tons of melons were exported. The main melon producing area is in the semiarid region of the northeast of Brazil which supplies all the melons for the internal market. Simultaneously in the last few years, the production of noble melons (*Cucumis melo* L., *cantalupensis* group) is growing steadily in the southeast region, already representing 15 to 20% of the market [2].

In the Southeast region, melon is cultivated mainly under protected conditions due to pluviosity in that region being frequently very high. The cultivation of net melon over the soil surface has met several phytosanitary problems so that alternative cultivation procedures have been sought. One of

the most important ones is that melon plants are cultivated in a substratum which, if properly managed, permits yields superior to those of the cultivation on soil [3].

Nowadays, cultivating in substrata has its management founded on fertigation and drainage of a certain percentage of the applied volume in order to keep substratum conditions adequate for the crop [4, 5]. On the other hand, this management causes a high residual volume, not used by the plants, which, when discarded, can contaminate the soil and water fountainheads [6, 7]. In addition to that, it is necessary to consider the direct influence on the production costs since what is being discarded is a part of the nutrients solution [8]. So, this management is to be used when the water for the preparation of the nutrients solution has a high electric conductivity. In Brazil, in the majority of the agricultural regions, the water is of excellent quality. So, the challenge to be overcome is the development of a management for the cultivation in coconut fiber in order to maintain the nutrients solution concentrations and the substratum chemical attributes at adequate levels. The lack or excess of the nutrients solution may affect the growth and the productivity of the plants [9].

K and N are the macronutrients that melon plants extract more—N, approximately 38% and K, approximately 45% of total of nutrients [10]. On the other hand, according to information in the literature, the concentrations of those two nutrients in the nutrients solution vary in function of cultivar, substratum, the climatic conditions, the forms and frequency with which water and nutrients are supplied, and the plant physiological stage among other factors which have influence on the growth and the mineral composition of the plant [11–13]. Studies concerning aspects of the nutrients solution in which melon plants are cultivated in coconut fiber without drainage were not found. Since there is no defined criterion for the use of fertigation in that system, the recommended N and K concentrations, when drainage is part of the management system, may not be the adequate concentrations when drainage is not part of that system.

So, the objective of this investigation was to evaluate the most efficient N and K concentrations in the nutrients solution when melon plants are cultivated in coconut fiber without drainage.

2. Materials and Methods

The experiment was carried out from July 1, 2011 to January 3, 2012, at the South Federal Institute of Minas Gerais, in Muzambinho (South latitude of $21°22'33''$, West longitude of $46°31'33''$, and at a mean altitude of 1000.75 m above sea level), in Muzambinho, Minas Gerais State, Brazil. The minimum, maximum, and mean temperatures during the experimental period were, respectively, 15.5, 35.6, and 25.0°C. A 30 × 14 m, 4 m high greenhouse, covered with a low density, 150 μm thick polyethylene film, with lateral and frontal shutting up to 3 m with a polypropylene screen which resulted in a shading of 30% was used.

Four N concentrations ($N_1 = 8$, $N_2 = 12$, $N_3 = 16$, and $N_4 = 20$ mmol L^{-1}) and four K concentrations ($K_1 = 4$, $K_2 = 6$, $K_3 = 8$, and $K_4 = 10$ mmol L^{-1}) were used; this resulted in 16 treatment combinations. These treatment combinations, each repeated 3 times, were arranged in the greenhouse according to a randomized complete block design. The melon genotype used in this experiment was the "Bonus no. 2," a F_1 hybrid belonging to the *reticulatus* botanical variety of the "net" type.

Seed sowing took place on September 21, 2011 in 128-celled expanded polystyrene trays, these cells being filled with coconut fiber powder. The trays were daily fertigated till the appearing of the first noncotyledonary leaf. The nutrients solution for this period of fertigation had half the concentrations of nutrients.

Fifteen days after sowing (DAS), the seedlings were transplanted to the cultivation channels at the spacing of 1.10 m between lines and 0.40 m between plants. Each channel was 0.2 m wide, 0.19 m high, and 3 m long. The channel was covered with a double-face polyethylene film and filled with 17.2 kg of the Golden-Mix 80 coconut fiber substratum,

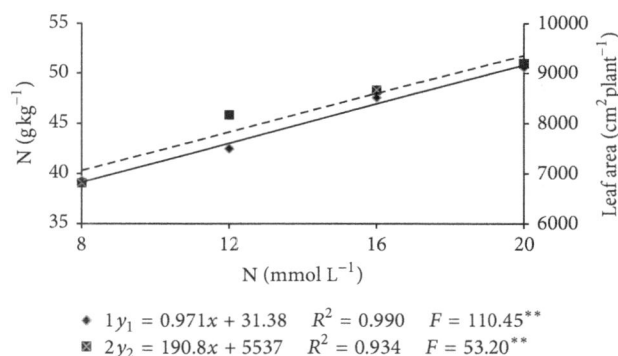

FIGURE 1: N content in the leaf used for the evaluation of the plant nutritional status 46 days after transplantation (DAT) (Y1) and leaf area 89 DAT (Y2) as influenced by the concentrations of N in the nutrients solution.

this amount meaning 0.01425 dm^{-3} per plant. This substratum results from a fifty-fifty mixture of coarse-texture and fine-texture substrata. The substratum on which the plants grew had the following chemical features: pH = 6.7, EC = 0.2 dS m^{-1}, and, in mg L^{-1}, 0.3 of N-NO$_3^-$, 1.2 of N-NH$_4^+$, 24.8 of K, 2.3 of P, 0.4 of Ca, 0.1 of Mg, and 0.03 of Zn. The maximum water retention capacity displayed by the substratum was 356 mL L^{-1}.

The substratum top was covered with a polyethylene film with the white side upwards. The cultivation channels were placed at the same level on top of the terrain with their extremities closed by the same film. The plants, supported by stalks, grew vertically. The basal secondary branches were pruned up to the tenth node, and, after that, the plants were allowed to grow freely to support the fruits. The pollination was carried out by bees with free access to the plants until there were four fruits per plant. After that, the plant buds were eliminated and two fruits suppressed so that there were two fruits per plant, one at the 11th node and the other at the 13th node. The apical buds of secondary branches, located after the third leaf, were eliminated and, if on the main branch, after the 22nd node. The first and the second fruits borne by each plant were harvested 83 and 89 days after transplantation (DAT), respectively. Preventive and curative measures were taken for phytosanitary reasons.

A drip irrigation system was adopted by the use of antidrainage and self-compensating emitters placed 0.4 m one from another with a uniformity of 99% and a water flow of 4.5 L h^{-1} under the operating conditions.

The water used to prepare the nutrients solution had the following characteristics: pH = 6.6 and, in mg L^{-1}, 0.47 of Zn, 0.0067 of Cu, 6.8 of chlorides, 0.30 of Fe, 0.0004 of N-NO$_2^-$, 0.5 of N-NO$_3^-$, 0.0019 of N-NH$_4^+$, 14.6 of S, 1.4 of B, and absence of chlorine, organic N, phosphorus, orthophosphate, potassium, sodium, calcium, and magnesium.

The characteristic curve of water retention by the substratum was determined for water tensions in the substratum up to 100 kPa. Fertigation was controlled by sensors of the Irrigas matricial tension type to be used in substrata capable of measuring tensions between 0 and 15 kPa. Fertigation

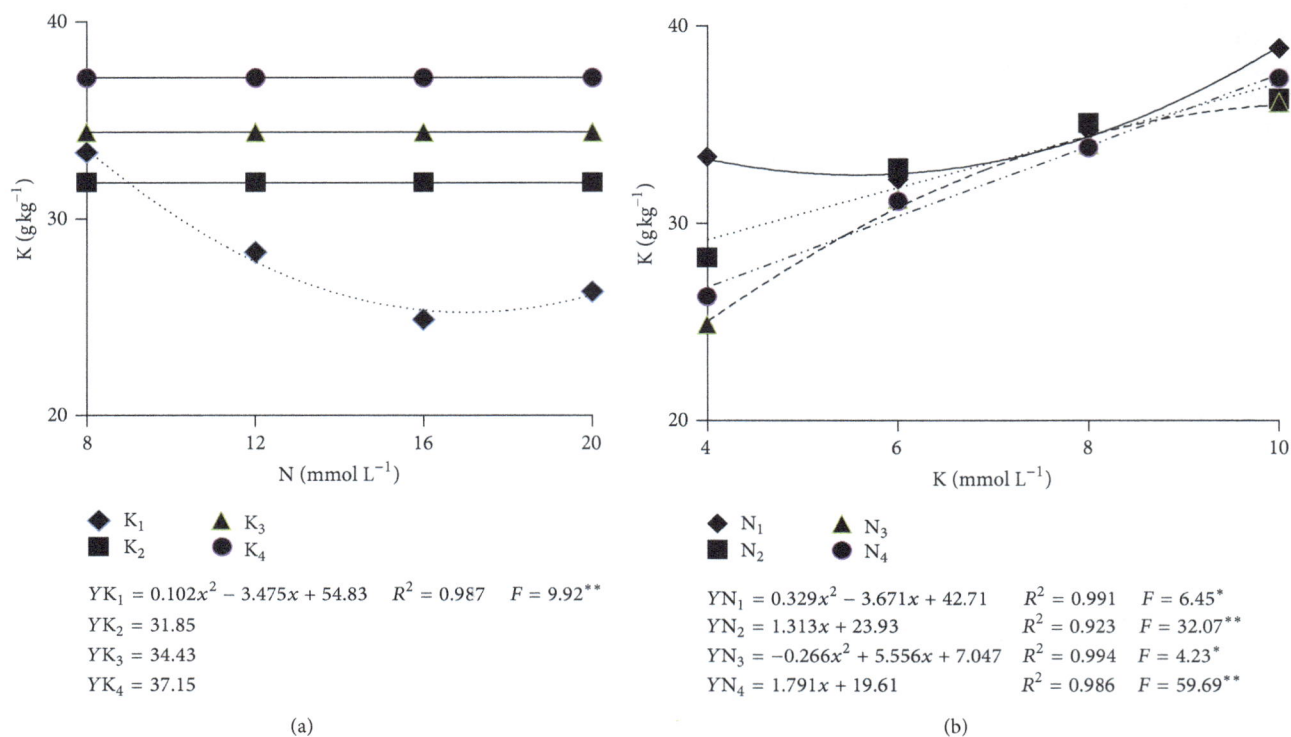

FIGURE 2: K content in the leaf used for the evaluation of the nutritional status of "Bonus no. 2" melon plants as influenced by N (a) and K (b) concentrations in the nutrients solution. **, *: Significant at the levels of 1 and 5%, respectively, according to the F test.

$YK_1 = 0.102x^2 - 3.475x + 54.83 \quad R^2 = 0.987 \quad F = 9.92^{**}$
$YK_2 = 31.85$
$YK_3 = 34.43$
$YK_4 = 37.15$

$YN_1 = 0.329x^2 - 3.671x + 42.71 \quad R^2 = 0.991 \quad F = 6.45^{*}$
$YN_2 = 1.313x + 23.93 \quad R^2 = 0.923 \quad F = 32.07^{**}$
$YN_3 = -0.266x^2 + 5.556x + 7.047 \quad R^2 = 0.994 \quad F = 4.23^{*}$
$YN_4 = 1.791x + 19.61 \quad R^2 = 0.986 \quad F = 59.69^{**}$

(a)   (b)

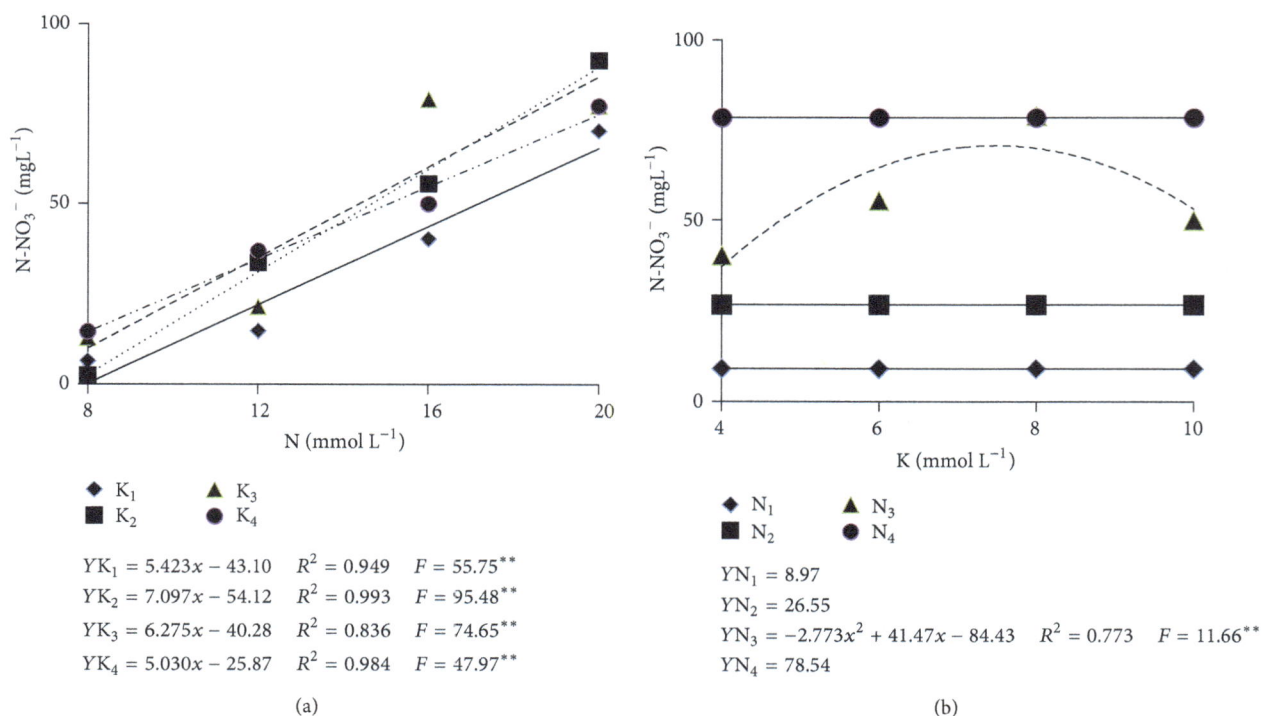

$YK_1 = 5.423x - 43.10 \quad R^2 = 0.949 \quad F = 55.75^{**}$
$YK_2 = 7.097x - 54.12 \quad R^2 = 0.993 \quad F = 95.48^{**}$
$YK_3 = 6.275x - 40.28 \quad R^2 = 0.836 \quad F = 74.65^{**}$
$YK_4 = 5.030x - 25.87 \quad R^2 = 0.984 \quad F = 47.97^{**}$

$YN_1 = 8.97$
$YN_2 = 26.55$
$YN_3 = -2.773x^2 + 41.47x - 84.43 \quad R^2 = 0.773 \quad F = 11.66^{**}$
$YN_4 = 78.54$

(a)   (b)

FIGURE 3: N-NO$_3^-$ levels (mg L^{-1}) in the substratum solution extract (method 1 : 1.5 v/v) 60 days after transplantation as influenced by the concentrations of N (a) and of K (b) in the nutrients solution. **: Significant at the level of 1%, according to the F test.

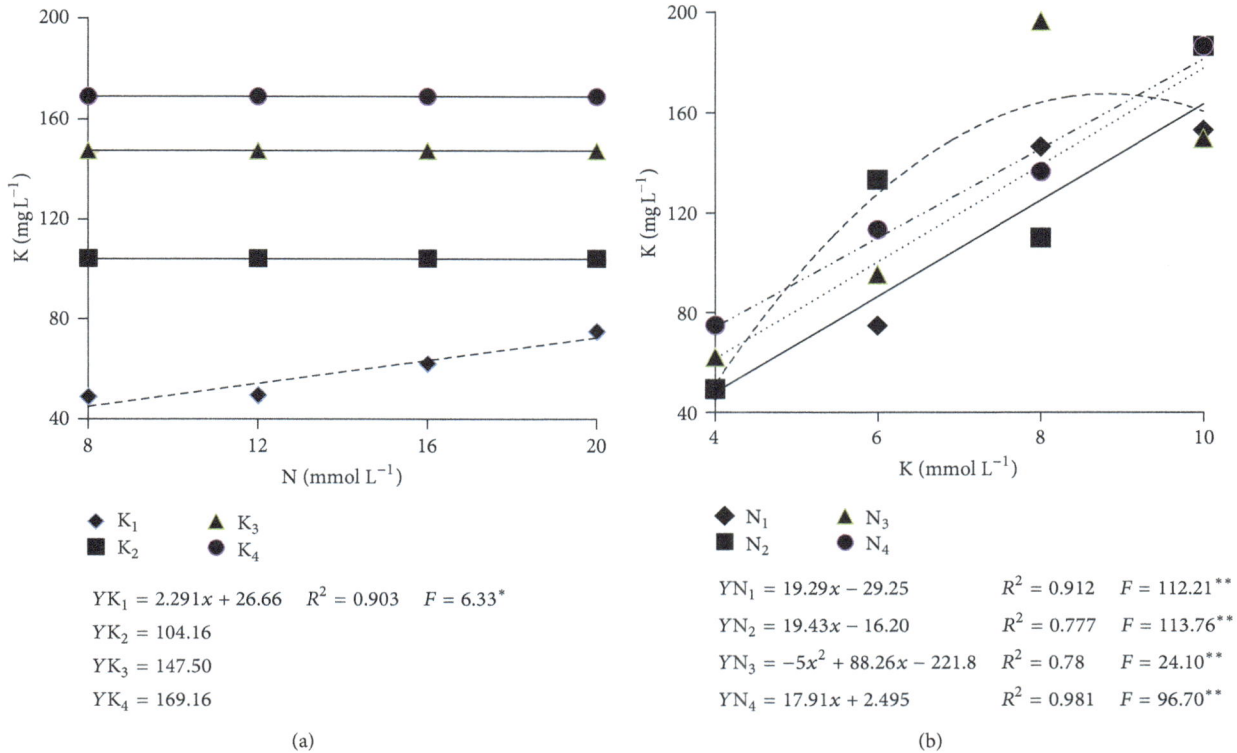

FIGURE 4: K levels ($mg\,L^{-1}$) in the substratum solution extract (method $1:1.5$ v/v) 60 days after transplantation as influenced by the concentrations of N (a) and of K (b) in the nutrients solution. $**$, $*$: Significant at the levels of 1 and 5%, respectively, according to the F test.

TABLE 1: Plant cycle in days after transplantation (DAT), maximum water tension in the substratum (MWT), and nutrients solution volume (SNV), as influenced by N concentration in the nutrients solution.

Cycle (DAT)	MWT (kPa)	SNV (L)
0–12	1.0	3.12
13–33	1.0	6.60
34–52	2.0	23.18
53–59	4.0	11.07
60–89	5.0	31.96
		Total = 75.95

was started soon after the plants were transplanted up to harvest with a frequency which was determined by the climatic conditions and the melon plant phonological stage. The irrigation duration was of 3 minutes and the substratum moisture tension to initiate fertigation was variable during the plant cycle (Table 1).

N and K levels in the leaves were evaluated making use of the 5th leaf starting from the tip of the branch, excluding the apical tuft, at the beginning of fruit set, which took place 46 DAT [14]; 60 DAT the pH (pH meter DIGIMED—model DM 21), the electrical conductivity (TECNAL—model TEC 4MP conductivimeter), the concentrations of K (Compaction ion meter C-131 Horiba Cardy), and N (distillation method)

were also determined. In order to make those measurements, four samples per plot, at a distance of 0.10 m from the plant stem, between 6:30 and 7:30 a.m., were taken. These samples were mixed to make a composed sample. The analyses were made using the $1:1.5$ extraction method [15]. Leaf area was measured at the end of the cycle (89 DAT) by measuring the width of all leaves of the plant and making use of the mathematical model ($AF = 0.826\,L^{1.89}$ ($R^2 = 0.97$)) to calculate it [16]. First and second fruits had their weight determined and also total yield in kilograms of fruits per plant.

The data were submitted to the analysis of variance by the F test and the polynomial regression analysis. The data related to the first and second fruits weight and the total yield per plant were analyzed by a regression study by the response surface methodology analysis.

3. Results and Discussion

Measurement made 46 DAT showed that leaf N content was significantly influenced only by the N concentration in the nutrients solution (Table 2) with the means showing adjustment to a first-degree equation (Figure 1). An increment of 30% in the N level (39.1 and 50.8 $g\,kg^{-1}$ of N) was verified when the nutrients solution used to fertigate the plants had the smallest and the highest N concentration, respectively.

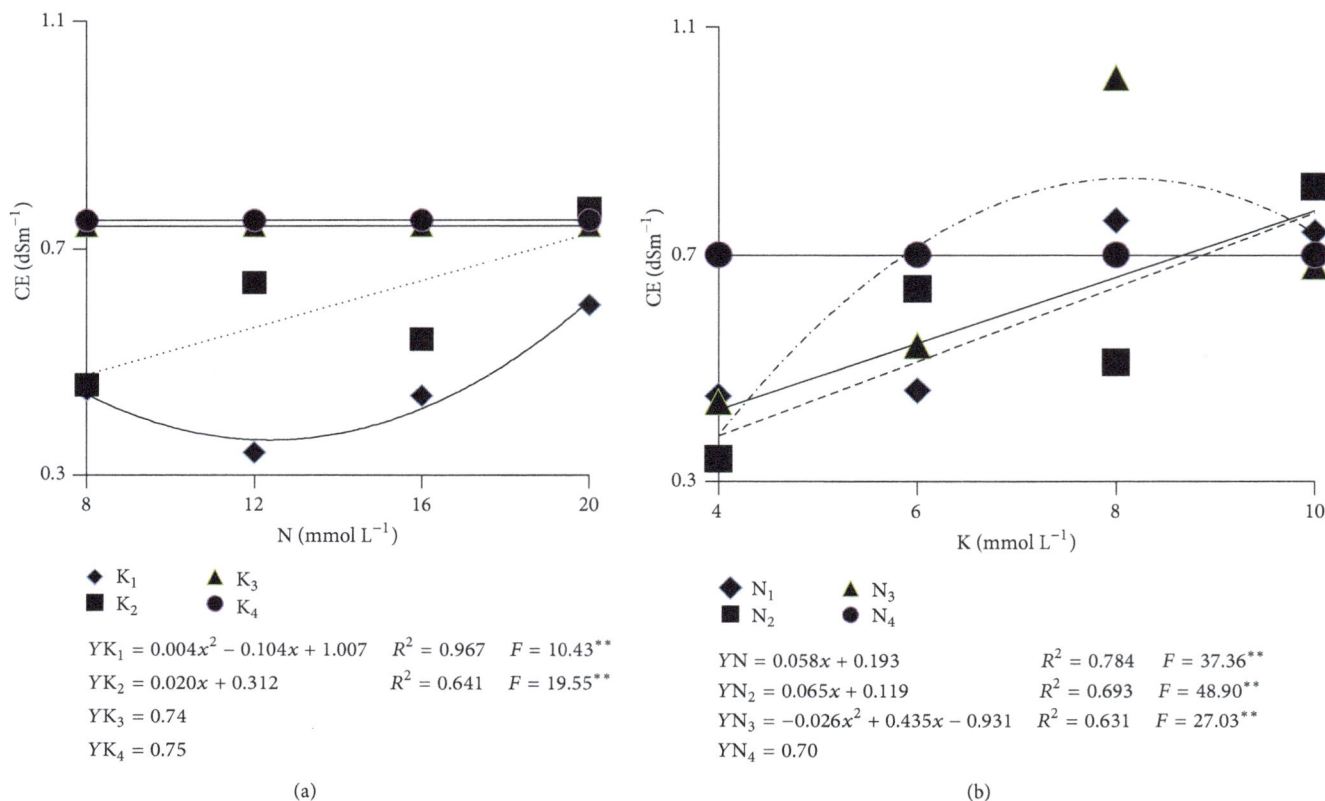

$YK_1 = 0.004x^2 - 0.104x + 1.007$ $R^2 = 0.967$ $F = 10.43^{**}$

$YK_2 = 0.020x + 0.312$ $R^2 = 0.641$ $F = 19.55^{**}$

$YK_3 = 0.74$

$YK_4 = 0.75$

(a)

$YN = 0.058x + 0.193$ $R^2 = 0.784$ $F = 37.36^{**}$

$YN_2 = 0.065x + 0.119$ $R^2 = 0.693$ $F = 48.90^{**}$

$YN_3 = -0.026x^2 + 0.435x - 0.931$ $R^2 = 0.631$ $F = 27.03^{**}$

$YN_4 = 0.70$

(b)

FIGURE 5: Electrical conductivity of the substratum solution extract (method 1 : 1.5 v/v) 60 days after transplantation as influenced by the concentrations of N (a) and K (b) in the nutrients solution. **: Significant at the level of 1%, according to the F test.

TABLE 2: Analysis of variance results for nitrogen leaf level (LN) and potassium leaf level (LK), $N\text{-}NO_3^-$, hydrogen ion potential (pH), electrical conductivity (CE), and N (NS) and K (KS) concentrations in the solution of substratum.

Treatments	LN	LK	pH	CE	NS	KS
	(g kg^{-1})		(dS m^{-1})		(mg L^{-1})	
N						
N_1	39.31	34.80	5.30	0.60	8.97	105.83
N_2	42.51	33.13	5.48	0.58	26.55	119.83
N_3	47.52	31.57	5.07	0.67	56.12	126.00
N_4	50.59	32.16	4.77	0.70	78.54	127.92
F	37.17**	7.43**	74.05**	7.29**	90.44**	6.01**
K						
K_1	45.12	28.22	4.96	0.46	32.82	58.75
K_2	45.03	31.86	5.14	0.60	45.24	104.17
K_3	44.12	34.43	5.23	0.74	47.57	147.50
K_4	45.67	37.15	5.29	0.75	44.56	169.17
F	0.60^{NS}	53.84**	16.44**	42.15**	4.15*	144.11**
$N \times K$	0.93^{NS}	2.63*	2.99*	11.86**	2.39*	10.58**
C.V. (%)	6.37	5.46	2.39	11.50	26.45	11.77

**, *, NS: Significant at the levels of 1 and 5% and non significant, respectively, according to the F test.

The K leaf content was influenced by the interaction of the factors (Table 2). Significant adjustments to first- and second-degree equations were verified in accordance with the N and K combination (Figure 2). While in the nutrients solution of 4 mmol L^{-1} of K the increase in N concentration resulted in a reduction in the level of K in the leaves. In the concentrations of 6, 8, and 10 mmol L^{-1} of K, no significant adjustment to a polynomial equation was observed and resulted in the levels of 31.9, 34.4, and 37.2 g kg^{-1} of K in the leaves, respectively (Figure 2(a)). On the other hand, in all N concentrations increments in K concentration were verified with each increment in K concentration, and the highest levels were observed when the nutrients solution had the lowest concentration of N (Figure 2(b)), this being explained by the lower growth of leaf area with the lowest concentrations of N (Figure 1).

Among the solutions with the lowest and the highest concentrations of K in the nutrients solution (4 and 10 mmol L^{-1}), an increment of 55.6% in the K leaf content (25.0 and 38.9 g kg^{-1}) was verified. Even with all the variation observed in the levels of K and N resulting from the different concentrations of these elements in the nutrients solution, all the leaf levels of N and K were verified to be within the range of values considered as adequate of 15 to 50 g kg^{-1} for N and of 25 to 40 g kg^{-1} for K [14, 17].

The N-NO$_3^-$ level in the substratum was significantly influenced by the interaction between the concentrations of N and K (Table 2). For all K concentrations, the higher the concentration of N in the nutrients solution, the higher the level of N-NO$_3^-$ in the substratum. The lowest level, that is, 0.28 mg L^{-1}, resulted from the lowest concentrations of N and K whereas the highest, that is, 87.8 mg L^{-1}, resulted from 20 of N and 6 mmol L^{-1} of K (Figure 3(a)). On the other hand, the breaking down of the interaction degrees of freedom for the level of N-NO$_3^-$ as a function of the N concentration in each K concentration, showed that a significant adjustment was verified only for the nutrients solution containing 16 mmol L^{-1} of N with the maximum level (70 g L^{-1}) resulting from 7.5 mmol L^{-1} of K. In the solutions with 8, 12, and 20 mmol L^{-1} of N, nitrate means in the substratum were 9.0, 26.6, and 78.5 mg L^{-1}, respectively (Figure 3(b)). The range from 4 to 6 mmol L^{-1} or 56 to 84 mg L^{-1} is adequate for the majority of crop species growing in organic substrata [18]. It is, thus, observed that only the nutrients solution with at least 16 mmol L^{-1} of N could result in N-NO$_3^-$ levels in the substratum within the range mentioned by that author.

The increment in N-NO$_3^-$ availability in the substratum may be explained by the concentration of N in the nutrients solution and by the demand of the nutrient by melon plants. The levels of N-NO$_3^-$ in the substratum solution extract were low when N and K concentrations were at their lowest (8 and 4 mmol L^{-1}). In this case, since the N concentration in the nutrients solution is low, the N added to the substratum via fertigation was used by the plant in almost its totality, this resulting in the lowest availability of N-NO$_3^-$ in the substratum. This same observation was reported by Gaion et al. [19] between 45 and 78 DAT, when the solution used had a lower concentration of N (12 mmol L^{-1}), and the reduction was attributed to the high N demand by fructifying melon plants.

The substratum K content was significantly influenced by the interaction between N and K concentrations (Table 2). The level of K as influenced by the N concentration only showed a significant equation adjustment for the solution with 4 mmol L^{-1} of K in which the K level in the substratum increases linearly with the increment in the concentration of N, reaching a maximum of 72.5 mg L^{-1}. In the nutrients solutions in which the K concentrations were of 6, 8, and 10 mmol L^{-1}, the mean K concentrations in the substratum were of 104.2, 147.5, and 169.2 mg L^{-1}, respectively (Figure 4(a)). This increment in K level in the substratum is observed in the significant adjustments of equations for each N concentration as influenced by K concentrations in the nutrients solution (Figure 4(b)).

The highest level of K in the substratum (181.6 mg L^{-1}) was observed with fertigation of the nutrients solution which had the highest concentrations of N and 10 mmol L^{-1} of K whereas the solution with 4 mmol L^{-1} of K resulted in K level in the substratum solution extract lower than the range considered as adequate by Baumgarten [18]. According to that author, for the majority of the crop species, the ideal range of K is between 1.9 and 3.5 mmol L^{-1} or 74.3 and 136.8 mg L^{-1} when the dilution extraction method 1 : 1.5 v/v is used. On the other hand, when the solutions with 8 and 10 mmol L^{-1} of K are used, the levels found in the substratum solution extract are higher than those of the optimum range.

The substratum electrical conductivity (CE) was significantly influenced by the interaction of the concentrations of N and K (Table 2). The mean CE values for the substratum fertigated with solutions with 8 and 10 mmol L^{-1} of K did not adjust to polynomial equations in response to the increments in N concentrations and showed very similar results—0.74 and 0.75 dS m^{-1}. Different from that, when the substratum was fertigated with nutrients solution with the lowest concentrations of K, an increment in the extract CE was verified as the concentrations of N increased (Figure 5(a)).

When the substratum solution extract CE study was undertaken as influenced by increments in K concentration in the nutrients solution, polynomial equations were not verified to describe the observed effect only when the N concentration was the highest (20 mmol L^{-1}) whereas the responses to the nutrients solutions with 8 and 12 mmol L^{-1} adjusted themselves linearly to increments of K in the nutrients solution. The solution with 16 mmol L^{-1} of N resulted in the highest substratum CE when containing 8 mmol L^{-1} (Figure 5(b)). Therefore, between the nutrients solution with higher or lower concentrations of N and K, the substratum solution extract saline index underwent a more than twofold increment—it went from 0.33 to 0.89 dS m^{-1} (Figure 5(b)). This increment can be partially attributed to the augments in the levels of nitrate and potassium in the substratum.

Only when nutrients solutions containing at least 8 mmol L^{-1} of K, independently of the N concentration, or solutions

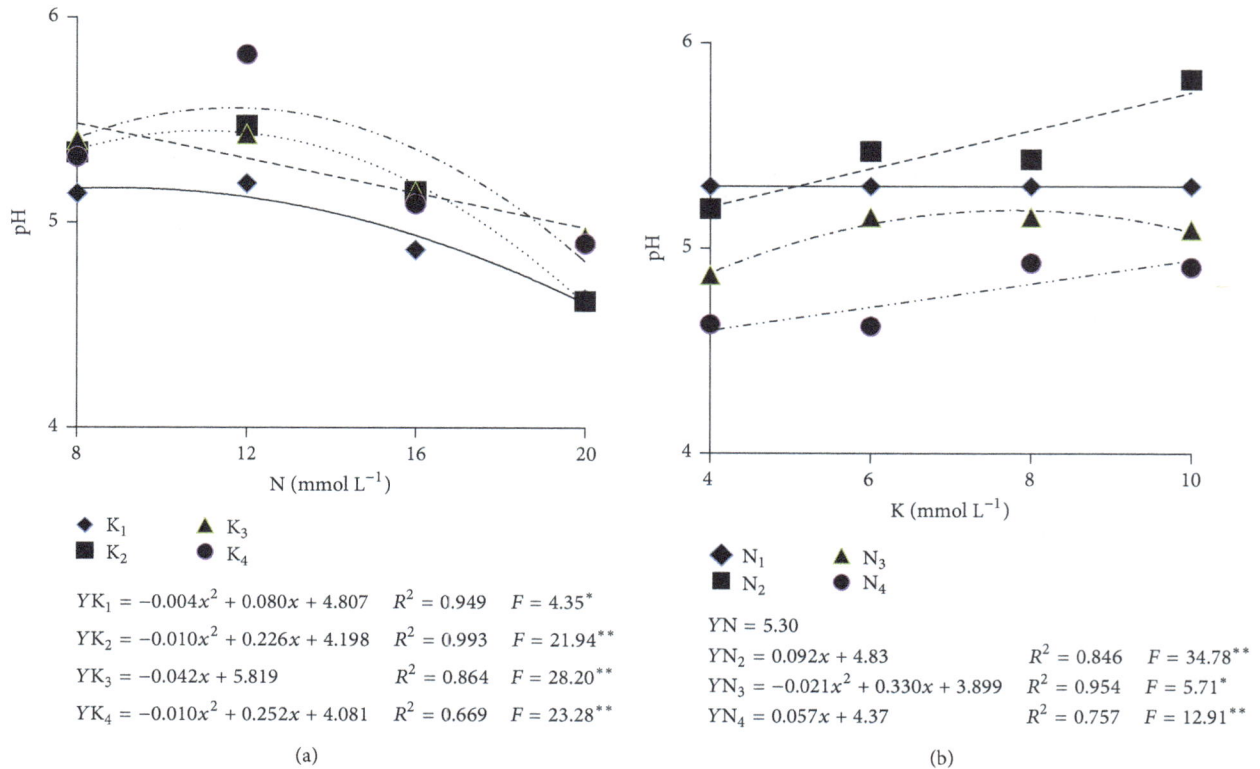

$YK_1 = -0.004x^2 + 0.080x + 4.807 \quad R^2 = 0.949 \quad F = 4.35^*$

$YK_2 = -0.010x^2 + 0.226x + 4.198 \quad R^2 = 0.993 \quad F = 21.94^{**}$

$YK_3 = -0.042x + 5.819 \quad\quad\quad R^2 = 0.864 \quad F = 28.20^{**}$

$YK_4 = -0.010x^2 + 0.252x + 4.081 \quad R^2 = 0.669 \quad F = 23.28^{**}$

(a)

$YN = 5.30$

$YN_2 = 0.092x + 4.83 \quad\quad\quad\quad R^2 = 0.846 \quad F = 34.78^{**}$

$YN_3 = -0.021x^2 + 0.330x + 3.899 \quad R^2 = 0.954 \quad F = 5.71^*$

$YN_4 = 0.057x + 4.37 \quad\quad\quad\quad R^2 = 0.757 \quad F = 12.91^{**}$

(b)

FIGURE 6: pH values of the substratum solution extract (method 1 : 1.5 v/v) 60 days after transplantation as influenced by the concentrations of N (a) and K (b) in the nutrients solution. **, *: Significant at the levels of 1 and 5%, respectively, according to the F test.

with 16 mmol L^{-1} of N and a minimum of 6 mmol L^{-1} of K or 20 mmol L^{-1} of N and any K concentration, the substratum CE reached values close to the inferior limit of the range considered adequate by Baumgarten [18], that is, from 0.8 to 1.5 dS m^{-1} and by Cavins et al. [20], that is, from 0.76 to 1.5 dS m^{-1} as determined by the extraction methods by dilution 1 : 1,5 v/v and 1.2 v/v, respectively.

Nutrients solutions with concentrations of N and K lower than those mentioned previously resulted in very low substratum solution extract CE values. Therefore, the adopted management of closed channels without drainage of part of the nutrients solution applied for the lixiviation of nutrients, did not cause the feared substratum salinization probably due to the water quality, which was very poor in ions.

The substratum pH was significantly influenced by the interaction between N and K concentrations (Table 2). For each K concentration in the nutrients solution as N concentration increased, there was a reduction in pH (Figure 6(a)).

In the study concerning pH values as influenced by augments in K for each N concentration, increments in pH values were verified with increasing K concentrations (Figure 6(b)). The lowest pH value was 4.7, which resulted from the fertigation with a solution containing 20 and 6 mmol L^{-1} of N and K, respectively. The highest pH value was 5.7, which resulted from nutrients solution containing 12.6 and 10 mmol L^{-1} of N and K, respectively.

It is probable that the augment in pH was a consequence of the fertilizers used to prepare the formulations. In the nutrients solution containing 8 and 12 mmol L^{-1} of N, the source of K was potassium chloride whereas in the 16 and 20 mmol L^{-1} of N, the K source was potassium nitrate. It is possible that the higher chloride concentration in the solutions with the lowest amount of N may have induced a larger absorption of this anion and, therefore, an increment in hydroxide (OH$^-$) concentration in the substratum, and in raising pH values [21]. According to data published by Martinez [22], the substratum pH values increase when the plants are in high demand of nitrate. In this case, however, the ion nitrate was of the same concentration in all the nutrients solutions.

N concentrations above 12 mmol L^{-1} caused the pH to decrease and reach the lowest value when the highest N concentration was used (20 mmol L^{-1}). pH values were 4.8, 4.7, 5.0, and 5.1 for the K concentrations of 4, 6, 8, and 10 mmol L^{-1}, respectively. Gaion et al. [19] reported pH values lowering from 6.3 to 5.0 during melon plants cycle in which the plants grew in a mixture of sand and peanut shell. Such reduction may be due to the plant extracting cations of basic nature (Ca, Mg, K, and Na), and this resulted in the increment of H$^+$ [23] or to the excessive use of ammoniacal fertilizers. On the other hand, this hypothesis is not applicable to this work since the concentration of N-NH$_4^+$ in the nutrients solution did not reach values above 10%. On the other hand,

$$MFF = 1028.1784 - 23.5779x - 36.9433y + 0.9974x^2$$
$$+ 1.0233xy + 1.7708y^2$$

$$MSF = 784.2265 - 15.4514x + 12.4346y + 0.8867x^2$$
$$+ 0.3804xy - 0,7969y^2$$

(g plant^{-1})

--- 957.236	·--·- 910	—— 870	--- 830
—— 940	--- 900	········· 860	······ 820
········· 930	······ 890	·--·- 850	--- 810
·····- 920	--- 880	·--·- 840	—— 800

(a)

(g plant^{-1})

--- 946	········· 880	······ 820
······ 933	·--·- 860	--- 800
--- 920	·--·- 850	—— 780
—— 900	--- 840	

(b)

$$MF = 1.8077 - 0.0387x - 0.0227y + 0.0018x^2 +$$
$$0.0018xy + 0.0006y^2$$

(kg plant^{-1})

·--·- 1.8947	—— 1.7894	--- 1.6841
--- 1.8684	········· 1.7631	······ 1.6578
········· 1.8421	·--·- 1.7368	--- 1.6315
--- 1.8157	·--·- 1.7104	—— 1.6051

(c)

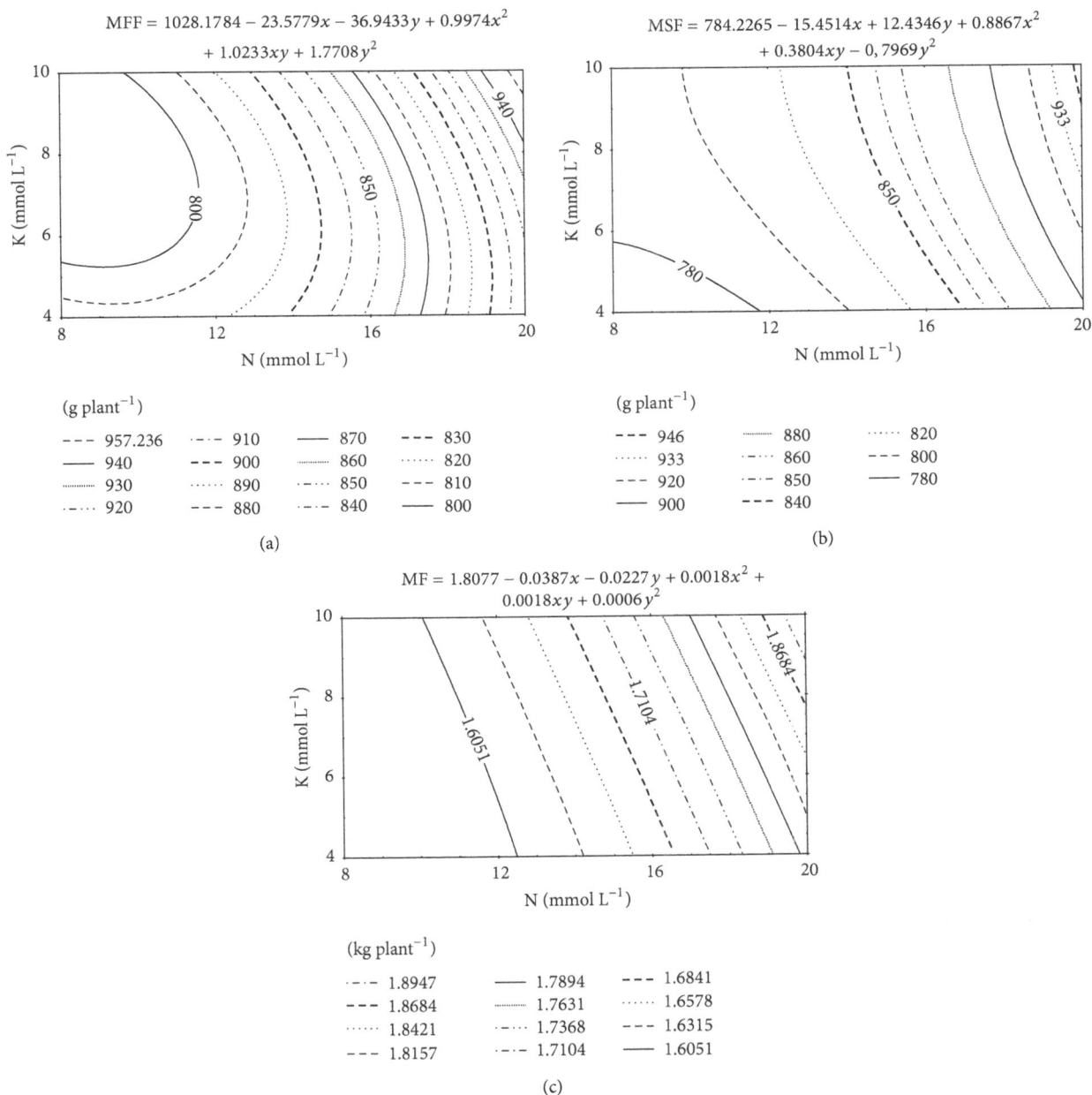

FIGURE 7: Mass of the first fruit (MFF) (a), of the second fruit (MSF) (b), and of fruits (MF) (c) as influenced by the concentrations of N and K in the nutrients solution.

the first hypothesis is more likely to explain the results at 38 DAT the first feminine flowers sprouted and at 41 DAT fructification started, and this caused an increment in nutrients demand by the plants, specially for potassium as reported by Lester et al. [24]. As a consequence, the roots excrete more H$^+$, thus reducing pH.

The pH values observed in this work are very close to the pH range recommended by several authors for organic substrata: 6.19 [18], 5.7 to 6.0 [23], 5.4 to 6.0 [25], 5.8 to 6.2 [26], and 5.4 to 6.4 [27]. Notwithstanding, the lowest value found (4.7) is below the values recommended by those

authors. pH values were not measured at the end of the cycle, and the analysis of the result at 60 DAT shows that the value at 89 DAT, when harvesting was made, may have reached critical values.

Melon plants leaf area (AF) was affected only by N concentration (Table 3). Leaf area measurements adjusted to first-degree equations and increased linearly with N concentration in the nutrients solution (Figure 1). Leaf areas resulting from N concentrations of 8 and 20 mmol L^{-1} in the nutrients solution were of 7,063 and 9,353 cm^2 per plant, respectively. The expressive effect of N on melon plant growth

TABLE 3: Analysis of variance results for leaf area (LA), mass of the first fruit (FFM), mass of the second fruit (SFM), and fruit productivity per plant (P) as influenced by the concentrations of N and K in the nutrients solution.

Treatments	LA ($cm^2 plant^{-1}$)	FFM (g)	SFM (g)	P (g)
N				
N_1	309.86	789.83	776.58	1568.33
N_2	371.11	835.50	820.75	1657.50
N_3	393.78	834.25	832.33	1668.33
N_4	417.99	943.75	933.25	1870.00
F	18.98**	39.33**	33.54**	47.01**
K				
K_1	379.56	847.33	813.00	1654.16
K_2	367.26	853.75	851.50	1705.00
K_3	374.83	833.75	836.33	1672.50
K_4	371.09	868.50	862.08	1732.50
F	0.24^{NS}	1.91^{NS}	3.47*	3.49*
N × K	0.57^{NS}	1.96^{NS}	1.34^{NS}	1.92^{NS}
C.V. (%)	9.87	4.25	4.71	3.81

**, *, NS: Significant at the levels of 1 and 5% and non significant, respectively, according to the F test.

has been intensively reported in the literature [13, 28, 29]. The highest melon plants AF were verified when the highest N and N-NO_3^- leaf levels were found in the substratum solution extract (Figure 2(a)).

The mass of the first melon fruit (MPF) was significantly affected by N concentration whereas the mass of the second fruit (MSF) and productivity (P) were influenced not only by N concentration but also by K concentration (Table 3).

According to a response surface methodology analysis, the lowest MPF (817 g, according to Figure 7(a)) and MSF (766 g, as shown in Figure 7(b)) values were verified when the melon plants were fertigated with a nutrients solution with the lowest concentrations of N and K (8 and $4 \, mmol \, L^{-1}$). As the concentrations of N and K were increased, higher MPF and MSF were attained. MPF and MSF increased by 18.5 and 24.0% reaching 968 and 951 g, respectively.

Productivity (Figure 7(c)) mirrored what happened with MPF and MSF. An increment in productivity of 22.5% (total fruit weight went from 1,589 g to 1,947 g per plant, or, 3.611 kg to $4.425 \, kg \, m^{-2}$) was observed when the solutions with the lowest and the highest concentrations in N and K were used.

Considering the K concentration of $10 \, mmol \, L^{-1}$, which was the highest of this element, the increment in N concentration from 8 to $20 \, mmol \, L^{-1}$ resulted in an increment of 356 g in productivity, that is, 29.7 g for each $1 \, mmol \, L^{-1}$ of N added to the nutrients solution.

On the other hand, keeping the concentration of $20 \, mmol \, L^{-1}$ of N and increasing the K concentration from 4 to $10 \, mmol \, L^{-1}$, the increment was of 131 g, or, 21.8 g for each $mmol \, L^{-1}$ of K added to the nutrients solution. This analysis makes clear that the effect of N on fruit productivity is larger than that of K. This is probably due to the effect N has on plant leaf area. The larger productivity did also correlate with the largest N and K leaf contents as well as with the levels of these nutrients in the substratum solution extract.

Therefore, higher concentrations of N and K in the nutrients solution in the interval from 8 to $20 \, mmol \, L^{-1}$ of N and from 4 to 10 of K cause increased levels of these nutrients in the plant leaves and in the substratum, of electrical conductivity, of leaf area, fruit mass, and productivity.

References

[1] Agrianual, *Anuário da Agricultura Brasileira*, FNP Consultoria e Comércio, São Paulo, Brasil, 2012.

[2] H. C. O. Charlo, R. Castoldi, P. F. Vargas, and L. T. Braz, "Desempenho de híbridos de melão-rendilhado cultivados em substrato," *Científica*, vol. 37, no. 1, pp. 16–21, 2009.

[3] J. L. Andriolo, G. L. Luz, O. C. Bortolotto, and R. S. Godoi, "Produtividade e qualidade de frutos de meloeiro cultivado em substrato com três doses de solução nutritiva," *Ciência Rural*, vol. 35, no. 4, pp. 781–787, 2005.

[4] B. Bar-Yosef, "Advances in fertigation," *Advances in Agronomy*, vol. 65, no. C, pp. 1–77, 1999.

[5] E. Gorbe and H. E. P. Calatayud, "Optimization of nutrition in soilless systems: a review," *Advances in Botanical Research*, vol. 53, no. 1, pp. 193–245, 2010.

[6] S. Jiménez, J. I. Alés, M. T. Lao, B. Plaza, and M. Pérez, "Evaluation of nitrate quick tests to improve fertigation management," *Communications in Soil Science and Plant Analysis*, vol. 37, no. 15–20, pp. 2461–2469, 2006.

[7] W. Bres, "Estimation of nutrient losses from open fertigation systems to soil during horticultural plant cultivation," *Polish Journal of Environmental Studies*, vol. 18, no. 3, pp. 341–345, 2009.

[8] A. O. Onanuga, P. Jiang, and A. Sina, "Residual level of phosphorus and potassium nutrients in hydroponically grown cotton

(*Gossypium hirsutum*)," *Journal of Agricultural Science*, vol. 4, no. 5, pp. 149–160, 2012.

[9] E. Dogan, H. Kirnak, K. Berekatoglu, L. Bilgel, and A. Surucu, "Water stress imposed on muskmelon (*Cucumis melo* L.) with subsurface and surface drip irrigation systems under semi-arid climatic conditions," *Irrigation Science*, vol. 26, no. 2, pp. 131–138, 2008.

[10] H. R. Silva, W. A. Marquelli, W. L. C. Silva et al., *Cultivo do Meloeiro Oara o Norte de Minas Gerais*, EMBRAPA- Hortaliças, Brasília, Brasil, 2000.

[11] M. A. Demiral and A. T. Köseoglu, "Effect of potassium on yield, fruit quality, and chemical composition of greenhouse-grown galia melon," *Journal of Plant Nutrition*, vol. 28, no. 1, pp. 93–100, 2005.

[12] A. Ferrante, A. Spinardi, T. Maggiore, A. Testoni, and P. M. Gallina, "Effect of nitrogen fertilisation levels on melon fruit quality at the harvest time and during storage," *Journal of the Science of Food and Agriculture*, vol. 88, no. 4, pp. 707–713, 2008.

[13] J. I. Contreras, B. M. Plaza, M. T. Lao, and M. L. Segura, "Growth and nutritional response of melon to water quality and nitrogen potassium fertigation levels under greenhouse mediterranean conditions," *Communications in Soil Science and Plant Analysis*, vol. 43, no. 1-2, pp. 434–444, 2012.

[14] P. E. Trani and B. Raij, "Hortaliças," in *Recomendações de Adubação e Calagem Para o Estado da São Paulo*, B. Raij, H. Cantarella, J. S. Quaggio, and A. M. C. Furlani, Eds., pp. 157–164, Instituto Agronômico de Campinas, Campinas, Brasil, 1997.

[15] C. J. Sonneveld, J. van der Ende, and P. van Dijk, "Analysis of growing media by means of 1:1, 5 volume extract," in *Communications in Soil Science and Plant Analysis*, vol. 5, pp. 183–200, 1974.

[16] I. B. Nascimento, C. H. A. Farias, M. C. C. Silva, J. F. Medeiros, J. E. Sobrinho, and M. Z. Negreiros, "Estimativa da área foliar do meloeiro," in *Horticultura Brasileira*, vol. 20, pp. 555–558, 2002.

[17] E. Malavolta, G. C. Vitti, and A. S. Oliveira, *Avaliação do Estado Nutricional das Plantas*, Associação Brasileira para Pesquisa da Potassa e do Fosfato, Piracicaba, Brasil, 1997.

[18] A. Baumgarten, "Methods of chemical and physical evaluation of substrates for plants," in *Encontro Nacional de Substratos Para Plantas*, Instituto Agronômico, Campinas, Brasil, 2002.

[19] L. A. Gaion, D. M. Melo, H. C. O. Charlo, R. Castoldi, and L. T. Braz, "Modificações químicas do substrato em função do cultivo com meloeiro rendilhado," *Horticultura Brasileira*, vol. 29, no. 2, pp. 1742–1748, 2011.

[20] T. J. Cavins, B. E. Whipker, W. C. Fonteno, B. Harden, I. Mccall, and J. L. Gibson, *Monitoring and Managing PH and EC Using the Pour-Thru Extraction Method*, North Carolina State University, Raleigh, NC, USA, 2000.

[21] A. O. Camargo, L. R. F. Alleoni, and J. C. Casagrande, "Reações dos micronutrientes e elementos tóxicos no solo," in *Micronutrientes e Elementos Tóxicos na Agricultura*, M. E. Ferreira, M. C. P. Cruz, B. Raij, and C. A. Abreu, Eds., pp. 89–124, CNPq/FAPESP/POTAFOS, Jaboticabal, Brasil, 2001.

[22] H. E. P. Martinez, *O Uso de Cultivo Hidropônico de Plantas em Pesquisa*, Universidade Federal de Viçosa, Viçosa, Brasil, 2002.

[23] R. F. Novais, V. V. H. Alvarez, N. F. Barros, R. L. F. Fontes, R. B. Cantarutti, and J. C. L. Neves, *Fertilidade do Solo*, Sociedade Brasileira de Ciência do Solo, Viçosa, Brasil, 2007.

[24] G. E. Lester, J. L. Jifon, and D. J. Makus, "Impact of potassium nutrition on postharvest fruit quality: melon (*Cucumis melo* L) case study," *Plant and Soil*, vol. 335, no. 1, pp. 117–131, 2010.

[25] W. C. Fonteno, "Growing media: types and physical/chemical properties," in *A Growers Guide to Water, Media, and Nutrition for Greenhouse Crops*, D. W. Reed, Ed., Ball, Madison, Wis, USA, 1996.

[26] M. Sailus and T. C. Weiler, *Water and Nutrient Management for Greenhouses*, Northeast Regional Agricultural Engineering Service, Cooperative Extension, New York, NY, USA, 1996.

[27] D. A. Bailey, P. V. Nelson, and W. C. Fonteno, "Substrate pH and water quality," North Caroline State University, 2000, http://www.ces.ncsu.edu/depts/hort/floriculture/plugs/ph.pdf.

[28] T. L. Pons and M. H. M. Westbeek, "Analysis of differences in photosynthetic nitrogen-use efficiency between four contrasting species," *Physiologia Plantarum*, vol. 122, no. 1, pp. 68–78, 2004.

[29] A. W. M. Verkroost and M. J. Wassen, "A simple model for nitrogen-limited plant growth and nitrogen allocation," *Annals of Botany*, vol. 96, no. 5, pp. 871–876, 2005.

24

Proximate Analysis of Five Wild Fruits of Mozambique

Telma Magaia,[1,2] Amália Uamusse,[3] Ingegerd Sjöholm,[4] and Kerstin Skog[2]

[1] Department of Biological Science, Science Faculty, Eduardo Mondlane University, Praca 25 de Junho,
P.O. Box 257, Maputo, Mozambique
[2] Division of Applied Nutrition and Food Chemistry, Lund University, P.O. Box 124, 221 00 Lund, Sweden
[3] Department of Chemistry, Science Faculty, Eduardo Mondlane University, Praca 25 de Junho,
P.O. Box 257, Maputo, Mozambique
[4] Division of Food Technology, Lund University, P.O. Box 124, 221 00 Lund, Sweden

Correspondence should be addressed to Telma Magaia; telma.magaia@food.lth.se

Academic Editors: L. F. Goulao and L. Kitinoja

Mozambique is rich in wild fruit trees, most of which produce fleshy fruits commonly consumed in rural communities, especially during dry seasons. However, information on their content of macronutrients is scarce. Five wild fruit species (*Adansonia digitata*, *Landolphia kirkii*, *Sclerocarya birrea*, *Salacia kraussii*, and *Vangueria infausta*) from different districts in Mozambique were selected for the study. The contents of dry matter, fat, protein, ash, sugars, pH, and titratable acidity were determined in the fruit pulps. Also kernels of *A. digitata* and *S. birrea* were included in the study. The protein content in the pulp was below 5 g/100 g of dry matter, but a daily intake of 100 g fresh wild fruits would provide up to 11% of the recommended daily intake for children from 4 to 8 years old. The sugar content varied between 2.3% and 14.4% fresh weight. The pH was below 3, except for *Salacia kraussii*, for which it was slightly below 7. Kernels of *A. digitata* contained, on average, 39.2% protein and 38.0% fat, and *S. birrea* kernels 32.6% protein and 60.7% fat. The collection of nutritional information may serve as a basis for increased consumption and utilization.

1. Introduction

In Mozambique, a large number of wild food plants are widely distributed throughout the country. The fruits and nuts are sold at informal markets during the harvest season and are consumed in various ways, and they are much appreciated by children [1–3]. The importance of wild fruits in the diet depends to a large extent on the availability of the fruits, since cultivated fruit trees are not particularly common in the dry regions of the country. Depending on the season, the fruits are eaten raw, pressed for juice, cooked with sugar, or used as flour to make porridge; the seeds or nuts are roasted to be eaten as snacks. The choice of fruit species varies according to region and cultural traditions [4].

Many wild fruits and nuts are good sources of carbohydrates, protein, fat, vitamins, and minerals that may be deficient in common diets [5]. There are some reports on the chemical composition of wild fruits from Southern African regions [5–9], but the literature data on the nutritional value of wild fruits in Mozambique is limited [4, 10]. People in many communities are not aware of the nutritional value of the fruits; for example, they often eat only the pulp of the fruits *Sclerocarya birrea* and *Adansonia digitata* while discarding the seeds, which contain a kernel with a higher protein and fat content than peanuts [1, 11].

The aims of this work were to perform a study on traditional utilization of wild fruits in Mozambique and to generate data on the proximate composition and other characteristics of five wild fruits as a basis for the selection of fruits suitable for processing. The long-term goal was to promote and increase the utilisation and consumption of indigenous fruits. The wild fruits selected for the present study were *Adansonia digitata*, *Landolphia kirkii*, *Salacia kraussii*, *Sclerocarya birrea*, and *Vangueria infausta*. These fruits are popular in Mozambique, and they play an important role in the diet, particularly in rural areas.

2. Materials and Methods

2.1. Species. Five wild fruit species were studied: *Adansonia digitata (A. digitata)* (family *Bombacaceae*, local name n'buyu or Malambe), *Landolphia kirkii (L. kirkii)* (family *Apocynaceae*, local name n'vhungwa), *Salacia kraussii (S. kraussii)* (family *Celastraceae*, local name n'phinsha), *Sclerocarya birrea (S. birrea)* (family *Anacardiaceae*, local name n'canhi), and *Vangueria infausta (V. infausta)* (family *Rubiaceae*, local name n'pfilwa). Ripe fruits were collected in 2008 and 2009, except for the fruits from *S. birrea*, which were collected only in 2009. *A. digitata* fruits, grown in the Tete district 1100 km from Maputo city, were bought at a local market in Maputo, and some fruits were collected in family orchards in the Vilankulos district, 700 km south of Maputo. *L. kirkii, S. kraussii,* and *V. infausta* fruits were collected in orchards in the Marracuene and Manhiça districts, 30 and 50 km south of Maputo. Fruits from *S. birrea* were obtained from a garden in Maputo city and *S. birrea* kernels, dried for 1–3 months, were obtained from a small family orchard in Manhiça. The fruits were collected in districts where there is an increased occurrence and consumption of them.

2.2. Sample Preparation. Unblemished fruits were selected and washed, the skin and seeds were removed, and the remaining parts were homogenized in a blender to obtain 100 g pulp of each type of fruit. Different numbers of fruit were used depending on fruit size and mass of pulp. The fruits from *A. digitata* had low moisture content and the pulp was ground into a fine powder and sieved (500 μm meshes). The seeds from *A. digitata* and *S. birrea* were crushed and the kernels inside were removed, milled, and sieved (500 μm meshes). Samples for determination of pH and titratable acidity were kept at room temperature and the analyses were performed on the day after collecting the fruit. The samples for the other analyses were vacuum-packed in plastic bags and stored at −18°C in a freezer.

2.3. Analysis. To determine the dry matter content, 2 g samples were dried in an oven at 105°C until constant weight [12]. The samples were weighed before and after drying and the contents of dry matter were calculated. The protein content was determined in an Elementar Analyzer (Flash EA 1112 Series, Thermo Fisher Scientific, Sweden), by means of combustion of 25 mg samples. Aspartic acid (Thermo Fisher Scientific, Delft, The Netherlands) was used as a standard. The amount of protein was calculated by converting the amount of nitrogen by a factor 6.25. The fat content was determined gravimetrically after extracting 1 g samples with petroleum ether (Sigma-Aldrich Chemicals Co., St. Louis, MO, USA) at 40–60°C for 1 hour using a Soxhlet equipment (SoxtecTM 2055, Foss, Höganäs-Helsingborg, Sweden) [13]; rapeseed oil was used as a standard. The ash content was determined by combustion of 2 g samples in silica crucibles in a muffle furnace (Carbolite, Sheffield, England) for 24 hours at 550°C [12]. The pH was determined using a pH meter (Carison GLP 21, serial no. 147012, Barcelona, Spain). Titratable acidity, expressed in percentage of citric acid, was determined after titration of 10 g samples, dissolved in 100 mL

water, with 0.1 M sodium hydroxide using phenolphthalein as indicator [12]. All determinations were performed at least in triplicate; the data are expressed as means ± standard deviations. Subsamples of fruits collected in 2009 were sent to an authorized laboratory for analysis of sugar content by high performance anion exchange chromatography (Dionex) with pulsed amperometric detection (HPAEC-PAD). The variation between duplicate determinations was below 15%.

2.4. Statistical Analysis. Statistical analyses were performed using SPSS (version 13). Significant differences were evaluated with one-way analysis of variance (ANOVA) followed by Tukey's multiple comparisons test. A value $P < 0.05$ was considered to be significant.

2.5. Interviews Regarding the Traditional Utilization of Wild Fruits. When the wild fruits were collected in the different districts, local people were interviewed about the traditional use of the fruits. For each fruit, 3–10 people were interviewed, both women and men of different ages, all of whom had grown up in rural areas. The aim of the interviews was to obtain information on knowledge concerning the occurrence of these wild fruit trees, their owners, and traditional habits regarding consumption, processing, and storage of wild fruits.

Examples of the interview questions are as follows. (i) What kind of wild fruits exist in this area? (ii) Who owns and takes care of these fruit trees? (iii) Who are the major consumers? (iv) How are the fruits eaten? (v) What is the typical amount harvested per day? (vi) How are the fruits stored? (vii) What kind of processing can be done?

3. Results and Discussion

The results of the determinations of dry matter content and proximate composition (protein, fat, and ash) for pulp and kernels are summarised in Table 1. The values differed somewhat between the years, but the difference was not significant for any of the fruits. The dry matter content of *A. digitata* pulp was very high, on average 88.5%, while for the other fruits, it varied between 16.7% and 34.8%. The high dry matter content of *A. digitata* was in the same range as has been reported in studies in other countries, 85–95% [5, 7, 8, 14–18]. There is little literature data on the dry matter content of the other fruits, and for *S. birrea* our data agree with one report [7], while one report is showing a lower value [8]. For *V. infausta* our results are somewhat higher than reported in fruits from Malawi [5], Botswana [8], and Tanzania [17]. For *L. kirkii*, the dry matter content was comparable with the literature data [1]. The average dry matter content of *A. digitata* kernels was 92.0%, which is in agreement with other results, ranging from 85% to 97% [1, 7, 9, 14, 15, 18–21]. The average dry matter content of *S. birrea* kernels was 94.3%, which is at the same level as the results from other reports [1, 11, 15, 22, 23].

The protein content of the pulp was low in general (below 5%) for all the fruits. This is in agreement with results of other reports, [1, 5, 7, 15–18]. For *A. digitata*, however, there

TABLE 1: Dry matter content, protein, fat, and ash expressed in g/100 g dry matter ($n = 3$).

Sample Location_year	Dry matter (%)	Protein	Fat	Ash
Adansonia digitata pulp				
Tete_2008	89.8 ± 0.1	2.4 ± 0.0	0.5 ± 0.1	5.5 ± 0.1
Tete_2009	89.1 ± 0.0	2.2 ± 0.1	0.7 ± 0.6	7.4 ± 0.0
Vilanculos_2009	86.5 ± 0.1	2.1 ± 0.0	0.5 ± 0.1	7.4 ± 0.0
Landolphia kirkii pulp				
Marracuene_2008	27.7 ± 0.2	2.1 ± 0.2	0.9 ± 0.1	2.9 ± 0.0
Marracuene_2009	23.9 ± 0.0	NA	0.4 ± 0.2	3.5 ± 0.0
Manhiça_2009	20.1 ± 0.3	1.7 ± 0.0	1.3 ± 0.1	3.0 ± 0.0
Salacia kraussii pulp				
Marracuene_2008	16.4 ± 0.3	3.7 ± 0.0	0.8 ± 0.5	4.8 ± 0.2
Manhiça_2008	17.3 ± 0.1	2.3 ± 0.0	1.5 ± 0.6	3.4 ± 0.0
Manhiça_2009	16.5 ± 0.1	2.1 ± 0.1	0.7 ± 0.2	3.6 ± 0.0
Sclerocarya birrea pulp				
Manhiça_2009	16.8 ± 0.1	1.4 ± 0.1	0.9 ± 0.3	3.0 ± 0.1
Vangueria infausta pulp				
Marracuene_2008	37.4 ± 0.5	2.9 ± 0.2	0.5 ± 0.8	7.8 ± 0.6
Marracuene_2009	30.0 ± 0.1	2.2 ± 0.3	0.7 ± 0.1	5.3 ± 0.0
Manhiça_2008	37.3 ± 0.9	4.7 ± 0.4	0.7 ± 0.2	5.7 ± 0.4
Manhiça_2009	34.5 ± 0.7	3.3 ± 0.8	0.2 ± 0.0	3.2 ± 0.2
Adansonia digitata kernel				
Tete_2008	93.6 ± 0.0	36.7 ± 0.9	35.0 ± 0.2	7.7 ± 0.0
Tete_2009	91.7 ± 0.1	38.6 ± 0.8	39.9 ± 6.2	7.2 ± 0.0
Vilanculos_2009	90.6 ± 0.0	42.7 ± 0.7	39.0 ± 7.0	8.5 ± 0.2
Sclerocarya birrea kernel				
Manhiça_2008	93.6 ± 0.2	30.1 ± 0.2	58.3 ± 1.3	3.8 ± 0.0
Manhiça_2009	95.0 ± 0.0	35.0 ± 0.1	63.1 ± 0.1	3.5 ± 0.0

is a report showing a much higher protein content, 15.3% [24]. The protein content in the pulp was significantly lower than in the corresponding kernels ($P < 0.05$). *A. digitata* kernels contained on average 39.3% protein and *S. birrea* kernels 32.6%. For *A. digitata*, our results are in accordance with other results [7, 18], but there are also reports showing lower protein content, 13–27% [14, 15, 19, 21], and one report showing higher protein content, 48.3% [20]. The protein content in the kernels of *S. birrea* was at the same level as found in another report [23]. The high protein content in the kernels is at the same levels as reported for soya beans, around 33% [11], which means that the kernels may be a potential source of protein and can be used to improve the diet in rural communities. For example, a daily intake of 100 g of fresh pulp from the wild fruits studied here would provide around 2–11% of the recommended daily intake (RDI) for children from 4 to 8 years old, while 20 g of *A. digitata* or *S. birrea* kernels would provide 32–39% of the RDI for children of the same age [25]. The protein quality of the kernels seems to be good since high amounts of the essential amino acid lysine as well as of arginine, glutamic acid, and aspartic acid have been reported for *A. digitata* kernels [15]. *S. birrea* kernels have

been shown to contain high amounts of the essential amino acids phenylalanine, lysine, and threonine [3, 6].

The fat content in the pulp was below 2% for all the fruits. The literature data on *A. digitata* and *S. birrea* pulp generally show fat contents below 1% [1, 7, 14–16, 18], while some reports show a higher fat content in *A. digitata*, 4% [5, 17, 24]. The pulp of wild fruits is typically low in fat and protein [5], while the kernels are good sources of fat and protein [10]. In the present study, the average fat content was 38.0% in kernels of *A. digitata* and 60.7% in kernels of *S. birrea*, and the fat content in the kernels was significantly higher than that in the pulp ($P < 0.05$). Our data on kernels are in agreement with results from other studies [1, 3, 22, 23], while the fat content was lower in two reports [20, 21] and higher in one [18]. The fat quality of the kernels is good according to the literature data: *A. digitata* kernels are rich in palmitic, oleic, and linoleic acid (essential fatty acid) [15], and *S. birrea* kernels are rich in oleic and linoleic acid [3].

The average ash content ranged from 3.0% to 7.8%. For *A. digitata* pulp and kernel, the results are at the same level as in other reports [7, 8, 14–17, 21]. The ash content was somewhat lower than that in some other reports for *S. birrea* [7, 8, 23] and *V. infausta* pulp [5]. The high ash content indicates that the fruits and kernels may be good sources of minerals.

The pH of the pulps showed an acidic character (around pH 3) except for *S. kraussii*, for which the pH was slightly above 6. The acidic character is in accordance with data on pulps from *A. digitata*, *S. birrea*, and *V. infausta* [8]. The pH of fruits generally varies between 2.5 and 4.5 due to their content of organic acids [26]; the low pH enhances the microbiological and physicochemical stability [27].

The titratable acidity of the pulp, which contributes to the acidity of the aroma, ranged from 0.6% to 1.7%. In another report, also using citric acid, the titratable acidity was 7.8% for *A. digitata*, 0.9% for *S. birrea*, and 1.7% for *V. infausta* [8]. Comparable data were 0.3% for mango pulp [28] and 0. 7% for orange juice [29].

Table 2 shows the sugar content of the investigated fruits expressed as g sugar/100 g pulp. The highest total sugar content was found in *A. digitata* and *L. kirkii*, 10.3 and 14.4 g/100 g, respectively. The value for *A. digitata* is much lower than that reported in another study, where the total sugar content was around 30% [16]. The highest sucrose content, 4.3 g/100 g, was found in *A. digitata*, while for the other fruits it was lower than 3 g/100 g. As expected, only very low amounts of maltose and lactose, below 0.04 g/100 g, were detected; the most abundant sugars in fruits, are glucose, fructose, and sucrose in various proportions, depending on species [30].

The sugar content, data on pH, and titratable acidity are essential characteristics, indicating the possibility for future use of these wild fruits. The sugar content is important for the development of the aroma and taste, and in product development it is important to find a good balance between pH, sugars, and titratable acidity to receive an optimal taste. The wild fruits in our study have different profiles regarding these characteristics but are in accordance with the literature data on some traditional fruits for juice production, for example, papaya, mango, pineapple, and orange [29–32].

Interviews. The interviews revealed that the majority of the fruits came from the forest and that wild fruits provide food for everyone, especially for children because they are more free to go into the forest to collect fruit. Some people said that in periods of hunger, the leaves, fruits, and seeds are used as food as well as for medical applications. Wild fruits can serve as a source of income for many families, depending on the amount of fruit that each family can reap in the forest or in the area surrounding their house. The fruit is usually sold early in the morning at informal markets to which people come to buy fruit to sell in the markets in the cities.

Different fruits are used in different ways; see Table 3. *A. digitata* fruits are sold in different forms, for example, whole fruit, pulp with seeds embedded, and fine powder made from the pulp packed in plastic bags. The seed-containing pulp is often consumed fresh and soaked in warm water to remove the seeds, and the remaining "milk" can be mixed with sugar to form juice, or boiled with maize flour or sorghum to make a porridge given to children before they go to school. Another way of using the pulp is to dilute it with warm water to prepare a juice, which is filtered, mixed with sugar and packed in small

TABLE 2: Sugar content of selected fruits obtained in 2009 from Manhiça expressed in g sugar/100 g pulp ($n = 2$).

Sample	Dry matter (%)	Glucose	Fructose	Sucrose
Adansonia digitata	88.0	3.0	3.0	4.3
Landolphia kirkii	18.0	7.5	5.7	1.2
Salacia kraussii	9.0	3.7	3.9	<0.1
Sclerocarya birrea	8.0	0.5	0.4	1.4
Vangueria infausta	31.0	1.4	1.4	2.7

The variation between duplicate determinations was below 15%.

TABLE 3: Traditional consumption and use of the studied fruits.

Species	Utilisation
Adansonia digitata	Pulp: fresh, diluted, and sweetened to juice, frozen to sweet ice, cooked to porridge, and fermented to an alcoholic drink Kernels: fresh or roasted, milled and boiled to sauce or porridge
Landolphia kirkii	Fresh or fermented to an alcoholic drink
Salacia kraussii	Eaten fresh
Sclerocarya birrea	Pulp: fresh, squeezed to juice, and fermented to an alcoholic drink Kernels: fresh or roasted, cooking oil can be extracted
Vangueria infausta	Fresh, soaked, squeezed to juice, mixed with sugar, water, or milk to a porridge, and fermented to an alcoholic drink

plastic bags and frozen. This sweet ice is commonly sold in informal markets, and it is served as refreshment consumed by both children and adults.

The seeds are crushed and the kernels inside can be consumed fresh or roasted. They can be milled to powder and mixed with a small amount of water and boiled with local plant food to make a sauce consumed with boiled maize flour. The seeds are also boiled with a small amount of water or milk to make porridge for children.

People in rural communities usually eat *L. kirkii* fruits fresh, but when large amounts of these fruits are available, they are squeezed and fermented to produce a local alcoholic drink that is consumed at social gatherings. Rural people of all ages eat *L. kirkii* fruit while walking to and working on their grassland and cattle farm plots, which are sometimes far away from their homes.

The fruits from the small *S. kraussii* bushes are more easily accessible to children. Many school children eat the fresh fruits on their way to and from school or while grazing cattle.

Fresh fruit from *S. birrea* is squeezed to make juice or fermented to produce a popular alcoholic drink. After juice extraction and fermentation, the juice may be stored in sealed clay pots or plastic containers for up to a year. The kernels can be eaten fresh or roasted or ground in a mortar together with water, boiled with local plant food to make a sauce. The kernels can also be used to produce oil for cooking.

In the southern part of Mozambique, *V. infausta* fruits are commonly consumed fresh, as juice, but they are also often fermented to produce alcoholic drinks. Fresh fruit is soaked in water, and the skin and seeds discarded before the preparation of a juice which is mixed with water and sugar or milk and served as porridge for children. Excess fruit is dried and stored for later use.

4. Conclusion

The wild fruits studied are consumed by people living in different districts of Mozambique and form a part of their normal diet. Our data on the proximate composition, pH, titratable acidity, and sugar content are consistent with the few reports available in the literature. We observed low but not significant variations between the growth locations and the harvest year, and these variations may be due to differences in climate, soil, and weather conditions. The findings of this study have shown that the analyzed fruits, and especially the kernels, are good sources of protein and fat. In Mozambique malnutrition is responsible for one-third of deaths in children under five years, and based on the above results, it may be concluded that promotion of consumption and processing of these fruits, to various products, may help to improve the diet and alleviate nutrient deficiencies.

Conflict of Interests

The authors declare that they have no conflict of interests.

Acknowledgments

This study was financially supported by the Swedish International Development Agency Project "Technology Processing of Natural Resources for Sustainable Development" at Eduardo Mondlane University, Maputo. The authors thank Mrs. Lisbeth Person and Mr. Dan Johnson for excellent technical assistance.

References

[1] P. M. Maundu, G. W. Ngugi, and C. H. S. Kakuye, *Traditional Food Plants of Kenya*, National Museums, Nairobi, Kenya, 1999.

[2] L. E. Grivetti and B. M. Ogle, "Value of traditional foods in meeting macro- and micronutrient needs: the wild plant connection," *Nutrition Research Reviews*, vol. 13, no. 1, pp. 31–46, 2000.

[3] R. S. Glew, D. J. VanderJagt, Y.-S. Huang, L.-T. Chuang, R. Bosse, and R. H. Glew, "Nutritional analysis of the edible pit of *Sclerocarya birrea* in the Republic of Niger (daniya, Hausa)," *Journal of Food Composition and Analysis*, vol. 17, no. 1, pp. 99–111, 2004.

[4] P. D. Mangue and M. N. Oreste, *Data Collection and Analysis for Sustainable Forest Management in ACP Countries—Linking National and International Efforts, EC-FAO Partnership Programme (1998–2000)*, Country Brief on Non-Wood Forest Products, Maputo, Mozambique, 1999.

[5] J. D. Kalenga Saka and J. D. Msonthi, "Nutritional value of edible fruits of indigenous wild trees in Malawi," *Forest Ecology and Management*, vol. 64, no. 2-3, pp. 245–248, 1994.

[6] R. H. Glew, D. J. Vanderjagt, C. Lockett et al., "Amino acid, fatty acid, and mineral composition of 24 indigenous plants of burkina faso," *Journal of Food Composition and Analysis*, vol. 10, no. 3, pp. 205–217, 1997.

[7] S. S. Murray, M. J. Schoeninger, H. T. Bunn, T. R. Pickering, and J. A. Marlett, "Nutritional composition of some wild plant foods and honey used by hadza foragers of Tanzania," *Journal of Food Composition and Analysis*, vol. 14, no. 1, pp. 3–13, 2001.

[8] J. O. Amarteifio and M. O. Mosase, "The chemical composition of selected indigenous fruits of botswana," *Journal of Applied Science and Environmental Management*, vol. 10, no. 2, pp. 43–47, 2006.

[9] I. I. Nkafamiya, S. A. Osemeahon, D. Dahiru, and H. A. Umaru, "Studies on the chemical composition and physicochemical properties of the seeds of baobab (*Adasonia digitata*)," *African Journal of Biotechnology*, vol. 6, no. 6, pp. 756–759, 2007.

[10] G. Saxon and C. Chidiamassamba, "Indigenous Knowledge of Edible Tree Products—the Mungomu Tree in Central Mozambique. Links project. Gender biodiversity and local knowledge systems for food security," FAO Report 40, 2005.

[11] O. Ogbobe, "Physico-chemical composition and characterisation of the seed and seed oil of Sclerocarya birrea," *Plant Foods for Human Nutrition*, vol. 42, no. 3, pp. 201–206, 1992.

[12] AOAC, *Official Methods of Analysis of the Association of Official Analytical Chemists*, 16th edition, 2000.

[13] AOAC, *Official Methods of Analysis of the Association of Official Analytical Chemists*, 17th edition, 1995.

[14] C. T. Lockett, C. C. Calvert, and L. E. Grivetti, "Energy and micronutrient composition of dietary and medicinal wild plants consumed during drought. Study of rural Fulani, Northeastern Nigeria," *International Journal of Food Sciences and Nutrition*, vol. 51, no. 3, pp. 195–208, 2000.

[15] M. A. Osman, "Chemical and nutrient analysis of baobab (*Adansonia digitata*) fruit and seed protein solubility," *Plant Foods for Human Nutrition*, vol. 59, no. 1, pp. 29–33, 2004.

[16] A. A. Abdalla, M. A. Mohammed, and H. A. Mudawi, "Production and quality assessment of instant baobab (*Adansonia digitata* L.)," *Advance Journal of Food Science and Technology*, vol. 2, no. 2, pp. 125–133, 2010.

[17] T. V. Emmanuel, J. T. Njoka, L. W. Catherine, and H. V. M. Lyaruu, "Nutritive and anti-nutritive qualities of mostly preferred edible woody plants in selected drylands of Iringa District, Tanzania," *Pakistan Journal of Nutrition*, vol. 10, no. 8, pp. 786–791, 2011.

[18] A. E. Assogbadjo, F. J. Chadare, R. G. Kakaï, B. Fandohan, and J. J. Baidu-Forson, "Variation in biochemical composition of baobab (*Adansonia digitata*) pulp, leaves and seeds in relation to soil types and tree provenances," *Agriculture, Ecosystems and Environment*, vol. 157, pp. 94–99, 2012.

[19] L. J. Proll, K. J. Petzke, I. E. Ezeagu, and C. C. Metges, "Low nutritional quality of unconventional tropical crop seeds in rats," *The Journal of Nutrition*, vol. 128, no. 11, pp. 2014–2022, 1998.

[20] H. O. Adubiaro, O. Olaofe, E. T. Akintayo, and O.-O. Babalola, "Chemical composition, cacium, zinc and phytate interrelationships in baobab (*Adansonia digitata*) seed flour," *Advance Journal of Food Science and Technology*, vol. 3, no. 4, pp. 228–232, 2011.

[21] C. Parkouda, H. Sanou, A. Tougiani et al., "Variability of Baobab (*Adansonia digitata* L.) fruits' physical characteristics and nutrient content in the West African Sahel," *Agroforestry Systems*, vol. 85, no. 3, pp. 455–463, 2011.

[22] I. C. Eromosele and C. O. Eromosele, "Studies on the chemical composition and physico-chemical properties of seeds of some wild plants," *Plant Foods for Human Nutrition*, vol. 43, no. 3, pp. 251–258, 1993.

[23] S. Muhammad, L. G. Hassan, S. M. Dangoggo, S. W. Hassan, K. J. Umar, and R. U. Aliyu, "Nutritional and antinutritional composition of Sclerocarya birrea seed kernels," *Studia Universitatis "Vasile Goldiş", Seria Ştiinţele Vieţii*, vol. 21, no. 4, pp. 693–699, 2011.

[24] I. C. Obizoba and N. A. Amaechi, "The effect of processing methods on the chemical composition of baobab (*Adansonia digitata* L.)," *Pulp and Seed. Ecology of Food and Nutrition*, vol. 29, pp. 199–205, 1993.

[25] The National Academies of Science, Dietary Reference intakes for Energy, Carbohydrates, Fiber, Fat, fatty acids, (Macronutrients), 2005, http://www.nap.edu/.

[26] A. L. Calvacanti, K. F. Oliveira, P. S. Paiva, M. V. Rebelo Dias, S. K. P. Costa, and F. F. Vieira, "Determination of total soluble solids content (Brix) and pH in milk drinks and industrialized fruit juices," *Pesquisa Brasileira em Odontopediatria e Clínica Integrada*, vol. 6, no. 1, pp. 57–64, 2006.

[27] P. R. Gayon, Y. Glories, A. Maujean, and D. Dubourdieu, "Organic acids in wine Handbook of Enology: the Chemistry of Wine and Stabilization and Treatments," vol. 2, pp. 9–10, John Wiley & Sons, England, UK, 2nd edition, 2006.

[28] N. I. Singh, C. Dhuique-Mayer, and Y. Lozano, "Physicochemical changes during enzymatic liquefaction of mango pulp (cv. Keitt)," *Journal of Food Processing and Preservation*, vol. 24, no. 1, pp. 73–85, 2000.

[29] B. K. Tiwari, K. Muthukumarappan, C. P. O'Donnell, and P. J. Cullen, "Kinetics of freshly squeezed orange juice quality changes during ozone processing," *Journal of Agricultural and Food Chemistry*, vol. 56, no. 15, pp. 6416–6422, 2008.

[30] F. J. Rambla, S. Garrigues, and M. De La Guardia, "PLS-NIR determination of total sugar, glucose, fructose and sucrose in aqueous solutions of fruit juices," *Analytica Chimica Acta*, vol. 344, no. 1-2, pp. 41–53, 1997.

[31] R. Shamsudin, W. R. W. Daud, M. S. Takriff, and O. Hassan, "Physicochemical properties of the Josapine variety of pineapple fruit," *International Journal of Food Engineering*, vol. 3, no. 5, article 9, 2007.

[32] L. E. Gayosso-García Sancho, E. M. Yahia, M. A. Martínez-Téllez, and G. A. González-Aguilar, "Effect of maturity stage of papaya maradol on physiological and biochemical parameters," *American Journal of Agricultural and Biological Science*, vol. 5, no. 2, pp. 194–203, 2010.

25

Critical Period of Weed Control in Aerobic Rice

M. P. Anwar,[1,2] **A. S. Juraimi,**[3] **B. Samedani,**[3] **A. Puteh,**[3] **and A. Man**[4]

[1] Institute of Tropical Agriculture, Universiti Putra Malaysia, Serdang 43400, Selangor, Malaysia
[2] Department of Agronomy, Bangladesh Agricultural University, Mymensingh 2202, Bangladesh
[3] Department of Crop Science, Universiti Putra Malaysia, Serdang 43400, Selangor, Malaysia
[4] Rice and Industrial Crops Research Centre, Malaysian Agricultural Research and Development Institute,
 Kuala Lumpur 50774, Malaysia

Correspondence should be addressed to A. S. Juraimi, ashukor@agri.upm.edu.my

Academic Editors: C. Dell, H. A. Torbert, and M. Tsubo

Critical period of weed control is the foundation of integrated weed management and, hence, can be considered the first step to design weed control strategy. To determine critical period of weed control of aerobic rice, field trials were conducted during 2010/2011 at Universiti Putra Malaysia. A quantitative series of treatments comprising two components, (a) increasing duration of weed interference and (b) increasing length of weed-free period, were imposed. Critical period was determined through Logistic and Gompertz equations. Critical period varied between seasons; in main season, it started earlier and lasted longer, as compared to off-season. The onset of the critical period was found relatively stable between seasons, while the end was more variable. Critical period was determined as 7–49 days after seeding in off-season and 7–53 days in main season to achieve 95% of weed-free yield, and 23–40 days in off-season and 21–43 days in main season to achieve 90% of weed-free yield. Since 5% yield loss level is not practical from economic view point, a 10% yield loss may be considered excellent from economic view point. Therefore, aerobic rice should be kept weed-free during 21–43 days for better yield and higher economic return.

1. Introduction

Critical period of weed control (CPWC) is an integral part of integrated weed management (IWM) and can be considered the first step to design weed control strategy [1]. The CPWC is the period of crop life cycle during which weeds must be controlled to prevent unacceptable or economic yield loss [2–4]. In theory, presence of weeds before or after CPWC will not pose a threat and should not cause significant yield loss [3, 5]. Thus, crop yield obtained by weeding during CPWC is almost similar to that obtained by the full season weed-free conditions. In general, one-third of the crop life cycle is considered as critical for weed control. A long critical period is the indication of less competitive crop or more competitive weeds and vice versa [6]. Studying CPWC could help identify residual action required for preemergence herbicide, improve timing and reduce the amount of postemergence herbicide applications [2], and thus may lessen potential environmental and ecological degradation [7]. Therefore,

CPWC has been the subject of extensive research in field crops for the last few decades [8, 9].

Critical period of weed control has got a beginning and an end as well. The beginning of CPWC is determined by estimating critical time for weed removal (CTWR) after which weed control must be initiated to ensure potential yield [7, 10]. The end of CPWC, on the other hand, is determined by estimating critical weed-free period (CWFP) required from planting to avoid irrevocable yield loss [3, 4]. The CPWC is determined by calculating the time interval between CTWR and CWFP. Critical period of weed control has commonly been reported as day after seeding (DAS), but due to differences in planting dates and environment, this may generate results with more variability among locations, seasons, and cultivars. Therefore, CPWC determination based on DAS has been criticized by many researchers [7]. In recent studies, CPWC has been reported as growing degree days (GDDs) because it is a biologically meaningful measure of time required for plant growth and development [11], and therefore, it would be applicable for comparing critical period across different agroclimatic conditions [4].

The CPWC is likely to be unique for every crop species because of their morphophysiological makeup [3], but it is not an inherent property of the crop rather a measurement of crop-weed-environment interaction [12]. Whether or not each crop has got CPWC is arguable [13]. Some crops have CPWC, while others may not. Since the introduction of CPWC concept in 1960s, countless studies have been conducted around the globe to determine the CPWC in rice [14–19]. In West Africa, Johnson et al. [15] estimated CPWC for lowland irrigated rice as 0–32 DAS in wet season, while 4–83 DAS in dry season to obtain 95% yield. In Malaysia, based on the 5% yield loss, Begum et al. [17] concluded that flood-irrigated rice must be kept weed-free from 14 to 28 DAS, while Juraimi et al. [18] suggested that direct-seeded rice should be kept weed-free for 2–71 DAS in saturated condition and 15–73 DAS in flooded condition. In the Philippines, Chauhan and Johnson [19] estimated CPWC of rice as between 18 and 52 DAS to obtain 95% of weed-free yield.

Aerobic rice, growing rice in nonpuddled and non-saturated soil like an upland crop, is gaining popularity day by day as a waterwise technology [20, 21]. But this technology is impeded by high weed pressure because of dry tillage and aerobic soil conditions [22, 23], and hence, weed management has been a challenge for this promising technology. The question remains unanswered is the following: will the aerobic rice system alter the CPWC? It is not unlikely that a switch from flood-irrigated to aerobic system will certainly bring a change in species composition, pressure, and emergence pattern of weeds resulting in a new dimension of weed-crop competition usually not existent in flood-irrigated rice system. Studies on single and multi-species weed interference effect on rice have been extensively conducted in different rice ecosystems, but a very little effort has been made so far to determine CPWC in aerobic rice system. Extrapolation of the findings obtained from flood-irrigated rice system to determine the CPWC in newly introduced aerobic rice system seems unwise which warrants urgent studies to determine CPWC in aerobic rice for its sustainability. The purpose of this study was to define and estimate the critical period of weed control for direct-seeded aerobic rice towards developing a less-herbicide-dependent weed management strategy. Other significant purpose was to evaluate the effect of different weed interference period on some agronomic and physiological traits of rice under aerobic soil conditions.

2. Materials and Methods

2.1. Experimental Site and Soil. The trial was carried out during off-season 2010 (May–July) and main season 2010/2011 (November–January) under field condition at the experimental farm, Universiti Putra Malaysia, Malaysia (3°00′ 21.34″ N, 101°42′ 15.06″ E, 37 m elevation). The local climate was hot humid tropic with high humidity and abundant rainfall. During the off-season, monthly

average maximum and minimum temperature and relative humidity ranged from 33.5 to 35.0°C, 23.6 to 24.6°C, and 93.3 to 93.9%, respectively, while rainfall, evaporation, and sunshine hours ranged from 4.2 to 10.8 mm/day, 3.65 to 4.90 mm/day, and 5.90 to 7.37 hrs/day, respectively. During the main season, monthly average maximum and minimum temperature and relative humidity ranged from 31.7 to 33.3°C, 22.9 to 23.8°C, and 94.1 to 94.6%, respectively, while rainfall, evaporation, and sunshine hours ranged from 6.1 to 9.9 mm/day, 2.64 to 4.66 mm/day, and 3.95 to 6.34 hrs/day, respectively. The experimental soil was sandy clay loam in texture (57.07% sand, 22.32% silt, and 20.61% clay) and acidic in reaction (pH 5.8) with 1.85% organic carbon, 1.43 g/cc bulk density, and 17.36 me/100 g soil CEC. Soil nutrient status was 0.33% total N, 21.2 ppm available P, 143 ppm available K, 794 ppm Ca, and 163 ppm Mg. At field capacity, soil water retention was 22.64% (wet basis) and 29.27% (dry basis).

2.2. Plant Material. An aerobic rice line AERON 1, sourced from International Rice Research Institute (IRRI), was used as the plant material in this study. The rice line AERON 1 was selected as the plant material because it performed well under aerobic soil conditions in previous study [21].

2.3. Experimental Treatments and Design. The experimental design was a randomized complete block with three replications. To determine critical period of weed control, a quantitative series of treatments comprising two components, (a) increasing duration of weed interference and (b) increasing length of weed-free period, were imposed. Timing of weed removal was based on the number of weeks after rice seeding. To determine the beginning of the CPWC, the first component, increasing duration of weed interference, was established by allowing the weeds to compete with the crops for 2, 4, 6, and 8 weeks after seeding (WAS) (referred to as weedy plots), after which plots were maintained weed-free until harvest. To evaluate the end of the critical period of the CPWC, the second component, increasing length of weed-free period, was established by maintaining weed-free condition for 2, 4, 6, and 8 WAS (referred to as weed-free plots) before allowing subsequent emerging weeds to compete for the remainder of the growing season. In addition, season long weedy check and weed-free check were included as control. No herbicide was used as weed control was accomplished by hand weeding. The experiment was conducted under naturally occurring population of mixed weed species.

2.4. Crop Husbandry. The experimental field was dry-ploughed and harrowed but not puddled during land preparation. Each plot was 5.0 m long and 3.0 m wide and accommodated 12 rows with 25 cm interrow spacing. Rice seeds were directly dry-seeded in rows with 15 cm intrarow spacing at the rate of eight seeds/hill. Each plot was fertilized with triple superphosphate (TSP) and muriate of potash (MP) at 100 kg P/ha and 100 kg K/ha, respectively, as basal application; urea was top-dressed thrice each at 50 kg N/ha at

2, 4, and 6 WAS. Field was maintained under nonsaturated aerobic conditions throughout the growing season. In both the seasons, the trial was primarily rain-fed, but supplemental sprinkler irrigation was applied when needed. Overflow canals were maintained to facilitate drainage whenever heavy rainfall resulted in ponding. Plant protection measures were taken, as needed, to avoid confounding effect of competition with insect and/or disease injury. Different intercultural operations and plant protection measures were conducted following standard practices.

2.5. *Data Collection.* At each weed removal time, a 25 cm × 25 cm quadrate was randomly placed lengthwise at four spots in each plot for recording weed data. Weeds were clipped to ground level, identified and counted by species, and separately oven dried at 70°C for 72 hours. Weed density (WD) and weed dry weight (WDW) were expressed as no./m^2 and g/m^2, respectively. Dominant weed species were identified using the summed dominance ratio (SDR) computed as follows [21]:

$$\text{SDR of a weed species} = \frac{[\text{Relative density (RD)} + \text{Relative dry weight (RDW)}]}{2}, \tag{1}$$

where RD (%) = (Density of a given weed species/Total weed density) × 100, RDW (%) = (Dry weight of a given weed species/Total weed dry weight) × 100.

Four central rows, excluding the harvesting area, were used for data recording. Aboveground crop biomass was recorded at panicle initiation, heading and harvesting stages from ten randomly selected hills of each plot by oven drying at 70°C for 72 hours. At maturity, yield components like panicle/m^2 and grains/panicle were recorded from ten randomly selected hills. Central 3 m^2 area of each plot was hand harvested to record grain yield and thousand-seed weight. Grain yield and thousand-seed weight were adjusted to 14% moisture content. Growing degree day (GDD) was accumulated from the date of seeding considering base temperature as 10°C [24]. The time of crop emergence was used as the reference point for the accumulation of GDD. The GDD was calculated as follows:

$$\text{GDD} = \frac{\sum(T_{max} + T_{min})}{2} - T_b, \tag{2}$$

where T_{max} and T_{min} are daily maximum and minimum air temperature (°C), and T_b is the base or threshold temperature below which physiological activities are inhibited.

2.6. *Statistical Analysis.* Statistical Analysis System (SAS 9.1) software was used to analyze the data, including analysis of variance (ANOVA) and comparison of means based on a protected LSD procedure at 5% level of probability [25]. Data for both the seasons were analyzed separately. For each season, mean yield across the three blocks was calculated for each treatment and converted to percentage values (relative yield, RY) of the season long weed-free control in each treatment group. To calculate CPWC, RY data for the weedy or weed-free treatments were regressed against the increasing duration of weed interference or increasing length of the weed-free period. A four-parameter logistic equation, proposed by Hall et al. [7] and modified by Knezevic et al. [3], was used to describe the effect of increasing duration of weed interference on relative yield of rice. The Gompertz model [3, 7] was used to provide a good fit to RY as it is influenced by increasing length of the weed-free period. The logistic equation was used to determine the beginning of the CPWC, and the Gompertz equation was used to determine the end of the CPWC for yield loss levels of 5 and 10% chosen arbitrarily [7, 26].

3. Results

3.1. *Weed Composition.* In the experiment, the naturally occurring weed community had a wide weed spectrum including broadlaf, sedges, and grasses. Weed population was mostly dominated by broadleaf weeds with very little contribution of grasses. The weed community composed of 23 species in off-season 2010 and 18 species in main season 2010/2011 representing 12 different families (Table 1). *Physalis heterophylla* Nees, *Scoparia dulcis* L., *Cleome rutidosperma* DC, and *Cyperus rotundus* L. were the most common species dominating the community in both the seasons. *Physalis heterophylla* Nees appeared as the most abundant species in both the seasons with about 58 and 71 plants/m^2 in off and main seasons, respectively. *Cyperus sphacelatus* Rottb., *Cyperus aromaticus*, *Axonopus compressus* (Sw.) Beauv, *Amaranthus viridis* L., and *Eclipta prostrata* (L.) L. were present in off-season but not found in main season.

3.2. *Weed Species Dominance Pattern.* Weed species dominance pattern was not found similar throughout the growing period (Table 2). In both the seasons, *Cyperus rotundus* L., representing sedge group, was the most dominant species at early growth stage of rice, But with the advancement of time, sedges were gradually replaced by broadleaf weeds. Among the grasses, only *Axonopus compressus* (Sw.) Beauv was recorded as one of the five most dominant species at early growth stages in both seasons and thereafter disappeared, whereas *Eleusine indica* (L.) Gaertn. and *Leptochloa chinensis* (L.) Nees were amongst the most dominant species in off-season and main season, respectively, in later growth stages. *Physalis heterophylla* Nees, *Scoparia dulcis* L., and *Cleome rutidosperma* DC were the broadleaf weeds started dominating the community from midgrowth stage of rice till maturity in both the seasons.

3.3. *Weed Density and Dry Weight.* Weed density and dry weight were recorded at the end of different weed competition periods. Due to aerobic soil conditions, weed density and dry matter were found very high (1241–1311 g/m^2 and 520–540 g/m^2, resp.) in both the seasons (Table 3). Weed pressure in terms of weed dry weight was higher in main season than in off-season. In weedy check, weed density and dry weight in main season were recorded as 520/m^2

TABLE 1: Weed composition in season long weedy plots of aerobic rice (Off-season 2010 and main season 2010/2011; summed dominance ratio (SDR ± SE)).

| Scientific name | Family name | Weed type | Summed dominance ratio | |
			Off-season 2010	Main season 2010/2011
Alternanthera sessilis (L.) R. Br. Ex DC.	Amaranthaceae	B	0.8 ± 0.19	0.6 ± 0.11
Amaranthus viridis L.	Amaranthaceae	B	0.8 ± 0.25	—
Axonopus compressus (Sw.) Beauv	Poaceae	G	1.1 ± 0.51	—
Cleome rutidosperma DC	Capparidaceae	B	10.3 ± 2.6	15.3 ± 5.1
Cyperus aromaticus	Cyperaceae	S	2.5 ± 0.81	—
Cyperus distans	Cyperaceae	S	4.1 ± 1.2	2.9 ± 0.86
Cyperus rotundus L.	Cyperaceae	S	9.2 ± 1.9	7.2 ± 2.8
Cyperus sphacelatus Rottb.	Cyperaceae	S	3.5 ± 0.98	—
Digitaria ciliaris (Retz.) Koel	Poaceae	G	1.5 ± 0.51	0.3 ± 0.11
Echinochloa colona (L.) Link	Poaceae	G	3.0 ± 0.76	2.2 ± 1.1
Eclipta prostrata (L.) L.	Asteraceae	B	0.3 ± 0.12	—
Eleusine indica (L.) Gaertn.	Poaceae	G	7.6 ± 2.2	1.7 ± 0.56
Emilia sonchifolia (L.) DC. Ex Wight	Asteraceae	B	0.6 ± 0.10	0.4 ± 0.10
Euphorbia hirta L.	Euphorbiaceae	B	0.7 ± 0.17	4.3 ± 1.34
Fimbristylis miliacea (L.) Vahl	Cyperaceae	S	6.4 ± 2.0	3.2 ± 1.2
Hedyotis corymbosa (L.) Lam.	Rubiaceae	B	1.4 ± 0.39	3.7 ± 1.2
Hyptis brevipes Poit	Lamiaceae	B	1.0 ± 0.22	0.7 ± 0.23
Jussiaea linifolia Vahl	Onagraceae	B	2.2 ± 0.25	6.6 ± 2.9
Leptochloa chinensis (L.) Nees	Poaceae	G	7.0 ± 1.3	1.1 ± 0.30
Mimosa pudica L.	Fabaceae	B	1.3 ± 0.36	6.4 ± 1.3
Phyllanthus niruri L.	Euphorbiaceae	B	1.9 ± 0.42	5.1 ± 0.84
Physalis heterophylla Nees	Solanaceae	B	20.8 ± 5.8	23.7 ± 7.2
Scoparia dulcis L.	Scrophulariaceae	B	12.0 ± 3.4	14.6 ± 4.6

B = broadleaf; S = sedge; G = grass.

and 431 g/m², respectively, while those in off-season were recorded as 471/m² and 390 g/m², respectively. Weed density and dry weight increased with the increasing duration of weed interference period up to 6 WAS and thereafter declined in both the seasons. In contrast, weed density and dry weight decreased with increasing duration of weed-free period.

3.4. Rice Biomass.
Aboveground crop biomass accumulation by rice variety AERON 1 was significantly influenced by weed interference period at all harvests in both off and main seasons (Table 4). AERON 1 accumulated more biomass in main season than in off-season. Increasing length of weed interference period caused lower biomass accumulation at all the growth stages except at panicle initiation stage, whereas weedy conditions for more than 2 WAS had no significance on biomass production in both season. Adverse effect of increasing weedy period on biomass production increased gradually with the advancement of growth stages. Season long weed competition encountered 63, 73, and 75% penalty in biomass accumulation at panicle initiation, heading, and harvesting stages, respectively, as compared with weed-free check in off-season. While in main season, biomass production was reduced by 62, 74, and 81% at panicle initiation, heading, and harvesting stages, respectively, as

a consequence of season long weed interference. Weeding after 8 WAS resulted in no significant increase in biomass accumulation of AERON 1.

3.5. Yield Components and Yield.
Yield components and grain yield of AERON 1 were significantly influenced by weed competition period in both off and main seasons (Table 5). Number of panicles/m², number of grains/panicle, and thousand-seed weight were increased with the increasing length of weed-free conditions and decreased with the increasing length of weedy conditions. In general, maintaining a weed-free condition beyond 8 WAS did not bring any improvement in the yield components like number of panicles/m² and thousand-seed weight. While, regarding number of grains/panicle, weed interference after 6 WAS had no adverse effect. Season long weed competition resulted in 47, 32, and 13% reduction in number of panicles/m², number of grains/panicle, and thousand-seed weight, respectively, as compared with season long weed-free conditions in off-season, while in main season, the respective values were 56, 27, and 13%. Grain yield of AERON 1 was significantly influenced by weed interference period in both the seasons; grain yield was increased with the increasing length of weed-free period up to 6 WAS after which no significant

TABLE 2: Five most dominant weed species at the end of different weedy periods in the two seasons (off-season 2010 and main season 2010/2011, with summed dominance ratios (SDRs) followed by standard error (SE)).

Weed species	Weedy 2 weeks	Weedy 4 weeks	Weedy 6 weeks	Weedy 8 weeks	Season long weedy
			Off-season 2010		
Axonopus compressus	13.7 ± 3.7	—	—	—	—
Cleome rutidosperma DC	—	11.4 ± 5.4	13.5 ± 3.2	15.7 ± 2.8	10.3 ± 2.6
Cyperus distans	15.5 ± 4.4	—	—	—	—
Cyperus rotundus L.	39.9 ± 16.2	31.4 ± 8.3	11.2 ± 4.0	9.9 ± 2.9	9.2 ± 1.9
Eleusine indica (L.) Gaertn.	—	—	—	—	7.6 ± 2.2
Fimbristylis miliacea (L.) Vahl	—	7.2 ± 1.8	6.1 ± 2.5	6.7 ± 3.4	—
Physalis heterophylla Nees	9.2 ± 2.4	17.6 ± 3.9	28.2 ± 7.5	26.5 ± 4.6	20.8 ± 5.8
Scoparia dulcis L.	7.1 ± 3.1	15.8 ± 5.1	15.9 ± 6.8	18.0 ± 5.2	12.0 ± 3.4
			Main season 2010/2011		
Axonopus compressus	5.7 ± 0.9	—	—	—	—
Cleome rutidosperma DC	16.4 ± 7.3	18.7 ± 5.7	20.1 ± 6.7	17.6 ± 7.9	15.3 ± 5.1
Cyperus rotundus L.	33.6 ± 8.6	26.2 ± 8.2	13.2 ± 2.3	8.3 ± 3.5	7.2 ± 2.8
Jussiaea linifolia Vahl	—	—	—	—	6.6 ± 2.9
Leptochloa chinensis (L.) Nees	—	—	3.6 ± 0.7	7.8 ± 2.7	—
Mimosa pudica L.	—	6.6 ± 2.5	—	—	
Physalis heterophylla Nees	22.3 ± 3.8	27.4 ± 5.9	29.5 ± 13.1	26.5 ± 6.8	23.7 ± 7.2
Scoparia dulcis L.	2.5 ± 0.4	6.3 ± 1.4	11.7 ± 3.5	10.4 ± 5.2	14.6 ± 4.6

TABLE 3: Effect of duration of weed competition on density and dry weight of weeds in both seasons (off-season 2010 and main season 2010/2011).

Weed competition period	Off-season 2010		Main season 2010/2011	
	Density (no./m^2)	Dry matter (g/m^2)	Density (no./m^2)	Dry matter (g/m^2)
Weedy until 2 WAS	879.67c	155.33de	950.33c	188.33e
Weedy until 4 WAS	1073.33b	215.00cd	1128.00b	239.0de
Weedy until 6 WAS	1241.00a	520.33a	1311.00a	548.33a
Weedy until 8 WAS	979.67b	503.00a	1033.33bc	532.33a
Weedy check	471.00d	390.00b	520.67d	431.67b
Weed-free until 2 WAS	473.00d	342.00b	506.00d	364.67c
Weed-free until 4 WAS	380.67d	256.67c	421.33d	286.0d
Weed-free until 6 WAS	206.00e	164.33de	224.33e	181.67e
Weed-free until 8 WAS	102.33f	107.33e	123.67e	101.33f

Data for weedy treatments were taken at the time of weed removal, whereas data for weed-free treatments were taken at the time of rice harvest.
Within a column for each factor, means sharing same alphabets are not significantly different at $P = 0.05$ probability level according to least significant difference test.
WAS = weeks after seeding.

improvement was observed. In contrast, grain yield was significantly decreased with the increasing span of weed interference period up to 8 WAS and thereafter remained unchanged. Apparently, grain yield was recorded slightly higher in main season than in off-season. Season long weed-free conditions produced a yield advantage of 115 and 122% over season long weedy conditions in off and main seasons, respectively.

3.6. Critical Period of Weed Control. Critical period of weed control (CPWC) was determined by using relative rice yield

(% of season long weed-free yield) and growing degree days (GDDs) as quantitative variables in the regression analysis. Rice seeding date was used as the reference point for accumulation of GDD for accounting the possibility of weeds emerging before the rice. The CPWC was determined based on arbitrarily chosen yield loss levels (AYLs) of 5% and 10%, which are judged to be acceptable considering the present economics of weed control. Predicted and observed relative rice yield as affected by weed interference and weed-free periods in off and main seasons are shown in Figure 1. Responses were highly significant as indicated by high R^2

TABLE 4: Effect of weed competition period on aboveground crop biomass production in aerobic rice variety AERON 1 at different growth stages in both seasons (off-season 2010 and main season 2010/2011 (g/m^2)).

Weed competition period	Off-season 2010			Main season 2010/2011		
	Panicle initiation stage	Heading stage	Harvest	Panicle initiation stage	Heading stage	Harvest
Weedy until 2 WAS	191.77a	364.35b	660.71ab	201.89a	328.23ab	669.60b
Weedy until 4 WAS	83.20c	293.42c	631.00b	81.40c	306.60b	652.00b
Weedy until 6 WAS	80.32c	215.00d	443.53d	76.00c	241.14c	431.30d
Weedy until 8 WAS	77.43c	119.42f	235.42e	75.28c	111.00e	205.55e
Weedy check	77.59c	108.29f	169.00f	75.67c	91.59e	130.49f
Weed-free until 2 WAS	165.30b	158.00e	247.48e	151.35b	169.47d	253.67e
Weed-free until 4 WAS	208.57a	302.66c	553.00c	185.48a	302.62b	589.07c
Weed-free until 6 WAS	208.98a	364.7b	664.36ab	197.73a	336.67ab	654.23b
Weed-free until 8 WAS	210.10a	381.85ab	673.00ab	204.16a	353.72a	687.58ab
Weed-free check	209.00a	406.00a	679.45a	202.31a	359.87a	729.33a

Within a column for each factor, means sharing same alphabets are not significantly different at $P = 0.05$ probability level according to least significant difference test.
WAS = weeks after seeding.

TABLE 5: Effect of weed competition period on yield components of aerobic rice variety AERON 1 in both seasons (off-season 2010 and main season 2010/2011).

Weed competition period	Off-season 2010				Main season 2010/2011			
	Panicles (no./m^2)	Grains/panicle (no.)	1000-seed weight (g)	Grain yield (t/ha)	Panicles (no./m^2)	Grains/panicle (no.)	1000-seed weight (g)	Grain yield (t/ha)
Weedy until 2 WAS	191.0ab	72.6ab	27.2ab	3.37ab	177.0a–c	72.3ab	27.5ab	3.46ab
Weedy until 4 WAS	163.6bc	68.3a–c	27.0b	2.99a–c	162.3bc	65.6a–c	27.0cd	3.07bc
Weedy until 6 WAS	150.3cd	59.0b–d	26.5bc	2.5c–e	148.0c	62.6a–c	26.2f	2.59d
Weedy until 8 WAS	134.0c–e	51.0d	26.1cd	1.98e–f	147.3cd	60.3bc	25.0g	2.01e
Weedy check	104.0e	50.3d	24.2e	1.68f	91.6e	53.3c	24.2h	1.73e
Weed-free until 2 WAS	118.3de	55.0cd	25.5d	2.22df	110.3de	59.6bc	26.4ef	2.43d
Weed-free until 4 WAS	135.3c–e	55.3cd	26.1cd	2.79b–d	140.6cd	61.6bc	26.6e	3.01c
Weed-free until 6 WAS	152.6cd	67.3a–c	26.5bc	3.22ab	165.3bc	63.6a–c	26.7de	3.51a
Weed-free until 8 WAS	193.3ab	72.3ab	27.1ab	3.49a	198.3a	67.3ab	27.1bc	3.76a
Weed-free check	199.0a	74.3a	27.8a	3.61a	205.6a	73.3a	27.8a	3.84a

Within a column for each factor, means sharing same alphabets are not significantly different at $P = 0.05$ probability level according to least significant difference test.
WAS = weeks after seeding.

values. In off-season, the beginning of CPWC based on 10% AYL occurred by 456 GDD corresponding to 23 days after seeding (DAS) (Table 6). In contrast, in main season at the same AYL, weeds required to be removed at 412 GDD, corresponding to 21 DAS. The end of the CPWC at 10% AYL occurred by 832 GDD or 40 DAS in the off-season and 847 GDD or 43 DAS in the main season. At 5% AYL, the onset of CPWC occurred at 137 GDD, relating to 7 DAS in off-season and 131 GDD or 7 DAS in main season. Weeds had to be controlled until 987 GDD, corresponding to 49 DAS in off-season at 5% AYL. In contrast for the main season and the same AYL, rice field should be kept weed-free until 1044 GDD, relating to 53 DAS. It is evident from our study that CPWC of AERON 1 was variable in length between seasons

and was a bit longer in main season than in off-season. The beginning of the critical period was relatively stable between seasons; the end, on the other hand, was more variable. The onset of CPWC became delayed and ended earlier as the predetermined AYL increased from 5% to 10%.

4. Discussion

Over the past few decades, weed management has been mostly herbicide dependent resulting raised public concern about the residual toxicity of herbicides, which necessitates the development of a less herbicide-dependent weed management system [27]. Study of critical period of weed competition (CPWC) is very crucial for sustainability view

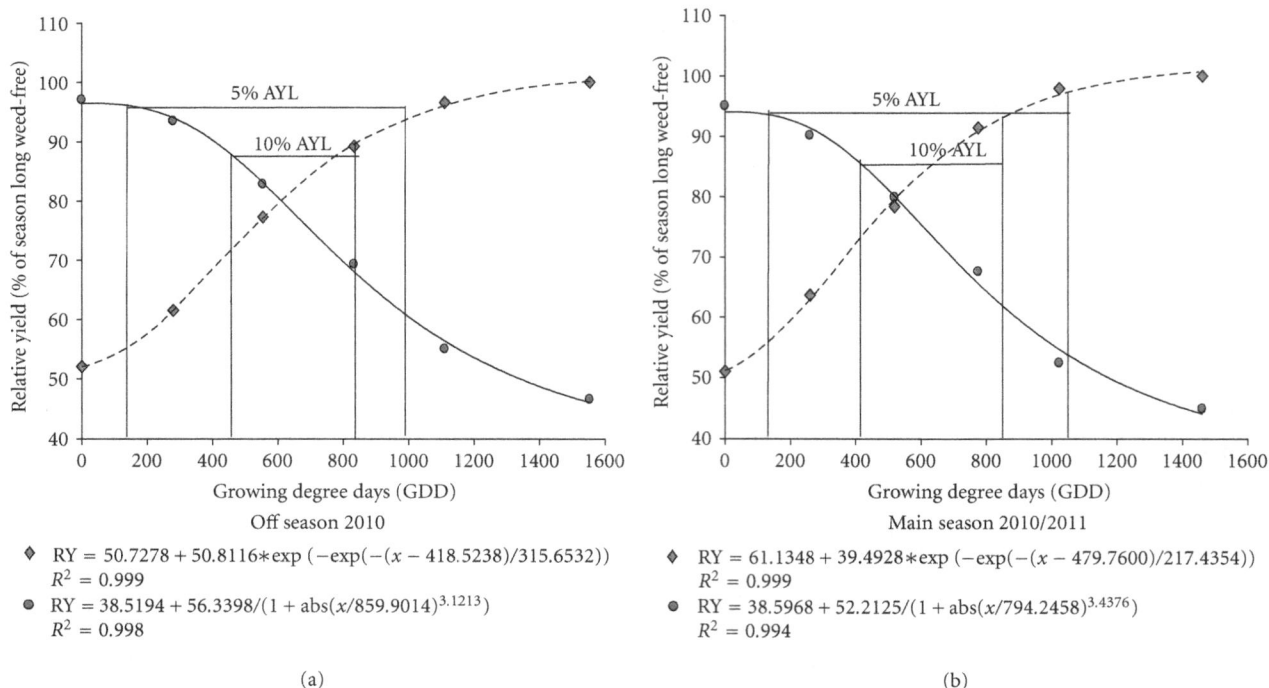

(a)

◆ RY = 50.7278 + 50.8116*exp (−exp(−(x − 418.5238)/315.6532))
$R^2 = 0.999$
● RY = 38.5194 + 56.3398/(1 + abs(x/859.9014)$^{3.1213}$)
$R^2 = 0.998$

(b)

◆ RY = 61.1348 + 39.4928*exp (−exp(−(x − 479.7600)/217.4354))
$R^2 = 0.999$
● RY = 38.5968 + 52.2125/(1 + abs(x/794.2458)$^{3.4376}$)
$R^2 = 0.994$

FIGURE 1: Influence of weed interference on relative yield of aerobic rice variety AERON 1 in the off-season of 2010 and main season of 2010/2011. Increasing duration of weed interference (●) data fitted to the logistic equation; increasing weed-free period (◆) data fitted to the Gompertz equation. The dots and the lines represent observed relative yield and fitted models, respectively. AYL = accepted yield loss; RY = relative yield.

TABLE 6: Estimated critical periods of weed control for two acceptable levels of crop losses in both seasons (off-season 2010 and main season 2010/2011).

Yield loss levels (%)	Critical period							
	Off-season 2010				Main season 2010/2011			
	Onset		End		Onset		End	
	Growing degree days	Days after seeding	Growing degree days	Days after seeding	Growing degree days	Days after seeding	Growing degree days	Days after seeding
5	137	7	987	49	131	7	1044	53
10	456	23	832	40	412	21	847	43

point since it optimizes time for implementing and maintaining weed control measures (e.g., timing of herbicide application) and thus reduces ecological risk and improves economics of herbicide application. Therefore, sustainable weed management in aerobic rice relies largely on the identification of CPWC.

The experiment was accomplished under naturally occurring weed population comprising 23 species in off-season and 18 species in main season. Based on summed dominance ratio (SDR), averaged over seasons, the most dominant weed species could be arranged in the order of *Physalis heterophylla > Scoparia dulcis > Cleome rutidosperma > Cyperus rotundus > Fimbristylis miliacea > Eleusine indica > Leptochloa chinensis*. Similarity in weed composition between seasons might be due to proximity of the experimental sites

and similarity of cropping pattern and weed management practices. The weed community was mostly dominated by broadleaf weeds followed by sedges and grasses, which is rather different from that of a typical aerobic rice field. In Karnataka, India, Gowda et al. [28] reported grasses and sedges as the predominant weed groups in aerobic rice field. Jaya Suria et al. [29] also accounted from their study with aerobic rice conducted at Penang, Malaysia that grassy weeds constituted about 80% of total weed community. The differences in the floristic composition and dominance pattern of weeds reported in different studies might be due to the variation in the agroecological conditions, cropping pattern, management practices, and weed seed bank composition among the experimental sites [30]. Weed species dominance pattern varied between seasons mostly due to the differences in soil moisture regimes; in main season, rice crop received a total rainfall of 228 mL throughout life cycle, while in off-season, the same received only 180 mL rainfall. Juraimi et al. [18] also reported that rice weed community is strongly influenced by soil moisture conditions. It was evident that broadleaf weeds are found to be more aggressive in main season (82% SDR) than in off-season (54% SDR), which might be due to higher moisture regimes in main season that favored broadleaf weeds more than the grasses or sedges. Abundance of broadleaf weeds under saturated conditions (higher-moisture regimes) has also been reported by Juraimi et al. [31]. Conflicting findings have been reported by Bhagat et al. [32], who recorded dominance of grasses under higher-moisture regimes.

Weed density and dry matter were recorded very high in both the seasons. The weed pressure was higher in main season than in off-season because of more favorable conditions in terms of soil moisture status. Differences in weed dry matter between rice seasons has also been reported by Johnson et al. [15] and Chauhan and Johnson [19]. The high weed pressure as observed in this study confirms the findings of many other researchers [29, 33, 34], who reported that weed pressure in aerobic rice is the highest among the rice ecosystems. In season long weedy check, weed dry matter in the previous study [19] with aerobic rice ranged from 458 to 692 g/m^2 in between seasons, whereas, in this study, weed dry matter ranged from 390 to 431 g/m^2, a 14–37% lower weed dry matter. The dissimilarity could be due to the contrasting rice variety, weed flora, soil moisture regimes, and agroclimatic conditions between experimental sites.

Biomass accumulation by AERON 1 was adversely affected by the increasing length of weed interference period and, conversely, favorably influenced by the increasing span of weed-free period up to 6 or 8 WAS. Maintaining weed-free conditions after 6 or 8 WAS failed to improve biomass accumulation of rice. Rice grown in a competitive advantage in comparison with weed will grow better than its counterpart and vice versa. At early crop growth stage, weeds may be better competitor than the crop, which is likely due to competitive advantages for the weeds in terms of preemption of resources. But with the advancement of times, crop start dominating over weeds, and after a certain stage, weed is no more a threat for crop growth, controlling weeds after that resulting no significant enhancement in crop growth. Similar findings have been reported by many researchers [17, 18, 35], who observed that weed interference up to a certain growth stage had negative impact on rice growth.

Rice grain yield decreased with prolonged delays in weed removal; conversely, grain yield increased with the increasing length of weed-free period in both the seasons. Weed competition throughout reduced crop yield by approximate 55% in both seasons as compared with season long weed-free period. These values are very close to those reported in a previous study, where season long weed competition reduced yield by approximately 50% [15]. Juraimi et al. [18] recorded 79 and 66% yield reduction in rice due to weed competition till harvest in flooded and saturated conditions, respectively. Chauhan and Johnson [19], on the contrary, reported as high as 95% yield reduction in aerobic rice due

to weed competition throughout the crop growing season. These contrasting findings might be due to differences in rice variety, agroclimatic conditions, soil moisture regimes, and weed flora among the experimental sites. Prolonged weed competition resulted in reduced biomass accumulation and lesser panicles/m^2, grains/panicle, and thousand-seed weight which ultimately translated into lower grain yield. Increased biomass accumulation by weeds with the increasing span of weed interference period might also be a plausible cause of yield reduction in rice. As Woolley et al. [36] stated, weed dry matter has been found to be highly correlated with crop yield loss.

Based on 5% AYL, our results imply that under similar experimental conditions rice can tolerate weed interference until 137 GDD or 7 DAS in off-season and 131 GDD or 7 DAS in main season, suggesting that control measures should start at that stage; the crop should be kept weed-free until 987 GDD corresponding to 49 DAS in off-season and 1044 GDD corresponding to 53 DAS in main season in order to prevent appreciable economic yield loss. Therefore, weed control measures must begin as soon as possible after rice emergence to prevent yield loss more than 5% in both seasons. At 10% AYL, the beginning of the CPWC was determined 456 GDD for off-season and 412 GDD in main season; to prevent more than acceptable yield loss, aerobic rice field should be maintained weed-free until 832 GDDs have been accumulated in off-season, while in main season, fields should continue to be scouted until 847 GDDs have been accumulated. This equates to controlling weeds from 23 to 40 DAS and 21 to 43 DAS in off and main seasons, respectively. This implies that keeping rice weed-free for that stipulated period is equivalent to keeping rice weed-free season long; presence of weeds before or after that period will not pose a threat [5], and hence, there may be very little benefit of subsequent weed control.

As Zimdahl [12] stated, the CPWC is "not an inherent property of the crop," and CPWC of a particular crop is weed species, site, and season specific [13]. Johnson et al. [15] estimated CPWC for lowland irrigated rice as 0–32 DAS in wet season and 4–83 DAS in dry season to obtain 95% yield in West Africa. In Malaysia, based on the 5% yield loss, Begum et al. [17] concluded that flood-irrigated rice must be kept free from weed competition from 14 to 28 DAS, while Juraimi et al. [37] suggested that direct-seeded rice should be kept weed-free for 2–71 DAS in saturated condition and 15–73 DAS in flooded condition. In the Philippines, Chauhan and Johnson [19] estimated CPWC of aerobic rice as between 18 and 52 DAS to obtain 95% of weed-free yield. Thus, CPWC is highly variable and is largely dependent on the relationship of crop seeding date to the emergence periodicity for the weed community of a particular site [26].

The onset of the critical period was found relatively stable between seasons, while the end was more variable. This phenomenon is supported by many researchers [4, 38, 39], who opined that the end of CPWC was variable and highly dependent on density, competitiveness, and emergence periodicity of the weed population. The CPWC in main season started earlier and lasted longer, as compared to off-season. A long critical period is the indication of less competitive crop or more competitive weeds and vice versa [1, 6]. A possible reason for starting earlier and lasting longer of CPWC in main season might be the conditions favorable for weed germination and growth. Main season received more rainfall than off-season which might provide weeds an advantage over the rice crop. Juraimi et al. [18] also estimated longer CPWC of rice in main season and shorter one in off-season in both flooded and saturated conditions. Johnson et al. [15], on the contrary, observed differences in CPWC between seasons in irrigated lowland rice.

The study portrays the significance of CPWC determination for sustainable weed management in aerobic rice. The practical implication of this study is that under the similar experimental conditions aerobic rice field should be kept weed-free during 7–49 DAS in off-season and 7–53 DAS in main season to achieve 95% of weed-free yield, and 23–40 DAS in off-season and 21–43 DAS in main season to achieve 90% weed-free yield. Since 5% yield loss level would not be practical from economic view point, a 10% yield loss may be considered excellent in terms of economic return, and this level can be achieved by early postemergence application of herbicide or weeding between 10 and 15 DAS followed by a post emergence application or weeding between 30 and 35 DAS. Since weeds emerge after this period are supposed to cause no substantial yield looses, the need for applying additional herbicides or weeding more than 2 times as practiced by most farmers would not be warranted and could lead to significant cost savings. Nevertheless, weed management can be extended beyond that period if the objective is not only to have higher yield but also to avoid weed seed rain to prevent buildup of the weed seed bank, which is of major concern for long-term sustainability of weed management.

Acknowledgment

The authors wish to acknowledge Universiti Putra Malaysia Research University Grant (01-04-08-0543RU) for conducting the research.

References

[1] M. D. Amador-Ramírez, "Critical period of weed control in transplanted chilli pepper," *Weed Research*, vol. 42, no. 3, pp. 203–209, 2002.

[2] R. L. Zimdahl, *Weed-Crop Competition: A Review*, International Plant Protection Control, Oregon State University, Corvallis, Ore, USA, 1980.

[3] S. Z. Knezevic, S. P. Evans, E. E. Blankenship, R. C. Van Acker, and J. L. Lindquist, "Critical period for weed control: the concept and data analysis," *Weed Science*, vol. 50, no. 6, pp. 773–776, 2002.

[4] S. P. Evans, S. Z. Knezevic, J. L. Lindquist, C. A. Shapiro, and E. E. Blankenship, "Nitrogen application influences the critical period for weed control in corn," *Weed Science*, vol. 51, no. 3, pp. 408–417, 2003.

[5] M. M. Williams II, "Planting date influences critical period of weed control in sweet corn," *Weed Science*, vol. 54, no. 5, pp. 928–933, 2006.

[6] H. Z. Ghosheh, D. L. Holshouser, and J. M. Chandler, "The critical period of johnsongrass (*Sorghum halepense*) control in field corn (*Zea mays*)," *Weed Science*, vol. 44, no. 4, pp. 944–947, 1996.

[7] M. R. Hall, C. J. Swanton, and G. W. Anderson, "The critical period of weed control in grain corn (Zea mays)," *Weed Science*, vol. 40, pp. 441–447, 1992.

[8] R. L. Zimdahl, *Weed-Crop Competition: A Review*, Blackwell, Oxford, UK, 2nd edition, 2004.

[9] B. L. Dillehay, W. S. Curran, and D. A. Mortensen, "Critical period for weed control in alfalfa," *Weed Science*, vol. 59, no. 1, pp. 68–75, 2011.

[10] S. Z. Knezevic, S. P. Evans, and M. Mainz, "Row spacing influences the critical timing for weed removal in soybean (*Glycine max*)," *Weed Technology*, vol. 17, no. 4, pp. 666–673, 2003.

[11] E. C. Gilmore and R. S. Rogers, "Heat units as a method of measuring maturity in corn," *Agronomy Journal*, vol. 50, pp. 611–615, 1958.

[12] R. L. Zimdahl, "The concept and application of the critical weed free period," in *Weed Management in Agroecosystems: Ecological Approaches*, M. A. Altieri and M. Liebman, Eds., pp. 145–155, CRC Press, Boca Raton, Fla, USA, 1988.

[13] S. E. Weaver, "Critical period of weed competition in three vegetable crops in relation to management practices," *Weed Research*, vol. 24, pp. 317–325, 1984.

[14] W. Zhang, E. P. Webster, D. Y. Lanclos, and J. P. Geaghan, "Effect of weed interference duration and weed-free period on glufosinate-resistant rice (*Oryza sativa*)," *Weed Technology*, vol. 17, no. 4, pp. 876–880, 2003.

[15] D. E. Johnson, M. C. S. Wopereis, D. Mbodj, S. Diallo, S. Powers, and S. M. Haefele, "Timing of weed management and yield losses due to weeds in irrigated rice in the Sahel," *Field Crops Research*, vol. 85, no. 1, pp. 31–42, 2004.

[16] M. Azmi, A. S. Juraimi, and M. Y. Mohammad Najib, "Critical period of weedy rice control in direct seeded rice," *Journal of Tropical Agriculture and Food Science*, vol. 35, no. 2, pp. 319–332, 2007.

[17] M. Begum, A. S. Juraimi, A. Rajan, S. R. S. Omar, and M. Azmi, "Critical period competition between *Fimbristylis miliacea* (L.) Vahl and rice (MR 220)," *Plant Protection Quarterly*, vol. 23, no. 4, pp. 153–157, 2008.

[18] A. S. Juraimi, M. Y. Mohamad Najib, M. Begum, A. R. Anuar, M. Azmi, and A. Puteh, "Critical period of weed competition in direct seeded rice under saturated and flooded conditions," *Pertanika Journal of Tropical Agricultural Science*, vol. 32, no. 2, pp. 305–316, 2009.

[19] B. S. Chauhan and D. E. Johnson, "Row spacing and weed control timing affect yield of aerobic rice," *Field Crops Research*, vol. 121, no. 2, pp. 226–231, 2011.

[20] T. P. Tuong and B. A. M. Bouman, "Rice production in water-scarce environments," in *Proceedings of the Water Productivity Workshop*, Colombo, Sri Lanka, November 2003.

[21] P. Anwar, A. S. Juraimi, A. Man, A. Puteh, A. Selamat, and M. Begum, "Weed suppressive ability of rice (*Oryza sativa* L.) germplasm under aerobic soil conditions," *Australian Journal of Crop Science*, vol. 4, no. 9, pp. 706–717, 2010.

[22] D. L. Zhao, G. N. Atlin, L. Bastiaans, and J. H. J. Spiertz, "Developing selection protocols for weed competitiveness in aerobic rice," *Field Crops Research*, vol. 97, no. 2-3, pp. 272–285, 2006.

[23] P. Anwar, A. S. Juraimi, A. Puteh, A. Selamat, A. Man, and A. Hakim, "Seeding method and rate influence on weed suppression in aerobic rice," *African Journal of Biotechnology*, vol. 10, no. 68, pp. 15259–15271, 2011.

[24] A. A. L. N. Sarma, T. V. L. Kumar, and K. Koteswararao, "Development of an agroclimatic model for the estimation of rice yield," *Journal of Indian Geophysics Union*, vol. 12, no. 2, pp. 89–96, 2008.

[25] SAS (Statistical Analysis System), *The SAS system for Windows, Version 9.1*, SAS Inst. Inc., Cary, NC, USA, 2003.

[26] S. G. Martin, R. C. Van Acker, and L. F. Friesen, "Critical period of weed control in spring canola," *Weed Science*, vol. 49, no. 3, pp. 326–333, 2001.

[27] C. J. Swanton, J. O'Sullivan, and D. E. Robinson, "The critical weed-free period in carrot," *Weed Science*, vol. 58, no. 3, pp. 229–233, 2010.

[28] P. T. Gowda, C. Shankaraiah, A. C. Jnanesh, M. Govindappa, and K. N. K. Murthy, "Studies on chemical weed control in aerobic rice (Oryza sativa L.)," *Journal of Crop and Weed*, vol. 5, no. 1, pp. 321–324, 2009.

[29] A. S. M. Jaya Suria, A. S. Juraimi, M. M. Rahman, A. B. Man, and A. Selamat, "Efficacy and economics of different herbicides in aerobic rice system," *African Journal of Biotechnology*, vol. 10, no. 41, pp. 8007–8022, 2011.

[30] A. S. Juraimi, M. Begum, M. N. Mohd Yusof, and A. Man, "Efficacy of herbicides on the control weeds and productivity of direct seeded rice under minimal water conditions," *Plant Protection Quarterly*, vol. 25, no. 1, pp. 19–25, 2010.

[31] A. S. Juraimi, A. H. Muhammad Saiful, M. Kamal Uddin, A. A. Rahim, and M. Azmi, "Diversity of weeds under different water regimes in irrigated direct seeded rice," *Australian Journal of Crop Science*, vol. 5, no. 5, pp. 595–564, 2011.

[32] R. M. Bhagat, S. I. Bhuiyan, K. Moody, and L. E. Estorninos, "Effect of water, tillage and herbicide on ecology of weed communities in intensive wet-seeded rice system," *Crop Protection*, vol. 18, no. 5, pp. 293–303, 1999.

[33] A. N. Rao, D. E. Johnson, B. Sivaprasad, J. K. Ladha, and A. M. Mortimer, "Weed management in direct seeded rice," *Advances in Agronomy*, vol. 93, pp. 153–255, 2007.

[34] G. Mahajan, B. S. Chauhan, and D. E. Johnson, "Weed management in aerobic rice in northwestern indo-gangetic plains," *Journal of Crop Improvement*, vol. 23, no. 4, pp. 366–382, 2009.

[35] A. Ali, M. A. Malik, R. M. Rehman, R. Sohail, and M. M. Akram, "Growth and yield response of wheat to different sowing times and weed competition durations," *Pakistan Journal of Biological Sciences*, vol. 3, no. 4, pp. 681–682, 2002.

[36] B. L. Woolley, T. E. Michaels, M. R. Hall, and C. J. Swanton, "The critical period of weed control in white bean (Phaseolus vulgaris)," *Weed Science*, vol. 41, pp. 180–184, 1993.

[37] A. S. Juraimi, M. Begum, A. M. Sherif, and A. Rajan, "Effects of sowing date and nutsedge removal time on plant growth and yield of tef [Eragrostis tef (Zucc.) Trotter]," *African Journal of Biotechnology*, vol. 8, no. 22, pp. 6162–6167, 2009.

[38] J. K. Norsworthy and M. J. Oliveira, "Comparison of the critical period for weed control in wide- and narrow-row corn," *Weed Science*, vol. 52, no. 5, pp. 802–807, 2004.

[39] I. Uremis, A. Uludag, A. C. Ulger, and B. Cakir, "Determination of critical period for weed control in the second crop corn under Mediterranean conditions," *African Journal of Biotechnology*, vol. 8, no. 18, pp. 4475–4480, 2009.

Permissions

The contributors of this book come from diverse backgrounds, making this book a truly international effort. This book will bring forth new frontiers with its revolutionizing research information and detailed analysis of the nascent developments around the world.

We would like to thank all the contributing authors for lending their expertise to make the book truly unique. They have played a crucial role in the development of this book. Without their invaluable contributions this book wouldn't have been possible. They have made vital efforts to compile up to date information on the varied aspects of this subject to make this book a valuable addition to the collection of many professionals and students.

This book was conceptualized with the vision of imparting up-to-date information and advanced data in this field. To ensure the same, a matchless editorial board was set up. Every individual on the board went through rigorous rounds of assessment to prove their worth. After which they invested a large part of their time researching and compiling the most relevant data for our readers. Conferences and sessions were held from time to time between the editorial board and the contributing authors to present the data in the most comprehensible form. The editorial team has worked tirelessly to provide valuable and valid information to help people across the globe.

Every chapter published in this book has been scrutinized by our experts. Their significance has been extensively debated. The topics covered herein carry significant findings which will fuel the growth of the discipline. They may even be implemented as practical applications or may be referred to as a beginning point for another development. Chapters in this book were first published by Hindawi Publishing Corporation; hereby published with permission under the Creative Commons Attribution License or equivalent.

The editorial board has been involved in producing this book since its inception. They have spent rigorous hours researching and exploring the diverse topics which have resulted in the successful publishing of this book. They have passed on their knowledge of decades through this book. To expedite this challenging task, the publisher supported the team at every step. A small team of assistant editors was also appointed to further simplify the editing procedure and attain best results for the readers.

Our editorial team has been hand-picked from every corner of the world. Their multi-ethnicity adds dynamic inputs to the discussions which result in innovative outcomes. These outcomes are then further discussed with the researchers and contributors who give their valuable feedback and opinion regarding the same. The feedback is then collaborated with the researches and they are edited in a comprehensive manner to aid the understanding of the subject.

Apart from the editorial board, the designing team has also invested a significant amount of their time in understanding the subject and creating the most relevant covers. They scrutinized every image to scout for the most suitable representation of the subject and create an appropriate cover for the book.

The publishing team has been involved in this book since its early stages. They were actively engaged in every process, be it collecting the data, connecting with the contributors or procuring relevant information. The team has been an ardent support to the editorial, designing and production team. Their endless efforts to recruit the best for this project, has resulted in the accomplishment of this book. They are a veteran in the field of academics and their pool of knowledge is as vast as their experience in printing. Their expertise and guidance has proved useful at every step. Their uncompromising quality standards have made this book an exceptional effort. Their encouragement from time to time has been an inspiration for everyone.

The publisher and the editorial board hope that this book will prove to be a valuable piece of knowledge for researchers, students, practitioners and scholars across the globe.

A. Dansi
Laboratory of Agricultural Biodiversity and Tropical Plant breeding (LAAPT), Faculty of Sciences and Technology (FAST-Dassa), University of Abomey-Calavi (UAC), P.O. Box 526, Cotonou, Benin
Crop, Aromatic and Medicinal Plant Biodiversity Research and Development Institute (IRDCAM), 071 BP 28 Cotonou, Benin

P. Assogba
Crop, Aromatic and Medicinal Plant Biodiversity Research and Development Institute (IRDCAM), 071 BP 28 Cotonou, Benin

M. Tamo
Institut International d Agriculture Tropicale (IITA), 08 BP 0932 Cotonou, Benin

A. H. Bokonon-Ganta
Service de la Protection des Vegetaux et du Controle Phytosanitaire, Direction de l Agriculture, BP 58 Porto-Novo, Benin

R. Vodouhè
Bioversity International, Office of West and Central Africa, 08 BP 0932 Cotonou, Benin

A. Akoegninou
National Herbarium, Department of Botany and Plant Biology, Faculty of Sciences and Technology (FAST), University of Abomey-Calavi (UAC), P.O. Box 526, Cotonou, Benin

A. Sanni
Laboratory of Biochemistry and Molecular Biology, Faculty of Sciences and Technology (FAST), University of Abomey-Calavi (UAC), P.O. Box 526, Cotonou, Benin

Santanu Sabhapondit, Lakshi Prasad Bhuyan and Mridul Hazarika
Department of Biochemistry, Tocklai Experimental Station, Tea Research Association, Assam, Jorhat 785008, India

Tanmoy Karak
Department of Soil, Tocklai Experimental Station, Tea Research Association, Assam, Jorhat 785008, India

Bhabesh Chandra Goswami
Department of Chemistry, Gauhati University, Assam, Guwahati 781014, India

Sondeep Singh, Anil K. Gupta and Narinder Kaur
Department of Biochemistry, Punjab Agricultural University, Ludhiana 141004, India

Jiancheng Wang
Seed Science Center, College of Agriculture and Biotechnology, Zhejiang University, Hangzhou 310058, China
Shandong Crop Germplasm Center, Shandong Academy of Agricultural Sciences, Jinan 250100, China

Yajing Guan, Yang Wang, Liwei Zhu, Qitian Wang, Qijuan Hu and Jin Hu
Seed Science Center, College of Agriculture and Biotechnology, Zhejiang University, Hangzhou 310058, China

Kristin R. Abney, Dean A. Kopsell and Carl E. Sams
Plant Sciences Department, The University of Tennessee, 2431 Joe Johnson Drive, Knoxville, TN 37996, USA

Svetlana Zivanovic
Department of Food Science and Technology, The University of Tennessee, 2605 River Drive, Knoxville, TN 37996, USA

David E. Kopsell
Department of Agriculture, Illinois State University, Normal, IL 61790, USA

Ilkay Yavas, Aydin Unay and Mehmet Aydin
Kocarli Vocational High School, Adnan Menderes University, 09100 Aydın, Turkey

Cinzia Benincasa, Enzo Perri and Caterina Briccoli Bati
Centro di Ricerca per 1 Olivicoltura e lIndustria Olearia, CRA, via Li Rocchi 111, 87036 Rende, Italy

Mohamed Ayadi, Moncen Khlif and Mariem Gharsallaoui
Olive Tree Institute, University of Sfax, Route de l'aeroport Km 1.5, BP 1087, 3000 Sfax, Tunisia

Slimane Gabsi
National School of Engineering, University of Gabes, Rue Omar Ibn El Khattab 6029, Tunisia

M. Bahari
Malaysian Agricultural Research and Development Institute, Bukit Tangga, 06050 Bukit Kayu Hitam, Kedah, Malaysia

G. B. Saleh
Department of Crop Science, Faculty of Agriculture, Universiti Putra Malaysia, 43400 UPM Serdang, Selangor, Malaysia

M. Y. Rafii
Department of Crop Science, Faculty of Agriculture, Universiti Putra Malaysia, 43400 UPM Serdang, Selangor, Malaysia
Institute of Tropical Agriculture, Universiti Putra Malaysia, 43400 UPM Serdang, Selangor, Malaysia

M. A. Latif
Department of Crop Science, Faculty of Agriculture, Universiti Putra Malaysia, 43400 UPM Serdang, Selangor, Malaysia
Plant Pathology Division, Bangladesh Rice Research Institute (BRRI), Gazipur-1701, Bangladesh

Luiz Augusto Gratieri
Instituto Federal de Educac ao, Ciencia e Tecnologia Sul de Minas Gerais (IFSULDEMINAS), Campus Muzambinho, Morro Preto s/n, 37890-000 Muzambinho, MG, Brazil

Arthur Bernardes Cecílio Filho, José Carlos Barbosa and Luiz Carlos Pavani
Universidade Estadual Paulista, Via de Acesso Prof. Paulo D. Castellane s/n, 14884-900 Jaboticabal, SP, Brazil

Telma Magaia
Department of Biological Science, Science Faculty, Eduardo Mondlane University, Praca 25 de Junho, P.O. Box 257, Maputo, Mozambique
Division of Applied Nutrition and Food Chemistry, Lund University, P.O. Box 124, 221 00 Lund, Sweden

Kerstin Skog
Division of Applied Nutrition and Food Chemistry, Lund University, P.O. Box 124, 221 00 Lund, Sweden

Amália Uamusse
Department of Chemistry, Science Faculty, Eduardo Mondlane University, Praca 25 de Junho, P.O. Box 257, Maputo, Mozambique

Ingegerd Sjöholm
Division of Food Technology, Lund University, P.O. Box 124, 221 00 Lund, Sweden

M. P. Anwar
Institute of Tropical Agriculture, Universiti Putra Malaysia, Serdang 43400, Selangor, Malaysia
Department of Agronomy, Bangladesh Agricultural University, Mymensingh 2202, Bangladesh

A. S. Juraimi, B. Samedani and A. Puteh
Department of Crop Science, Universiti Putra Malaysia, Serdang 43400, Selangor, Malaysia

A. Man
Rice and Industrial Crops Research Centre, Malaysian Agricultural Research and Development Institute, Kuala Lumpur 50774, Malaysia

www.ingramcontent.com/pod-product-compliance
Lightning Source LLC
Chambersburg PA
CBHW080635200326
41458CB00013B/4639